Chinmedomics

Chinmedomics

The Integration of Serum Pharmacochemistry and Metabolomics to Elucidate the Scientific Value of Traditional Chinese Medicine

Editor

Xijun Wang

Associate Editors

Aihua Zhang

Hui Sun

ELSEVIER

AMSTERDAM • BOSTON • HEIDELBERG • LONDON • NEW YORK
OXFORD • PARIS • SAN DIEGO • SAN FRANCISCO • SINGAPORE
SYDNEY • TOKYO

Academic Press is an Imprint of Elsevier

Academic Press is an imprint of Elsevier
125, London Wall, EC2Y 5AS, UK
525 B Street, Suite 1800, San Diego, CA 92101-4495, USA
225 Wyman Street, Waltham, MA 02451, USA
The Boulevard, Langford Lane, Kidlington, Oxford OX5 1GB, UK

British Library Cataloguing-in-Publication Data
A catalogue record for this book is available from the British Library

Library of Congress Cataloging-in-Publication Data
A catalog record for this book is available from the Library of Congress

ISBN: 978-0-12-803117-9

For information on all Academic Press publications
visit our website at http://store.elsevier.com/

Publisher: Mica Haley
Acquisition Editor: Kristine Jones
Editorial Project Manager: Molly McLaughlin
Production Project Manager: Julia Haynes
Designer: Mark Rogers

Typeset by Thomson Digital

Contents

List of Contributors

All contributors are located at: National TCM Key Laboratory of Serum Pharmacochemistry, Laboratory of Metabolomics and Chinmedomics, Department of Pharmaceutical Analysis, Heilongjiang University of Chinese Medicine, Harbin, Heilongjiang, China

Yang Bo
Ying Han
Qi Liu
Haitao Lv
Hui Sun
Huiyu Wang
Ping Wang
Xijun Wang
Wenjun Sun
Xin Tong
Guangli Yan
Aihua Zhang
Shuxiang Zhang
Xiaohang Zhou

About the Editor

Xijun Wang, PhD, is a Professor of Pharmacology of Traditional Chinese Medicine (TCM) and Vice President at Heilongjiang University of Chinese Medicine in China. Dr. Wang has published more than 150 papers in peer-reviewed journals, received the National Science and Technology Progress Award and National Technological Invention Award and holds 20 patents. His research focuses on serum pharmacochemistry of TCMs. By integrating metabolomics with serum pharmacochemistry of TCM, Dr. Wang has developed and established the theory and research methods of Chinmedomics, whereby the clinical efficacy of TCM formulae can be measured and analyzed.

Dr. Wang can be contacted at xijunw@sina.com

Preface

Chinmedomics, proposed by our team, is the theory and research method of mining the chemical biological characters of Traditional Chinese Medicine (TCM) Syndrome and evaluating the efficacy of the TCM formulae (prescription) which consists of multiple herbal drugs, and consequently to shed new light on the efficacy of the formulae to discover the effective constituents, and finally to elucidate the scientific value of TCM.

Syndrome and formulae are two key issues in TCM, the vagueness of syndrome and complexity of formulae make the diagnosis and evaluation difficult, and greatly limit the locating of the effective constituents of TCMs related with clinical efficacy. Therefore, the chemical biological characters of syndrome and efficacy of formulae will be the key scientific problems for discovery of the TCM effective material basis. Our team has engaged in the innovation and research methods of TCM for a long time, aiming to solve the key scientific issues such as the chemical biological essence of syndrome and ascertain the effective substances of the TCM formulae. In the beginning of the 1990s, we first established the theory and method of the serum pharmacochemistry of TCMs, providing a methodology for the discovery of active constituents *in vivo* from TCMs, exploring their efficacy, and safety issues. In the twenty-first century, we integrated the serum pharmacochemistry of TCMs with metabolomics, and developed a system method called *chinmedomics* for elucidating the chemical biological essence of syndrome and the efficacy of TCMs, as well as providing an effective material basis. Briefly, using metabolomic technology to clarify the biomarkers of a syndrome, and to evaluate the efficacy of TCM formulae. And to further discover the potential active constituents in the serum post-oral administration of TCM formulae by using the serum pharmacochemistry of TCMs under the acting condition of the efficacy, and further analyzing the correlation between the exogenous constituents of formulae *in vivo* and the endogenous biomarker of syndrome to find the constituents highly associated with efficacy of formulae, as the effective substance. By clarifying the activities of the effective substances we may discover the lead compounds and make innovative drug discoveries based on TCM clinical experiences. By way of variation of the effective substances in the process of formulae compatibility, we can elucidate the principles of the combination therapy of TCM formulae.

The connotation and methods of *chinmedomics* has been published in *OMICS*. Presently, we have published 126 papers related to *chinmedomics*, for example, *Scientific Reports*, *Mol Cell Proteomics*, *Anal Chem*, etc. Up to now, the theory and research methods of *chinmedomics* not only deepens in our laboratory, but has also been widely cited and used by international scholars. Thus, we systematically arranged the theory and the research methods of *chinmedomics*, and combined and compiled these with the results of our research work into this book. We sincerely expect that *chinmedomics* will be more widely used to elucidate the scientific value of TCM.

Xijun Wang
Professor, PhD
General Director, National TCM Key Lab
Vice President, Heilongjiang University of Chinese Medicine
No. 24, Heping Road, Harbin 150040, China
December 26, 2014
Harbin, China
Email: xijunw@sina.com

CHAPTER 1

Origin of Chinmedomics

Xijun Wang, Aihua Zhang, Hui Sun, Guangli Yan

CONTENTS

1.1 INTRODUCTION

Traditional Chinese Medicine (TCM) has been practiced for thousands of years and is well integrated in the Chinese health care system as one of the mainstream medical practices, rapidly gaining attention in China for improving the health condition of human beings and preventing or healing diseases (Wang et al., 2012a). TCM attempts to bring the body, mind, and spirit into harmony and adjusts humans to remain in a healthy status. For improving health, TCM applies multiple natural therapeutic approaches including Chinese medical formulae (prescription), and herbal medicine, etc. As one of the most important parts in TCM, formulae have been accepted by the academic community, and patients in the treatment of multiple organ system disorders. To enhance the therapeutic effects of TCM, practitioners advocated combinatorial therapeutic strategies and often prescribe a combination of herbs, called formulae, that work together harmoniously to achieve an ideal therapeutic outcome (Zhang et al., 2014). It is believed that multiple herbs in the formulae could hit multiple targets and exert synergistic therapeutic efficacies. Differing from the single component strategy of Western drugs – the formulae is a complex system consisting of multiple compounds, which makes the scientific evaluation of TCM more difficult. Therefore, it is essential to strengthen the formulae research. There have been many recent attempts to address these issues but most of them were still based on the "reductionism" philosophy, whereas TCM is based on "holism" philosophy rather than "reductionism."

Syndrome and formulae are two key issues in TCM and the premise finding for effective constituents of TCM (Zhang et al., 2012a). However, vagueness of syndrome and complexity of formulae greatly limited the evaluation, to syndromes and effective substance basis of formulae. Therefore, the question posed is how to solve the evaluation of syndrome, and discover the efficacy material basis in formulae – these are the current hot issues of international concern. In the early 1990s, we first established the theory and method on the Serum Pharmacochemistry of TCM, providing a methodology

Chinmedomics. http://dx.doi.org/10.1016/B978-0-12-803117-9.00001-9

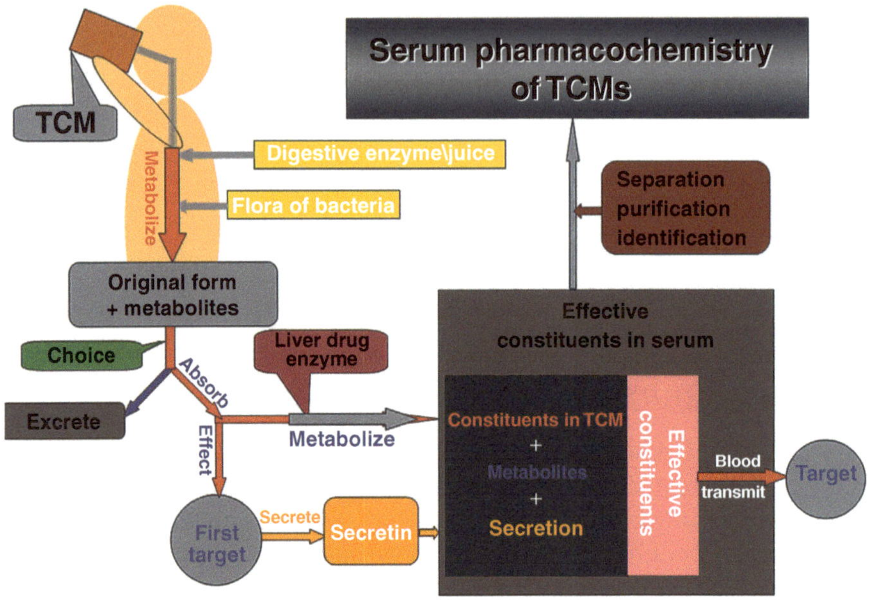

FIGURE 1.1 The procedure for active compounds from TCM illustrates the effectiveness and general content of the Serum Pharmacochemistry of TCM.

for the discovery of active constituents *in vivo* from TCM; thus solving the efficacy and safety issues of TCM. The Serum Pharmacochemistry of TCM as a preliminary was defined as isolated and identified the efficacy material basis from serum containing active constituents (after oral formulae) so that the findings reflect the *in vivo* state of Chinese medicine treatment, access to the therapeutic substance of formulae, and effectively guiding modern innovative drug design (Fig. 1.1).

Syndrome is a basic description of disease in TCM, and the formulae are the corresponding drugs used to act against the syndrome. Syndrome diagnosis is the foundation and core of differential treatment. Due to lack of objective criteria for syndrome diagnostics – it is difficult to properly evaluate the efficacy of formulae. From the point of modern system medicine, the TCM syndrome is a functional state that is caused by the body's metabolic imbalances. Metabolomics and TCM have some similar characteristics such as entirety, comprehensiveness, and dynamic changes and its property is in concert with the holistic efficacy of TCM. Metabolomics can capture comprehensive analysis of small molecule metabolites and provides a powerful approach to establish metabolic profiling, to discover biomarkers and related pathways for exploring the TCM syndrome. A typical metabolomics study is depicted schematically in Fig. 1.2.

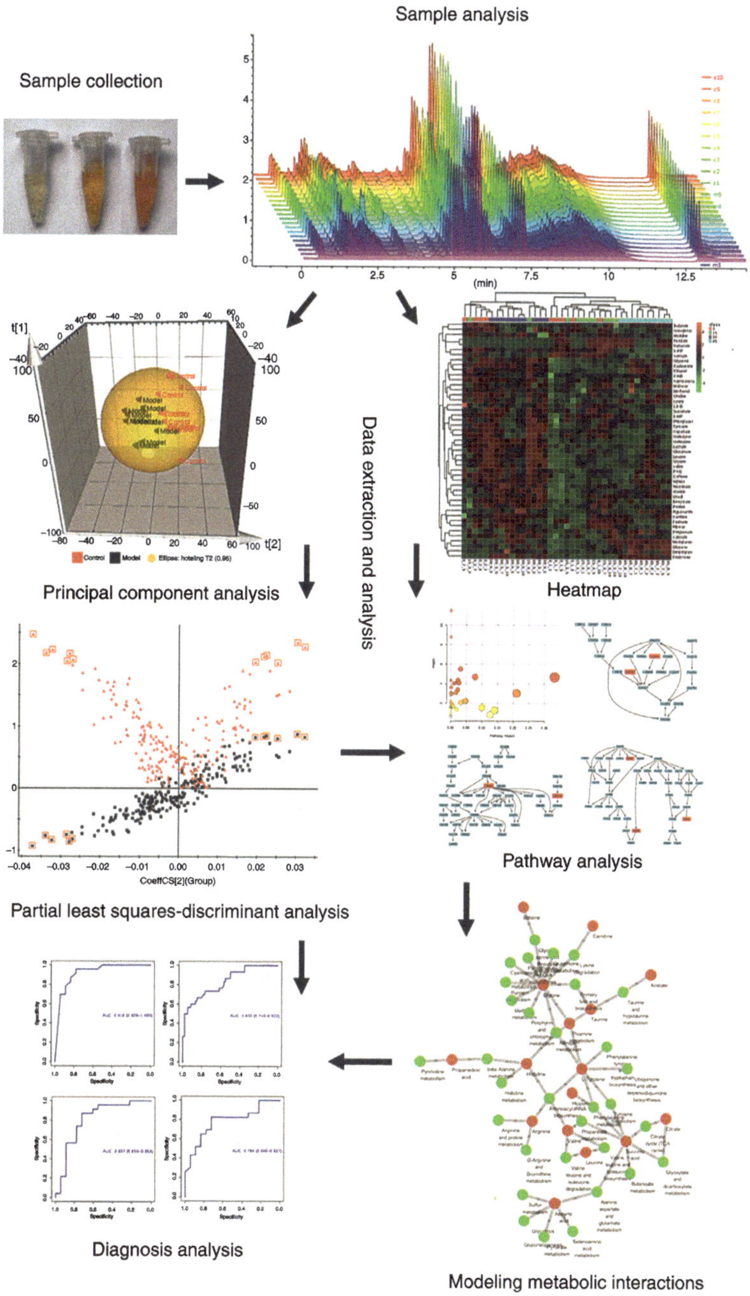

FIGURE 1.2 Typical metabolomics experiment flow diagram from large biological data sets.

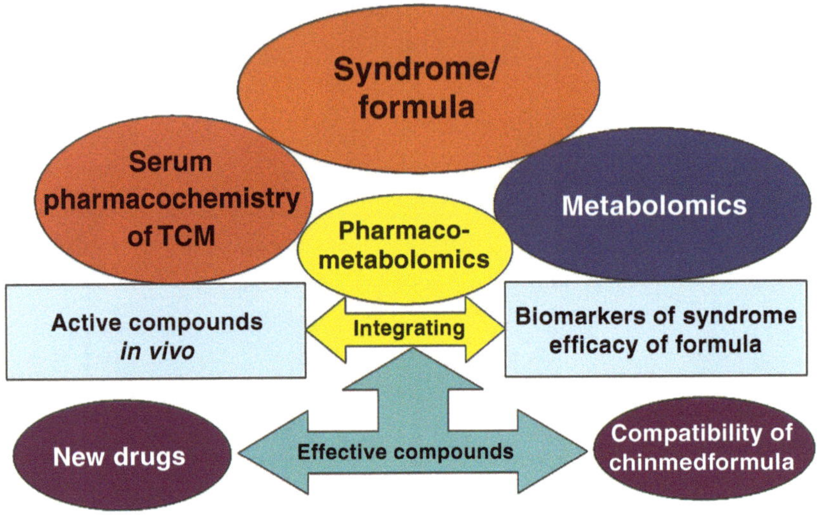

FIGURE 1.3 Flow chart of the research strategy of chinmedomics.

In the twenty-first century, the Serum Pharmacochemistry of TCM has become integrated with metabolomics, developed a new integrated method called Chinmedomics that is a unique method of TCM research, and has made outstanding contributions in solving international concerns; such as the effectiveness and safety aspects of TCM (Wang et al., 2012b). The connotation and methods of chinmedomics was published in *OMICS*. The overall procedure of the chinmedomics analysis is shown in Fig. 1.3. Metabolomics technology is used to clarify the biological nature of the syndrome; the use of Serum Pharmacochemistry is to locate the chemical substances in the body after oral TCM. Under the premise of the efficacy and effectiveness of formulae, correlation analysis between the exogenous ingredients *in vivo* after oral formulae and endogenous marker metabolites *in vivo* are used to clarify the active ingredients and their synergistic properties. On the basis of the biological characterization of syndrome, we will establish metabolic profiling of the animal models of the TCM syndrome, and related metabolic fingerprints as well as metabolic biomarkers, to evaluate the overall effects of the TCM formulae and the corresponding relationship between syndrome and formulae.

The serum pharmacochemistry of TCM is an essential approach to ensure optimal drug efficacy and assess therapeutic treatment; the advances in metabolomics provide new tools to the understanding of formulae action and mechanism. As a systemic approach, metabolomics adopts a "top-down" strategy to reflect the function of organisms from the terminal symptoms of the metabolic network and to understand the metabolic changes of a complete system caused by interventions in an holistic context. This property is in concert with the holistic efficacy of TCM; metabolomics has the potential to impact our understanding of the theory behind the evidence-based Chinese medicine (Zhang et al., 2010). This project integrates metabolomics with the Serum Pharmacochemistry of TCM, elucidating the therapeutic effects, synergistic properties and the law of formulae compatibility, based on the metabolic profiling, fingerprinting, metabolic markers and pathways of the animal model of the TCM syndrome, to further discover the active constituents *in vivo* from formulae. As a new research approach, chinmedomics may be the best method to be suitable for the holistic concept of TCM, and it provides an effective research tool for the nature of the TCM syndrome, modern formulae research, and opens the door for the modernization of the TCM industry.

1.2 APPLICATION AND CHALLENGES OF TCM

TCM developed based on ancient Chinese philosophy, and is considered as a complementary or alternative medical system in most of the Western countries whilst remaining as a form of primary care throughout most Asian countries. Its characteristics include Yin-Yang theories, five-element concept, subjective diagnostic methods, holism, treatment of differential syndrome, and dynamic functional aspect, etc. There are three major advantages of TCM treatments. First, is holism (concept of global and system-thinking) which considers the various parts of the human body as an organic whole, and the various organs closely relate to each other physiologically and pathologically, with emphasis on the close relationship between the human and the natural environments. Second, is treatment based on the treatment of differential syndrome (TDS) (diagnosis and treatment based on over-all symptoms and signs). Consistent with the modern view of homeostasis, the clinical diagnosis, and treatment in TCM is based primarily on the diagnosis and TDS. Third, TCM concerns the disease from various dynamic functional aspects as outlined from a nonlinear point of view. TCM is a comprehensive system for the assessment and treatment of acute and chronic disorders, as well as preventative health care and maintenance. TCM focuses on the patient rather than the disease, promoting health and enhancing the quality of life, with therapeutic holistic strategies. TCM advocates combinatory therapeutic strategies by formulae, typically, formulae are comprised by several types of crud herbs or minerals, in which one represents the key component, and others serve as adjuvant ones to assist the effects or facilitate the delivery of the key component. More specifically, according to the rules of TCM theory, the famous formulae include four elements: the monarch (which plays the most important role in the formulae); the minister (which increases the effectiveness of the monarch herb); the assistant (which helps the monarch and minister herbs reach their target positions); and the servant (which can reduce the adverse effects and/or increase the potency of the whole formulae). The therapeutic efficacy of TCM usually attributed to the synergistic property, of minimizing adverse reactions or improving the therapeutic efficacy, among multiple herbs and constituents which were named as the "formulae compatibility" of TCM (Zhang et al., 2014). It is recognized that there are existing principles of formulae compatibility which also have the objective material basis that could be revealed by certain established methods.

At least 11,146 plant species representing 2,309 genera and 383 families, 1,581 animal species, and 80 minerals and substances, ranging from precious stones to mineralized fossils, are used as TCM medicines. Many TCM derived products developed in China or the USA (PHY906W as an adjuvant cancer remedy) have been approved for clinical trials by the US Food and Drug Administration (US FDA). TCM is a complex system that contains lots of components with diversities in chemical structures, biological activities and interactions among compounds, and the content for each component varies greatly. In the past few years, the pharmaceutical industry has a shift from the search of disease treatments with the "1 disease-1 target-1drug" and the "1-drug-fits-all" concepts to the pursuit of combination therapy that comprises of more than one active ingredient. Combination medicine with multi-effect-pathways-target have better efficacy than a single drug alone, and can change the current predicament state in the therapy. A single drug acting on an individual molecular target usually exerts unsatisfying therapeutic effects when used in diseases of complicated causes, such as tumor, diabetes, and inflammation. Modern medicine has acknowledged for quite some time the usefulness of combination therapies of formulae which regulate multiple nodes of the disease network simultaneously and show synergistic effect in the treatment of multifactorial diseases. Compared to chemical drugs, TCM have multitarget and multicomponent characteristics when treating a disease. That means to say, the healing efficacy of a TCM depends on the combined action of multiple components because it usually contains a lot of ingredients. The combination application of TCM can achieve a synergistic interaction capable of yielding a sufficient effect at low doses that has increased significantly in recent years (Zhang et al., 2011). Many Chinese therapeutic herbs traditionally used in the cotreatment, but not monotreatment series, demonstrate significantly better pharmacological effects. Like almost all other traditional medicines, at present, formulae are still practicing in its original form and although it has been effective in treating many conditions, it lacks necessary scientific research. Differing from the single component orientation of Western drugs – formulae is a complex system consisting of multiple compounds, and can cause many difficulties to control the quality, and in the search for the effective substance(s) of TCM, which make scientific evaluation of TCM more difficult. There have been many recent attempts to address these issues but most of them were still based on the "reductionism" philosophy. There is a lack of a comprehensive understanding of the relationship between syndrome and formulae; this poses a difficulty for the advancement of TCM. Fortunately, the chinmedomics approach may beneficially contribute to explain the biological mechanism of syndrome and evaluating the efficacy of formulae, and thus shed new light on the effectiveness of formulae to find effective substances, and subsequently to elucidate the scientific value of TCM.

1.3 METABOLOMICS: AN OVERVIEW

Metabolomics and the related metabonomics approach is a specific "top-down" systems biology manifestation of metabolic profiling and is defined as "the quantitative measurement of the dynamic multiparametric metabolic response of living systems to pathophysiological stimuli or genetic modification" (Nicholson and Lindon, 2008). The small-molecule metabolites including lipids, amino acids, peptides, nucleic acids, and organic acids, vitamins, thiols, carbohydrates, etc., have an important role in biological systems and represent attractive candidates to understand disease phenotypes. Metabolomics has recently demonstrated significant potential in many fields such as responses to environmental stress, toxicology, nutrition, studying global effects of genetic manipulation, cancer, comparing different growth stages, diabetes, disease diagnosis, and natural product discovery (Wang et al., 2011). It is now finding applications that span almost the full length of the drug discovery and development pipeline, from lead compound discovery to postapproval drug surveillance (Clayton et al., 2006). In contrast to classical biochemical approaches that often focus on single metabolites, single metabolic reactions and their kinetic properties, and/or defined sets of linked reactions, metabolomics involves the collection of quantitative data on a broad series of metabolites in an attempt to gain an overall understanding of metabolism, and/or metabolic dynamics. Metabolomics promises to have the potential to contribute significantly to biomedical research and, ultimately, to clinical medical practice, because it can capture information with regard to mechanisms of disease and of drug action. The rapid development of metabolomics technology platforms could provide more rapid, direct, concise, and effective methods for disease research, especially for exploring the essence of Chinese medicine syndromes. It provides a methodological basis for a deep understanding of the essence of Chinese medicine syndromes from the aspect of systems biology. Presently, metabolomics has been used to explore the particular metabolites, potential biomarkers and pathways of TCM syndromes, and these findings show the robust metabolomics platform opens new perspectives to resolve special TCM issues (Zhang et al., 2013).

A typical metabolomics study is depicted schematically in Fig. 1.2. Samples of interest (e.g., plasma, cerebral spinal fluid, or tissue biopsies) are collected. Blood samples are the most commonly collected biological fluid, providing a representation of the global "metabolome" at the time of collection. Urine collection offers a low cost alternative (or complementary) biological sampling strategy to blood; it has the advantage of providing a time-averaged pattern of the end products of metabolism as well as capturing the excretion of exogenous compounds. Unlike earlier analytical methods, metabolomics utilizes instruments that can simultaneously quantitate thousands of small molecules in a biological sample. The application of analytical technologies to the measurement of low-molecular-weight metabolites that could offer deeper insights into mechanisms of the specific diseases, has helped illuminate the effects of drug perturbations. Those analytical techniques include: gas chromatography (GC), liquid chromatography (LC), and capillary electrophoresis (CE) coupled to mass spectrometry (MS) and nuclear magnetic resonance (NMR) spectroscopy, that could enable separation, detection, characterization, and quantification of such metabolites and related metabolic pathways (Zhang et al., 2012b). No single analytical method is capable of obtaining a complete molecular fingerprint of the collected biofluid specimen; instead, integrated platforms have been frequently used to provide the sensitive and reproducible detection of thousands of metabolites in a biofluid sample. Owing to the complexity of metabolites and the large number of metabolites therein, combinations of different analytical technologies can be used to gain a broad perspective of the metabolome and augment separations and/or to expand the analyte information collected. Continued development of these analytical platforms will also accelerate widespread use of metabolomics into systems biomedicine.

Analytical platforms for metabolomic studies must then be joined to sophisticated mathematical tools that can identify a molecular signal from biological data sets. However, automated identification and quantification is complicated by the large number of peaks. Using pattern recognition approaches such as "unsupervised" principal component analysis (PCA) and "supervised" partial least squares-discriminant analysis (PLS-DA), metabolomics could provide a comprehensive assessment of the metabolites. Figure 1.2 gives an overview of the data analysis workflow and salient questions that must be considered. Potential biomarkers which are very useful for diagnosing and monitoring disease progress, were extracted from loading-plots constructed following the PLS-DA analysis, and biomarkers were chosen based on their contribution to the variation and correlation within the data set. The development of analytical technologies will aid the discovery of novel biomarkers, hopefully leading to the early detection of various diseases. Metabolite identification, biologic interpretation, and clinical validation are key components of metabolomic studies. Furthermore, public pathway resources such as HMDB (http://www.hmdb.ca/), KEGG (http://www.genome.jp/kegg/), METLIN (http://metlin.scripps.edu/), Massbank (http://www.massbank.jp), LIPIDMAPS (http://www.lipidmaps.org/) and Chemspider (http://www.chemspider.com), MetaboAnalyst (http://www.metaboanalyst.ca) are useful for exploring the link between identified metabolites, and functional and regulatory processes within known metabolic pathways.

1.4 PROCESS OF THE SERUM PHARMACOCHEMISTRY OF TCM

TCM has its own unique medical system with the significant characteristics of pursuing an overall therapeutic effect with a multitarget treatment. It consists of multiple medicinal plants (called "formulae:") that can regulate balance and homeostasis of the body in a holistic fashion, and exert synergistic effects via the interaction of individual metabolites in the formulae. Due to the complexity of the chemical compositions and many endogenous intervening substances, it is challenging for the identification of chemical constituents and metabolites related to TCM. Conventional, bioactivity method guided extraction and separation fails to reflect the overall efficacy of formulae. Because, it has a low collecting rate, activity losing, only structure elucidation; does not reflect the biotransformation and express the drug interaction in the formulae *in vivo*. This approach ignores the fact that in the vast majority of cases the clinical administration of TCM is through oral administration. After being metabolized by the gastrointestinal tract and liver, the components that finally reach the blood are usually not the original drug ingredients. There are also some metabolism ingredients. Many ingredients in TCM can only produce an effect after absorption into the bloodstream. Theses blood contained ingredients may then be the substances playing a direct role in the body. Research on the components in the serum could be an effective approach to identify the effective substance(s). To resolve these problems, in early 1990s, we first established the Serum Pharmacochemistry of the TCM method that can reflect drug action, absorption, distribution, and interaction in the body. The Serum Pharmacochemistry of TCM was as a preliminary defined as isolated, and identified the efficacy material basis from serum containing active constituents after oral formulae, reflecting the *in vivo* state of the therapeutic substance of formulae, effectively guiding modern, and innovative drug design. With development of more than 20 years, it has become the key technology for analysis and screening active ingredients of TCM *in vivo*. It provides a methodology for the discovery of active constituents *in vivo* from TCM, solving the efficacy and effectiveness of TCM.

An ideal assay of Serum Pharmacochemistry of TCM for herbal medicines should simultaneously qualitatively identify and characterize "what are absorbed (the bioavailable phytochemical compounds)" and "what are produced (the new compounds produced through biotransformation)" and present their quantitative or semiquantitative time courses *in vivo*. In the mean time, a time-dependent response of biochemical substances, to the multicomponent intervention should be identified. In recent years, many analytical methods have been successfully applied in the Serum Pharmacochemistry of TCM for analysis of the TCM formulae. Liquid chromatography coupled with tandem mass spectrometry (LC–MS) was widely applied for the analysis of herbal constituents *in vivo*. Especially, ultraperformance liquid chromatography with minor particles (sub 1.8 μm) has greatly improved the sensitivity, resolution and speed, integrating with MS to form a high-performance platform for TCM. MetaboLynx™ data processing approach is concerned with the analysis and interpretation of complex data variables. It is of significance to systematically analyze the chemical constituents and *in vivo* metabolites in TCM formulae so as to interpret its material basis for pharmacological effects. The advent of high throughput biochemical analyses coupled with chemo-informatics has allowed us to address many technical issues involved in the *in vivo* evaluation of multi components. This proposed Serum Pharmacochemistry of TCM approach has been successfully applied on searching multicomponent and screening active ingredients from Chinese medicinal formulae such as Yinchenhao Tang, Liuweidihuang Wan, Shengmai San, and Kaixin San, etc. For the establishment and implementation of Serum Pharmacochemistry of TCM, we had won the 2002 National Science and Technology Progress Second Award of China. The research achievements in this area have been increasing in the last 15 years, with a lot of articles pertaining to research on TCM having been published in the last 10 years in Science Citation Index-cited journals such as *J Ethnopharmacol, J Sep Sci., Analyst,* and *Biomed Chromatogr, etc.* In the field of TCM research, there are approximately 108 projects related with the Serum Pharmacochemistry of TCM funded by the China National Natural Science Foundation from the year 2002.

1.5 ESTABLISHMENT AND EVOLUTION OF CHINMEDOMICS

Syndrome and formulae are two key issues in TCM and the premise finding for effective constituents of TCM. However, vagueness of syndromes and complexity of formulae make the evaluation to syndromes and efficacy of TCM formulae somewhat difficult, and as a consequence greatly limits the research regarding the effective substances of TCM. Therefore, the biological essence of syndrome and effect of formulae will be the key scientific problems for the discovery of the effective substance of TCM. The Serum Pharmacochemistry of TCM only considers the chemical composition qualitatively or quantitatively, while the efficacy is not taken into account. Only in respect of the premise of the curative effect of TCM, the characteristic chemicals of a TCM *in vivo* will be related to its biological activity to better reflect the effective materials of TCM. Therefore, we designed the Serum Pharmacochemistry of TCM to be integrated with metabolomics, and developed a new chinmedomics method that may provide the possibility to compensate for the abovementioned shortcomings of the Serum Pharmacochemistry of TCM. It could reflect the real and comprehensive pharmacological information of active constituents. Metabolomics is used to clarify the biological mechanism of the TCM syndrome or disease, and the Serum Pharmacochemistry of TCM is to found the chemical substances in the body after oral formulae. On the premise of the curative effect of formulae, correlation analysis between the exogenous ingredients *in vivo* after oral formulae and endogenous marker metabolites *in vivo* are used to clarify the effective substances and their synergistic properties.

As a hotspot of TCM studies, the chinmedomics method research has gradually drawn extensive attention and has been supposed to provide a way to clarify the active materials of TCM. There are a variety of methods to establish the chemical fingerprint of the TCM and the serum fingerprint of the TCM, including thin layer chromatography, high-performance liquid chromatography, gas chromatography, high speed counter current chromatography, mass spectroscopy and capillary electrophoresis. To overcome the shortcomings of fingerprints, it uses data-processing technology to analyze the weight coefficients of components, monitoring efficacy. In recent years, scholars have been constantly improving the data processing methods applied in establishing the chinmedomics. The processing methods used are not uniform, mainly including correlation analysis, cluster analysis, regression analysis, pattern-recognition analysis (principal component analysis, partial least-squares-discriminant analysis, and orthogonal partial least-squared discriminant analysis, etc). Various mathematical processing methods using combined can be a complement to each other, insuring information maximization and improving accuracy. Through the analytic statistics method, chinmedomics could establish a mathematical model of a TCM's chemical composition and efficacy; the chinmedomics technique uses informatics methods to find the alignment of key active compounds; it optimizes the compatibility of TCM through the forecasting model of multi-component effects; it also explores the synergy or antagonism mechanisms of the variety of aligned TCM active compounds.

On the basis of the biological characterization of syndrome, we will establish metabolic profiling of animal models of the TCM syndrome, and the related metabolic fingerprints as well as metabolic biomarkers, to evaluate the overall effects of the TCM formulae and the corresponding relationship between syndrome and formulae. Under the premise of therapeutic efficacy, an optimal plotting of a correlation model between the marker metabolites and chemical composition (PCMS) was used to screen the constituents or metabolites absorbed into the blood postoral administration of the TCM formulae. To determine which constituents contributed to the mainly therapeutic effect of the TCM formulae, for example, PCMS was performed using Pearson's correlation, and the correlation coefficient (r) described the degree of correlation between two variables: all the biomarkers and the chemical composition of the TCM formulae *in vivo*. The r values were between −1 and +1; $r > 0$ indicating that two variables were positively correlated, whereas $r < 0$ indicates a negative correlation. The larger the absolute value, the higher the correlation; therefore r = 0.90–1.00 was defined as extremely related and r = 0.80–0.89 as highly related. According to the correlation coefficient, the absorbed chemical components that were positively, highly, and negatively correlated to marker metabolites, contributed most of the therapeutic effect. This can reflect not only the pharmacological effects of the components coming into the body and their metabolites, but also the role played by any formulae-induced endogenous components. The PCMS model should follow the principle of reasonability, effectiveness, and accuracy and has been chosen carefully so as to evaluate the pharmacological activity of the components more precisely; the weight coefficients of the components contributing to the efficacy were analyzed and then effective ingredient groups with synergism or antagonism effects can be sought out to further reveal the TCM effective material basis. Appropriate multi-objective optimization strategy of the components combination from formulae should be established to search potential optimal combinations of multicomponent drug candidates. Being an interdisciplinary and cutting-edge science, the chinmedomics method is a technology integrating Chinese medicinal chemistry, analytical chemistry, the Serum Pharmacochemistry of TCM, the metabolomics of TCM and chemometrics, so it is considered a more accurate way to study TCM's active materials. Nowadays, it has thus become a hotspot in TCM studies. Overall, chinmedomics based on the metabolic profiling, fingerprinting, biomarkers and pathways of the TCM syndrome; further provide for the discovery of active constituents *in vivo* from formulae, and open the door for the modernization of the TCM industry.

1.6 POTENTIAL VALUE OF CHINMEDOMICS

As a systemic approach, chinmedomics also reflects the function of organisms from the terminal symptoms of the metabolic changes of a complete system caused by formulae interventions in a holistic context. Its property consists within the holistic thinking of TCM, suggesting that chinmedomics has the potential to impact our understanding of the action mode of formulae. Consequently, the development of robust chinmedomics platforms will greatly facilitate various applications for formulae. The opportunities presented by chinmedomics will be a new revolution in TCM. Chinmedomics can be applied at any stage in the following settings: (1) predictive biomarkers for a syndrome in animal models; (2) understanding of the biochemical mechanisms of formulae action to target organ, or to target organ pathologies from animal to man; (3) developing biomarkers for formulae in nonclinical development; and (4) predictive biomarkers for formulae-related effects in humans during clinical trials. Chinmedomics performs studies focussed on the complex interactions of system components of formulae, emphasizing the whole system rather than the individual parts.

Technological advances have opened a new chapter in TCM by using chinmedomics as an approach to study the metabolism of the formulae. Recent advances in the two analytical platforms of MS and NMR have driven forward the discipline of chinmedomics. NMR is one of the most commonly used technologies in omics research, providing detailed information on the probing metabolites. The high selectivity of MS with low-detection limits makes MS an ideal tool for metabolomic applications. It requires separation of the metabolic components using either GC or LC. Ultra high performance liquid chromatography/mass spectrometry (UPLC/MS) is often used to obtain the largest possible biochemical profile information subset. It is a sensitive tool that can be used to characterize, identify, and quantify a large number of compounds in a biological sample where metabolite concentrations might cover a broad range of information with regard to disease pathophysiology. All chinmedomics studies result in complex multivariate datasets that require visualization software and chemometric and bioinformatic methods for interpretation. The application of software tools for the analysis of the information contained in a database can identify the signature of the syndrome.

TCM performs treatment based on TDS, which could be identified in clinical special phenotypes by patients' symptoms (Sun et al., 2012). With syndrome as the core of diagnosis and therapy in TCM, it has the advantage of collecting macroscopic information on patients for diagnosis. The effectiveness of TCM treatment depends on the accuracy of TDS. However, it lacks objectivity and repeatability and has caused controversy, because syndrome classification only depends on observation and the clinical experience of TCM doctors. Chinmedomics could provide useful tools for exploring the essence of syndromes. Following development of chinmedomics technologies, it has been more and more applied in the study of objectively differentiating TCM syndrome and understanding its biological mechanisms. The uniqueness and effectiveness of TCM in disease treatment would have a significant impact on people's health worldwide. The application of the chinmedomics approach, constituting a novel platform of TCM study, will provide novel insights into the action mechanism of TCM. This process will not only improve our understanding of TCM, but further explore the application of formulae in the discovery and development of new drugs and therapies for diseases responsive to TCM treatment.

Chinmedomics would thus provide a novel tool for the validation of TCM therapeutic effects and the enrichment of modern medicine. It is believed that chinmedomics will greatly promote Chinese medicine research and be beneficial to the modernization of formulae in terms of extending the application of modern means. In the holistic and systemic context, chinmedomics has a convergence with TCM, which could overcome the one-sidedness of TCM, and avoid both complications and tedious processes. Chinmedomics opens up "dialogue" between Chinese and Western medicine, it is conceivable that the application of chinmedomics technologies will eventually lead to the reconciliation and integration of Chinese and Western medicine. Overall, integration of chinmedomics into TCM would make it possible to explore drug discovery and development of TCM, the molecular mechanism of therapeutic effects of herbal medicine with multiple targets.

1.7 FUTURE PERSPECTIVES

Although many TCM have been recognized for the treatment of various diseases, more information is needed, including action mechanisms, therapeutic effects, toxicological evaluation, and the safety profile of TCM. The actual value of TCM has not been fully recognized worldwide due to the lack of scientific approaches. Progress in chinmedomics has opened new gateways in therapeutics and drug discovery and development. As a new research approach, chinmedomics, cannot only interpret the essence of the syndrome but also elucidates the scientific connotation of formulae. With the global application of chinmedomics, more and more natural resources and products from formulae are being studied and recognized. The chinmedomics approach as pillars of the potential bridge – may both beneficially influence and provide an opportunity to explain, the theoretical meaning of evidence-based TCM. Chinese medicine has a wealth of experience and chinmedomics has a substantial research potential, the integration of both in the systemic context will bring a greater and deeper understanding of TCM. The combining chinmedomics approach with TCM into modern health care systems may lead to a revolution in TCM therapy. Although the subject is only at an early stage and requires further exemplification and validation, the approach has major implications for TCM efficiency improvements.

References

Clayton, T.A., Lindon, J.C., Cloarec, O., Antti, H., Charuel, C., Hanton, G., Provost, J.P., Le Net, J.L., Baker, D., Walley, R.J., Everett, J.R., Nicholson, J.K., 2006. Pharmaco-metabonomic phenotyping and personalized drug treatment. Nature 440 (7087), 1073–1077.

Nicholson, J.K., Lindon, J.C., 2008. Systems biology: metabonomics. Nature 455 (7216), 1054–1056.

Sun, H., Zhang, A., Wang, X., 2012. Potential role of metabolomic approaches for Chinese medicine syndromes and herbal medicine. Phytother. Res. 26 (10), 1466–1471.

Wang, X., Sun, H., Zhang, A., Sun, W., Wang, P., Wang, Z., 2011. Potential role of metabolomics apporoaches in the area of traditional Chinese medicine: as pillars of the bridge between Chinese and Western medicine. J. Pharm. Biomed. Anal. 55 (5), 859–868.

Wang, X., Zhang, A., Sun, H., Wang, P., 2012a. Systems biology technologies enable personalized traditional Chinese medicine: a systematic review. Am. J. Chin. Med. 40 (6), 1109–1122.

Wang, X., Zhang, A., Sun, H., 2012b. Future perspectives of Chinese medical formulae: chinmedomics as an effector. OMICS 16 (7–8), 414–421.

Zhang, A., Sun, H., Wang, Z., Sun, W., Wang, P., Wang, X., 2010. Metabolomics: towards understanding traditional Chinese medicine. Planta Med. 76 (17), 2026–2035.

Zhang, A., Sun, H., Yuan, Y., Sun, W., Jiao, G., Wang, X., 2011. An *in vivo* analysis of the therapeutic and synergistic properties of Chinese medicinal formula Yin-Chen-Hao-Tang based on its active constituents. Fitoterapia 82 (8), 1160–1168.

Zhang, A., Sun, H., Wang, P., Han, Y., Wang, X., 2012a. Future perspectives of personalized medicine in traditional Chinese medicine: a systems biology approach. Complement. Ther. Med. 20 (1–2), 93–99.

Zhang, A., Sun, H., Wang, P., Han, Y., Wang, X., 2012b. Modern analytical techniques in metabolomics analysis. Analyst 137 (2), 293–300.

Zhang, A.H., Sun, H., Qiu, S., Wang, X.J., 2013. Recent highlights of metabolomics in chinese medicine syndrome research. Evid Based Complement. Alternat. Med. 2013, 402159.

Zhang, A., Sun, H., Wang, X., 2014. Potentiating therapeutic effects by enhancing synergism based on active constituents from traditional medicine. Phytother. Res. 28 (4), 526–533.

CHAPTER 2

Methods and Protocols of Chinmedomics

Xijun Wang, Aihua Zhang, Hui Sun, Ying Han

2.1 INTRODUCTION

Chinmedomics, the primary definition is an elucidation of the therapeutic, synergistic properties, and the metabolism of Traditional Chinese Medicine (TCM) formulae and related metabolic pathways using modern analytical techniques, through the integration of Serum Pharmacochemistry and Metabolomics to elucidate the scientific value of TCM; this elucidation has recently showed significant potential in the TCM field (Wang et al., 2012a). Briefly, metabolomics technology is used to clarify the biological mechanism of syndrome, and the Serum Pharmacochemistry of TCM is to discover the potential active substances in the body after oral TCM. Under the premise of the efficacy and effectiveness of formulae, correlation analysis between the exogenous ingredients *in vivo* after oral formulae and endogenous marker metabolites is to clarify the effective substances that are highly associated with the overall effect of the formulae. Adoption of the chinmedomics approach would assist in exploring the scientific connotation of the TCM syndrome and the function of formulae, and forming linking bridge between TCM and modern medicine. The Chinmedomics approach is considered to have the potential to revolutionize TCM research, and to advance the development of formulae. It holds the promise of a comprehensive, noninvasive analysis of metabolic biomarkers that could help to monitor treatment response and drug discovery of formulae. Some new molecular targets and signaling pathways can be identified through the chinmedomics analysis, which could then become a platform for the development of new therapies or drugs.

The introduction of the concept of chinmedomics, enables the study of low-molecular-weight metabolites in living systems from a holistic perspective based on the metabolic profiling of syndrome. Metabolites are important indicators of physiological or pathological states, and can provide information for the identification of biomarkers for syndrome, and assist in the understanding of its occurrence and progression (Wang et al., 2012b). These small molecules, including peptides, amino acids, nucleic acids, carbohydrates, organic acids,

CONTENTS

Chinmedomics. http://dx.doi.org/10.1016/B978-0-12-803117-9.00002-0

vitamins, polyphenols, alkaloids and inorganic, are small-molecule biomarkers that represent the functional phenotype (Wang et al., 2013a). In recent years, the hyphenated analytical techniques such as liquid chromatography-mass spectrometry (LC–MS), gas chromatography-mass spectrometry (GC–MS), capillary electrophoresis-mass spectrometry (CE–MS), and matrix assisted laser desorption ionization-mass spectrometry, etc. often leads to better analyte coverage of the metabolome. Bioinformation derived from these techniques could provide accurate and clinically useful diagnostic capability for the management of diseases at the metabolic level.

2.2 ANALYTICAL TECHNOLOGIES

Separation and identification of the small molecules has been made possible by technological advances. These innovational technologies including accurate measurement of high-resolution MS and NMR technology, can accomplish detection of metabolites within a few minutes. Measuring such low-molecular-weight metabolites could offer deeper insights into mechanisms of the specific diseases. Metabolites are biological characteristics that are objectively measured and evaluated as indicators of normal biological, and pathological processes or pharmacologic responses to a therapeutic intervention, widely used in clinical practice. The application of analytical technologies to the measurement of metabolites has helped illuminate the effects of drug perturbations. Owing to the complexity of metabolites, and the large number of metabolites therein, advanced and high-throughput separation techniques have been coupled to high-resolution MS platforms that are frequently used to provide the sensitive, and reproducible detection of hundreds to thousands of metabolites in biofluids.

Every technique has its advantages and drawbacks, no existing analytical technique can be versatile. NMR has many advantages, but the sensitivity of NMR is relatively poor compared with MS methods, and concentrations of potential biomarkers may be below the detection limit. GC/MS requires sample derivatization to create volatile compounds. Nonvolatile compounds that do not derivatize and large or thermo-labile compounds will not be observed in the GC/MS analysis. The recent introduction of UPLC, employs porous particles with internal diameters smaller than 2 μm, in combination with MS, resulting in higher peak capacity, improved resolution, and increased sensitivity compared to conventional LC columns, therefore, making it even more suitable for a metabolomics approach (Wang et al., 2011). In view of the recent developments in the separation sciences, leading to the advent of UPLC and MS-based technologies showing the ever-improving resolution of metabolite species and the precision of mass measurements. Each technique has associated advantages and disadvantages. Clearly, no single analytical methodology is ideal for all of the many thousands of metabolites in a biological system; instead, integrated platforms have been frequently used to provide the sensitive and reproducible detection of thousands of metabolites in a biofluid sample. Thus, a combination of different analytical technologies must be used to gain a broad perspective of the metabolome.

2.3 SAMPLING

Metabolomics represents a comprehensive method for metabolite assessment that involves measuring the overall metabolites of biological samples. It enables the parallel assessment of the levels of a broad range of endogenous and exogenous metabolites, and has been shown to have a great impact in the investigation of physiological status, diagnosing diseases, discovering biomarkers, and identifying perturbed pathways. The small-molecule metabolites have an important role in biological systems and represent attractive candidates to understand disease phenotypes. It represent a diverse group of low-molecular-weight structures: including lipids, amino acids, peptides, nucleic acids, organic acids, vitamins, thiols, and carbohydrates. Metabolite changes that are observed in diseased individuals as a primary indicator; have become possible, and hence the measurement of metabolites has become an important part of clinical practice. Employing a range of biofluids including: plasma and whole blood, serum, urine, saliva, cerebrospinal fluid, synovial fluid, semen, and tissue homogenates has a number of advantages, that have ensured the widespread use of metabolites as a diagnostic tool in clinical practice (Zhang et al., 2012a). Metabolic profiling of biofluids and tissues can provide a panoramic view of abundance changes in endogenous metabolites when monitoring cellular responses to perturbations, such as diseases and drug treatments. Precise identification and the accurate quantification of metabolites facilitates downstream pathway and network analysis for the further discovery of clinically accessible and minimally invasive biomarkers of TCM efficacy and toxicity. On the other hand, chinmedomics offers potential advantages that classical diagnostics approaches do not, based on following the discovery of a suite of clinically relevant biomarkers that are simultaneously affected by the disease.

2.4 DATA EXTRACTION AND ANALYSIS

Owing to the complexity and volume of the data generated by today's advanced technologies, pattern recognition methods have become predominant in the medical sciences, and may be suitable for certain diagnostics medical applications. Metabolomics data preprocessing workflow (see Fig 2.1). By applying advanced analytical and statistical tools, the metabolome is mined for biomarkers that are associated with the state of diseases. Multivariate statistical

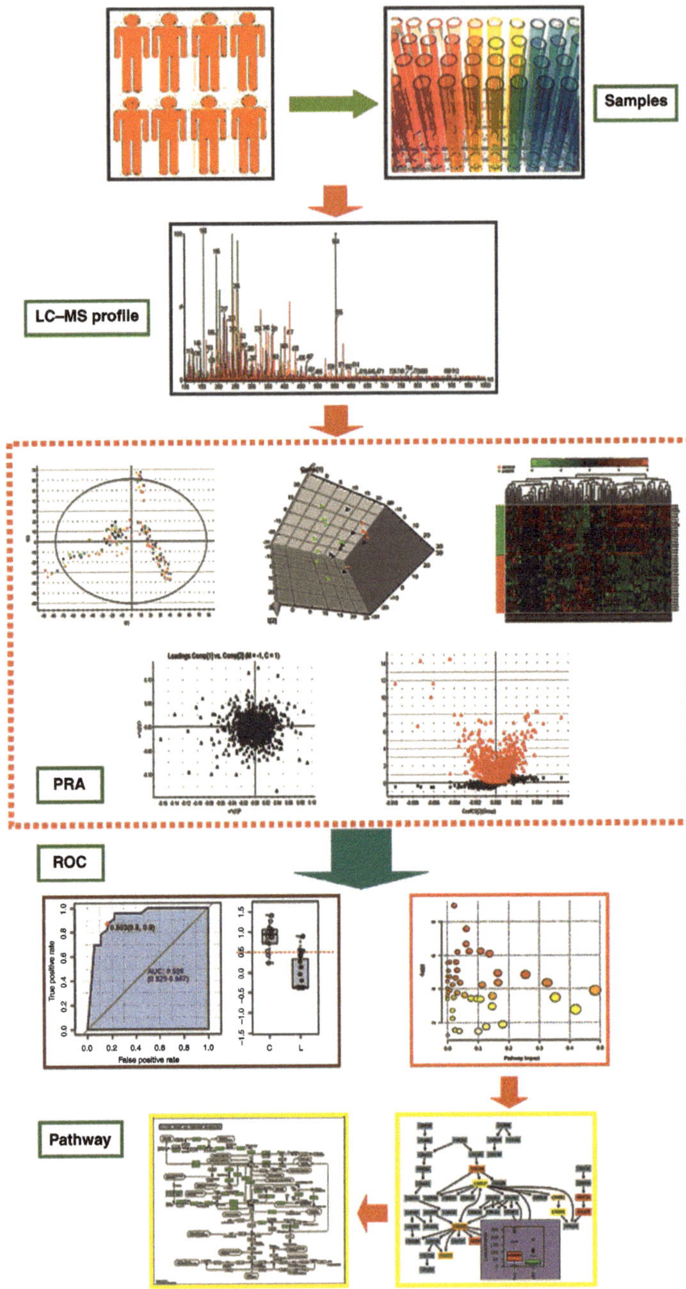

FIGURE 2.1

Metabolomics data preprocessing workflow. Samples collected in a uniform manner to minimize variability, are analysed by full scan MS profile to collect data on all metabolites potentially present in the sample. Pattern recognition approach (PRA) includes PCA, PLS-DA, OPLS, Heatmap, support vector machines and random forests method and other modes aiming to highlight underlying trends and visualization tools such as contribution, trend and box plots that are used to further evaluate these. Receiver operating characteristic curves (ROC) are generally considered the method of choice for evaluating the performance of potential biomarkers. The markers are eventually placed in a metabolic pathway to provide insight into the biochemical phenomena.

methods and pattern-recognition programs were developed to handle the acquired data, and to search for the discriminating features between data acquired from two sample sets: healthy and diseased. Using pattern recognition approaches, such as "unsupervised" principal component analysis (PCA) and "supervised" partial least squares-discriminant analysis (PLS–DA), metabolomics could provide a comprehensive assessment of the metabolites (Wang et al., 2012c). The most commonly used unsupervised technique for identifying patterns and trends in metabolomics data is PCA, that is a data analysis method that can deal with large volumes of data, and this is often used as a starting point for analysis. It attempts to reduce multidimensional data so that it can be plotted in a two- or three-dimensional system. In the PCA scores, each point represents an individual sample. The PCA results are displayed as score plots indicating the scatter of the samples, which indicate similar metabolomic compositions when clustered together and a compositionally different metabolome when dispersed. PCA is usually followed by a supervised learning technique, such as PLS–DA. It attempts to maximize the covariance between the independent and dependent variables to discriminate amongst samples, and this method often discriminates between classes more effectively than PCA. The purpose of the PLS–DA analysis was to calculate models that differentiate among groups. In the PLS–DA modeling, the samples from the different groups were sorted into different classes using score plots, and endogenous metabolites that contribute to the classification were identified in loading plots, which showed the importance of each variable to the classification. Potential biomarkers were extracted from loading-plots constructed following the PLS–DA analysis, and biomarkers were chosen based on their contribution to the variation and correlation within the data set. A specialized form of PLS–DA is orthogonal projections to latent structures (OPLS) in which noncorrelated systematic variance is removed from the model. Many other supervised methods have been used to analyze metabolomics data such as artificial neural networks, and classification and regression trees, support vector machines, and random forests method, etc.

2.5 MARKER IDENTIFICATION AND VALIDATION

Biomarkers are very useful for diagnosing and monitoring disease progression and are important for patient selection, aiding selection of appropriate patient treatment. High-resolution MS is highly discriminatory for a range of pathological processes affecting the diseases. The use of the novel MS tools to study metabolism provides a unique insight into disease progression. For the clinical studies of disease therapeutics, there is an increasing need for diagnostic markers to ensure that therapies are targeted at the right patient. Metabolomics is the study of a unique fingerprint of small molecules present in biological systems under healthy and disease conditions. One of the major challenges in metabolomics is validation of fingerprint molecules to identify specifically perturbed pathways in metabolic aberrations. The identification and interpretation of metabolic biomarkers is a challenging task. Another one is certainly lab-to-lab reproducibility. To address it, protocols should be as simple and adaptable as possible. Most of the identified biomarkers have failed to replace existing clinical tests. While the past decade has seen great advancements in metabolomics research producing potential biomarkers, to become a clinically approved test, a potential biomarker should be confirmed and validated using hundreds of specimens, and should be reproducible, specific, and sensitive (Zhang et al., 2012b). The quest for an ideal biomarker in disease is still underway. Since at present we cannot apply the application of the perfect biomarkers, it is possible that combining different molecules will provide information that compensates for the shortcoming of individual tests. The development of analytical technologies will aid the discovery of novel biomarkers, hopefully leading to the early detection of various diseases.

2.6 CORRELATION ANALYSIS BETWEEN MARKER METABOLITES AND ABSORBED CHEMICAL COMPONENTS

Under the premise of therapeutic efficacy, a plotting of a correlation model between the marker metabolites, and chemical composition (PCMS) was used to screen the constituents or metabolites absorbed into blood after oral administration of the TCM formulae. To determine which constituents contributed to the mainly therapeutic effect, PCMS was performed using Pearson's correlation, and the correlation coefficient (r) described the degree of correlation between two variables: all the biomarkers and the chemical composition of the TCM formulae *in vivo*. The r values were between -1 and $+1$; $r > 0$ indicating that the two variables were positively correlated, whereas $r < 0$ indicates a negative correlation. The larger the absolute value, the higher the correlation; therefore $r = 0.90–1.00$ was defined as extremely related, and $r = 0.80–0.89$ as highly related. According to the correlation coefficient, the absorbed chemical components that were positively highly and negatively correlated to marker metabolites, contributed most of the therapeutic effect. The analysis can identify the best drug combination to display a synergy or attenuation effect. It can also probes into the active substance or the main drugs contained in a formula.

2.7 GOALS FOR CHINMEDOMICS

Through a series of studies on chinmedomics, we have improved the methodology for formulae and syndrome. The use of metabolomics technology aims to find biomarkers and the metabolic network of a syndrome or disease, and to objectively understand and evaluate the biological essence of the syndrome or disease. – Under the premise that the validity of biological effects of formulae and further integration research of the Serum Pharmacochemistry of TCM may determine the effective substances *in vivo* from formulae, and the correlation analysis of pharmaceutical (exogenous) ingredients, and marker metabolites (endogenous) in the syndrome or disease, and thereby finally locate the active constituents from TCM.

2.7.1 Biological Mechanisms of the TCM Syndrome

The essence of the research into the TCM syndrome is based on the essential premise for a standardized foundation of TCM, its indepth exploration of syndrome diagnosis or evaluation, and has practical clinical value that is a key way to establish TCM standardization and an evaluation system of the efficacy of TCM. The research on the biological mechanisms of the TCM syndrome using chinmedomics includes the following two aspects: (1) the use of metabolomics technology to identify biomarkers, and fully understand the mechanisms of syndromes; a biomarker as a bridge to establish the animal model associate with syndrome; and (2) application of high-throughput and high-resolution analytical instruments combined with the pattern recognition method and bioinformatics, to establish characteristic metabolic pathways of the TCM syndromes.

2.7.2 Pharmacological Effects of Formulae

TCM often uses formulae tailored to an individual's condition based on subjective diagnosis methods. With the wave of "returning to nature," people from all over the world are becoming more and more interested in the magical effect of the formulae. Increasing evidence demonstrates that, in treating illnesses, formulae can usually amplify the therapeutic efficacies of each agent, leading to maximal therapeutic efficacy with minimal adverse effects. The rapid development of chinmedomics, considered as one of the available and ideal means in this process, especially the advances in the high-throughput and comprehensive research technologies provide new strategies in the analysis of active components of the formulae. Research on the effective substance of formulae includes the following three aspects: (1) the establishment of the key technologies of the chinmedomics study on TCM formulae, carried out to determine the biomarkers on therapeutic effect (deeply into their biological pathways), and multidimensionally to characterize the overall efficacy of the formulae; (2) overall evaluation on effects of formulae, with the overall biomarkers as indicators, as the dynamic concentration levels to evaluate the effects of formulae; and (3) establishment of regulatory networks of TCM formulae – syndrome – the overall effects, indepth exploration of synergistic properties of formulae, and to promote the standardization and scientific dosing regimen.

2.7.3 Correlation Analysis of Formulae and Syndrome for Effective Substances

The establishment and implementation of chinmedomics assists in the exploration of the intrinsic link between the essence of syndrome and formulae efficacy: (1) identification and characterization of multicomponent *in vivo* postoral formulae, by the use of modern multidimensional chromatography technology; (2) based on the form of TCM, formulae as a starting point to analyze the therapeutic effect of TCM, and under the premise of the efficacy and effectiveness of formulae, used the Serum Pharmacochemistry of TCM combined with metabolomics to establish the PCMS correlation analysis method to screen the potentially effective substances with the highly associated overall effect of formulae; (3) through the biological validation on the cell or animal assay, determining the TCM effective material basis or key components; (4) the dynamic changes under the different compatibilities and establishing the relationship between the pharmacokinetics and pharmacodynamics of the key components, clarifying the compatibility principle and scientific connotation; and (5) building syndrome-key components-biomarkers' association networks, integrally clarifying the action mechanism of formulae, achieving the modern and innovative drug discovery based on TCM clinical experience.

2.8 CHINMEDOMICS FOR YINCHENHAO TANG: A CASE STUDY

Although the formulae were designed by ancient doctors, its mode of function can be revealed using the modern chinmedomics tool. One example is a famous TCM herbal formulae Yinchenhao Tang (YCHT), (Fig. 2.2A) whose efficacy in treating Jaundice syndrome (Huangdan Zheng, in Chinese). YCHT was recorded in "*Shanghanlun*" – a classic resource on TCM written by Zhongjing Zhang (150–215 AD). It consists of *Artemisia* annua L., *Gardenia* jasminoides Ellis, and *Rheum* Palmatum L. (Fig. 2.2B) and they work together harmoniously to achieve an ideally therapeutic effect for liver disorders. However, its precise mechanism and drug candidates remain very complex and poorly understood. The therapeutic and synergistic effects of YCHT were conducted using biochemistry, pharmacokinetics, and systems biology tools, to exploring the key molecular mechanisms. A completely validated method based on HPLC coupled with photodiode array detector was used for controlling quality of medical formula YCHT (Fig. 2.2C). As a result, fingerprint chromatogram determined 15 representative general fingerprint peaks. The fingerprinting method was used to identify the chemical constituents of YCHT *in vivo*. Forty-five compounds in YCHT (Fig. 2.2D) and 21 compounds *in vivo* (Fig. 2.2E) were identified by the established method on the Serum Pharmacochemistry of TCM (Wang et al., 2008). Metabolomics was incorporated to analyze the pharmacological

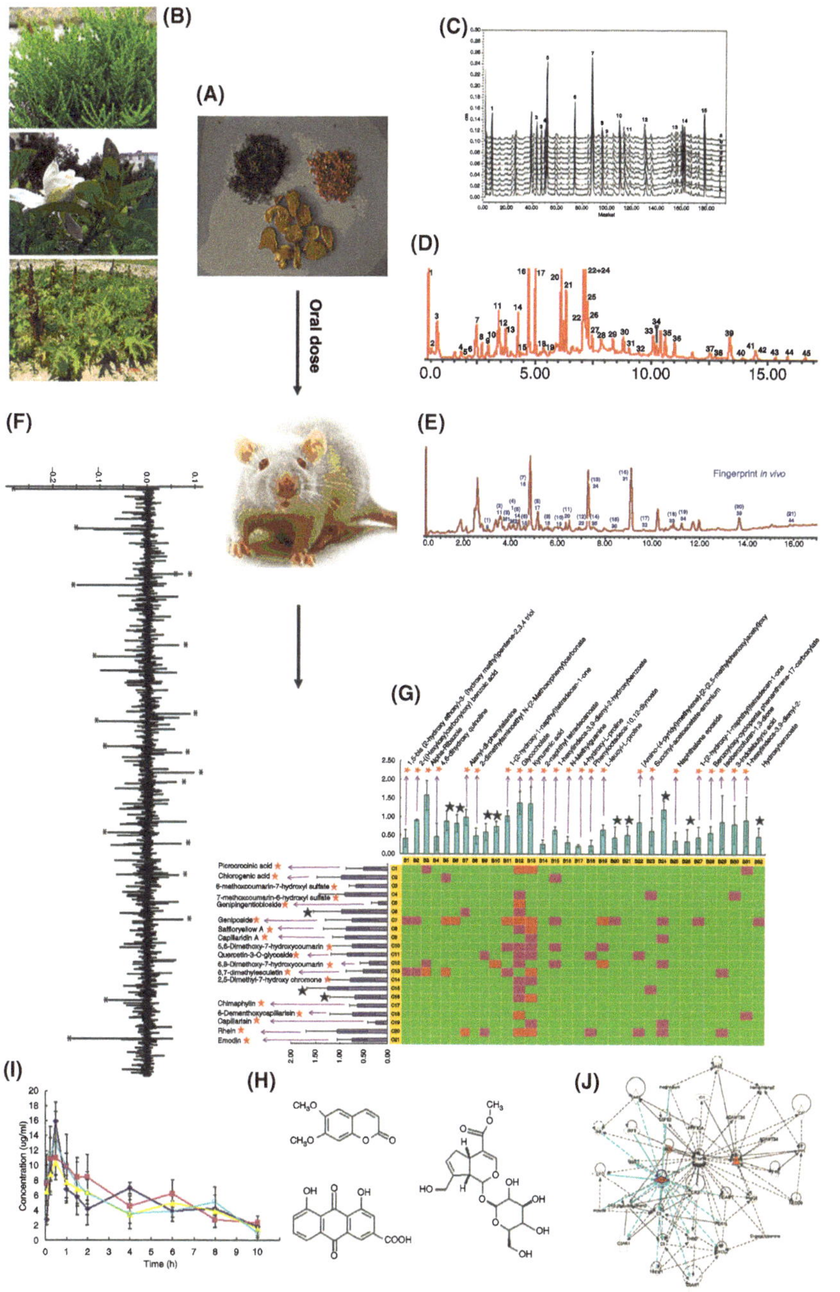

FIGURE 2.2

Advancing drug discovery and development from ancient Yinchenhao Tang, a famous TCM practitioner. (A) the flowers of *Artemisia capillaries* Thunb, the fruits of *Gardenia jasminoides* Ellis, and roots of *Rheum officinale* Baill; (B) prepared herbal medicines of Yinchenhao Tang; (C) quality control of YCHT samples by HPLC-UV fingerprint chromatogram method; (D) compounds derived from YCHT; (E) compounds *in vivo*; (F) marker metabolites; (G) correlation analysis of compounds *in vivo* and marker metabolites; (H) chemical structures of three drug candidates; (I) pharmacokinetic characteristics of major active components; (J) merged network combining major signaling networks.

effects of YCHT, and found 32 marker metabolites *in vivo* (Fig. 2.2F). By correlation analysis of these compounds *in vivo* and marker metabolites (Fig. 2.2G), three components (6,7-dimethylesculetin (D), geniposide (G) and rhein (R)) were found as drug candidates (Fig. 2.2H).

Continuing work was to elucidate the drug-interaction property of marker compound D, G, and R, based on the pharmacokinetic characteristics (Fig. 2.2I). The DGR combination could significantly increase the plasma level, and slow the elimination rate (more than any one or two of the three individual compounds) which may be an indication of a synergism (Wang et al., 2012b). Mapping merged major signaling networks, DGR combinations exert a more robust synergistic effect than any one or two of the three individual compounds by hitting multiple targets (Fig. 2.2J). They can regulate molecular networks through activating both intrinsic and extrinsic pathways to intensify therapeutic effects (Wang et al., 2013b). To test whether compounds D, G, and R and their combination had an effect equal to that of YCHT, the combination of these individual components was administered to primary hepatocytes alcohol-induced apoptosis. Interestingly, the antiapoptotic effect of the combination of D and G with R was markedly stronger than the efficacy of monotherapy with each of the three components individually of its constituents, and better than the original YCHT. The biochemical analysis and immunohistochemical assay further supports the cooperation of D and G with R in up-regulating BCL-2 and down-regulating FAS, with the strongest effect occurring with the DGR combination. At the levels of the metabolome, the underlining molecular mechanisms were explored, and it was found that DGR activated an array of factors that are involved in energy, amino acid, nucleotide, fatty acid, cofactor, and vitamin metabolism. The DGR targets not only immunity and metabolism, but also targets key regulatory pathways, thereby helping to restore the normal function.

2.9 CONCLUSIONS

Millennia-old TCM treats disease with many combination therapies involving ingredients used in clinical practice. Warranted or not, enthusiasm for the Chinmedomics techniques is irrepressible. Chinmedomics can be considered a useful pilot trial in the effort to evaluate traditional formulae on a larger scale, and to bridge Western and Eastern medicines in this era of systems biology. It may provide a promising design in natural products and new combination medicine pattern derived from TCM. We believe that the integration of chinmedomics principles into TCM would make it possible to interpret and explore the essence of the TCM syndrome, and may be the direction to enable a revolution for future health care. Perhaps, also it is time to embrace the arrival of the "TCM-OMICS" era in TCM research.

References

Wang, X., Sun, W., Sun, H., Lv, H., Wu, Z., Wang, P., Liu, L., Cao, H., 2008. Analysis of the constituents in the rat plasma after oral administration of Yin Chen Hao Tang by UPLC/Q-TOF-MS/MS. J. Pharm. Biomed. Anal. 46 (3), 477–490.

Wang, X., Sun, H., Zhang, A., Wang, P., Han, Y., 2011. Ultra-performance liquid chromatography coupled to mass spectrometry as a sensitive and powerful technology for metabolomic studies. J. Sep. Sci. 34 (24), 3451–3459.

Wang, X., Zhang, A., Sun, H., 2012a. Future perspectives of Chinese medical formulae: chinmedomics as an effector. OMICS 16 (7–8), 414–421.

Wang, X., Zhang, A., Han, Y., Wang, P., Sun, H., Song, G., Dong, T., Yuan, Y., Yuan, X., Zhang, M., Xie, N., Zhang, H., Dong, H., Dong, W., 2012b. Urine metabolomics analysis for biomarker discovery and detection of jaundice syndrome in patients with liver disease. Mol. Cell. Proteomics 11 (8), 370–380.

Wang, X., Yang, B., Sun, H., Zhang, A., 2012c. Pattern recognition approaches and computational systems tools for ultra performance liquid chromatography-mass spectrometry-based comprehensive metabolomic profiling and pathways analysis of biological data sets. Anal. Chem. 84 (1), 428–439.

Wang, X., Zhang, A., Sun, H., 2013a. Power of metabolomics in diagnosis and biomarker discovery of hepatocellular carcinoma. Hepatology 57 (5), 2072–2077.

Wang, X., Zhang, A., Wang, P., Sun, H., Wu, G., Sun, W., Lv, H., Jiao, G., Xu, H., Yuan, Y., Liu, L., Zou, D., Wu, Z., Han, Y., Yan, G., Dong, W., Wu, F., Dong, T., Yu, Y., Zhang, S., Wu, X., Tong, X., Meng, X., 2013b. Metabolomics coupled with proteomics advancing drug discovery toward more agile development of targeted combination therapies. Mol. Cell. Proteomics 12 (5), 1226–1238.

Zhang, A., Sun, H., Wang, P., Han, Y., Wang, X., 2012a. Recent and potential developments of biofluid analyses in metabolomics. J. Proteomics 75 (4), 1079–1088.

Zhang, A., Sun, H., Wang, X., Jiao, G., Yuan, Y., Sun, W., 2012b. Simultaneous in vivo RP-HPLC-DAD quantification of multiple-component and drug-drug interaction by pharmacokinetics, using 6,7-dimethylesculetin, geniposide and rhein as examples. Biomed. Chromatogr. 26 (7), 844–850.

CHAPTER 3

Chinmedomics Advancing Drug Discovery and Development from Yinchenhao Tang

Xijun Wang, Aihua Zhang, Hui Sun

3.1 INTRODUCTION

Yinchenhao Tang (YCHT), a well-known Chinese herbal formulae, was recorded in *Shanghanlun'* – a classic resource on Traditional Chinese Medicine (TCM) written by Zhongjing Zhang (150–215 A.D.) and was officially listed in the Chinese Pharmacopoeia (China Pharmacopoeia Committee, 2010). It consists of *Artemisia* annua L. (the monarch herb), *Gardenia* jasminoides Ellis (the minister herb), and *Rheum* Palmatum L. (the assistant and servant herb) and has been used for thousands of years to treat jaundice and liver disorders. It possesses many medical applications, such as the treatment of cholestasis, liver fibrosis, hepatitis C, biliary cirrhosis, and cholestatic diseases.

Chemical studies on the single herbs of YCHT have been conducted. However, there has been no fully integrated study of the constituents in the global formulae, which is not simply combinations of the individual herbs, but a prepared decoction. The pharmacological effects of the TCM prescription are demonstrated to express their effects through multiactive chemical constituents, which refer to the total compounds that have active functions on disease. The sources of these effective compounds are complicated, including original compounds from the prescription and the products of processed decoction. For the screening and analysis of effective compounds in TCM prescription, conventional phytochemical approaches that isolate and identify individual components one by one could be used to search the lead compounds in herbal medicine, but this strategy is both time consuming and labor intensive. It is essential to develop novel and available methods that could overcome all these limitations, to reveal the complex compounds in a TCM prescription. Since only the compounds in the blood have the probability to become the effective constituents to show the therapeutic effects, the serum pharmacochemistry of TCM can reflect with regard to drug action, absorption, distribution, and interaction in the body.

From the point of modern systems medicine, the TCM syndrome is a functional state that is caused by the body's metabolic imbalances. Metabolomics

CONTENTS

Chinmedomics. http://dx.doi.org/10.1016/B978-0-12-803117-9.00003-2

and TCM have some similar characteristics such as entirety, comprehensiveness, and dynamic changes, and its property is in concert with the holistic efficacy of TCM. Metabolomics can capture comprehensive analysis of small molecule metabolites and provides a powerful approach to establish metabolic profiling, to discover biomarkers, and the related pathways for exploring the TCM syndrome. To determine which constituents contributed to the mainly therapeutic effect, statistical analyses were performed using Pearson's correlation, and the correlation coefficient (r) described the degree of correlation between all the biomarkers and the chemical composition of the YCHT *in vivo*. Correlation analysis between metabolite markers with serum constituents originating from TCM is used for the discovery of the efficacy of the material basis of YCHT. Therefore, this study was undertaken to investigate the chemical constituents and the metabolites *in vivo*, and further to screen the absorbed bioactive components in YCHT.

3.2 EXPERIMENTAL DETAIL

3.2.1 Chemicals and Materials

HPLC grade acetonitrile was purchased from Dikma Technology Inc. (Richmond Hill, ON, Canada). Deionized water was purified by the Milli-Q system (Millipore, Bedford, MA, USA). Formic acid and phosphoric acid was of an analytical grade purchased from Beijing Reagent Company (Beijing, P.R.C.). Leucine enkephalin was purchased from Sigma-Aldrich (MO, UK), and carbon tetrachloride (CCl_4) was purchased from the Chemicals Factory (Tianjin, P. R.C.). Other chemicals, except as noted, were analytical grade. The Flos Artemisiae was purchased from Xinneitian Pharmaceutical Inc. (China branch office, Japan); Fructus Gardeniae Jasminoidis and Radix et Rhizoma Rhei were purchased from Harbin Tongrentang Drug Company (Harbin, China). All crude drugs were authenticated by pharmacognosist, Prof. Xijun Wang, Department of Pharmacognosy of Heilongjiang University of Chinese Medicine. YCHT was prepared according to the method recorded in *Shanghanlun* and the extract was freeze-dried.

3.2.2 Metabolomics Study

3.2.2.1 Animals Handling

Male Wistar rats were maintained in a specific pathogen-free environment, and allowed to acclimatize in metabolic cages for 1 week prior to treatment. The animals were randomly assigned to three groups of six rats each, as follows: control, hepatic injury (HI model), and YCHT groups. The rats in the control group were given distilled water for 8 consecutive days and anesthetized via an intraperitoneal injection of 1% pentobarbital sodium (0.15 mL/100 g body weight) on day 9. The rats in the HI group were orally administrated 20%

CCL_4 (1 mL/kg body weight) for 8 consecutive days (at 6:00 p.m.), and on day 9, the rats were anesthetized as in the control group. Simultaneously, the YCHT group was administrated CCL_4 and treatment for 9 consecutive days, then anesthetized as in the control group. Blood was collected from the hepatic portal vein, and plasma was separated via centrifugation at 4500 rpm for 5 min at 4°C, flash frozen in liquid nitrogen and stored at −80°C until the metabolomics analysis were performed. Urine was collected daily (at 6:00 a.m.) from the metabolic cages at ambient temperature throughout the entire procedure and centrifuged at 10,000 rpm at 4°C for 5 min; the supernatants were then stored frozen at −20°C for subsequent analysis.

3.2.2.2 Metabolomics Analysis Platform

Serum and urine were collected for the UPLC–HDMS analysis. For the reversed-phase UPLC analysis, the ACQUITY UPLC BEH C_{18} column (50 × 2.1 mm i.d.,1.7 μm, Waters Corp., Milford, MA, USA) was used. The column temperature was maintained at 35°C, the flow rate during the mobile phase was 0.50 mL/min, and the injection volume was fixed at 2.0 μL. Mobile phase A involved the use of 0.1% formic acid in acetonitrile, whereas mobile phase B involved the use of 0.1% formic acid in water. For the urine UPLC–HDMS (Waters Corp., Milford, USA) analysis, the optimal conditions for positive ion mode were as follows: capillary voltage of 2500 V, desolvation temperature of 350°C, sample cone voltage of 35 V, extraction cone voltage of 3 V, microchannel plate voltage of 1600 V, collision energy of 4 eV, source temperature of 110°C, cone gas flow of 50 L/h, and desolvation gas flow of 600 L/h. The mass spectrometry full-scan data were acquired in the positive ion mode from 100 Da to 1000 Da with a 0.1 s scan time. For the plasma UPLC–HDMS analysis, the desolvation gas flow was 600 L/h, whereas the other parameters were the same as for the urine. The MS data were generated recorded using MassLynx V4.1 (Waters Corp., Milford, USA), MarkerLynx Application Manager (Waters Corp., Milford, USA) was used for peak detection, and EZinfo 2.0 software which is included in the MarkerLynx Application Manager was used for the PCA, PLS-DA, and OPLS. Putative markers were extracted from S-plots that were constructed following the analysis using OPLS, and markers were chosen based on their contribution to the variation and correlation within the data set. The potential biomarkers were matched with the structure message of metabolites acquired from available biochemical databases.

3.2.3 Constituents Analysis Using Serum Pharmacochemistry

3.2.3.1 Animals

Male Wistar rats (200 ± 20 g) were obtained from the Laboratory Animal Center of Heilongjiang University of Chinese Medicine (Harbin, China). Animals were bred in a breeding room with temperature of 24 ± 2°C, humidity of

60 ± 5%, and 12 h dark–light cycle. They were given tap water and fed normal food *ad libitum*. All the experiment animals were housed under the above conditions for 3 days acclimation, and were fasted overnight before the experiments. The animal facilities and protocols were approved by the Institutional Animal Care and Use Committee, Heilongjiang University of Chinese Medicine.

3.2.3.2 Preparation of YCHT Samples for LC/MS Analysis *In Vitro* and *In Vivo*

According to the original composition and preparation method of YCHT recorded in *Shanghan Lun*, YCHT was prepared in the following procedure. Flos Artemisiae (18 g) was boiled with 1400 mL distilled water until the volume of water was reduced to about 400 mL, then 9 g Fructus Gardeniae Jasminoidis and 6 g Radix et Rhizoma Rhei were added, and kept boiling for 10 more minutes. The extracted solution was filtered through five-layer gauzes and made to a concentration of 1 g crude drug per milliliter, and finally the solution was freeze-dried. Two hundred milligrams of the freeze-dried powder were extracted with 50 mL methanol for 30 min under ultrasonic. The methanol extraction was centrifuged at 13,000 rpm for 10 min at 4°C, and the supernatant was filtered through a 0.20 μm filter, the filtrate was used as UPLC sample.

Freeze-dried powder of YCHT was dissolved with distilled water as stock solution (0.8 g/mL). The above solution was orally administrated to male Wistar rats (1 mL/100 g body weight). Fifteen minutes after drug administration, the animals were anesthetized by an intraperitoneal injection of 1% pentobarbital sodium (0.15 mL/100 g body weight). The blood was collected from the hepatic portal vein and then centrifuged at 13,000 rpm for 5 min at 4°C. The supernatant obtained were frozen immediately and stored at −26°C, and thawed before analysis. Phosphoric acid (10 μL) and some internal standard kaempferide solution were added to 500 μL of the above supernatant and ultrasonic for 1 min, and vortexed for 30 s. The mixed solution was applied to a preactivited OASIS HLB solid phase extraction C_{18} column (30 μm, 60 mg, Waters Corp., USA). The column was washed with 1 mL of water, 1 mL of 2% methanol, and 2 mL of 100% methanol. The 100% methanol elutes were collected and dried under nitrogen gas at 45°C. The residues were dissolved in 350 μL of methanol and filtered through a 0.20 μm filter, the filtrate was used as an UPLC sample.

3.2.3.3 Instrumentation and Conditions

3.2.3.3.1 Q-TOF/MS analysis

Waters Micromass Q-TOF-micro™ (Manchester, UK) equipped with an electrospray ion source operating in either positive ion or negative ion mode, and with Masslynx data analysis system. The ion source temperature was set at 110°C with a cone gas flow of 100 L/h, a desolvation gas temperature of 300°C, and a desolvation gas flow of 600 L/h. The capillary voltage was set at 3.1 kV for positive ion mode and 2.8 kV for negative ion mode, the cone voltage was up to 35 V. A scan time of 0.48 s with an interscan delay of 0.1 s was used throughout, with collision energy of 3 eV and a collision gas pressure of ~2.8×10^{-3} mbar argon. A lock-mass of leucine enkephalin at a concentration of 0.5 ng μL^{-1}, in 50:50 acetonitrile: water (0.1% formic acid) for positive ion mode ($[M + H]^+ = 556.2771$) and 1 ng μ/L, in 50:50 acetonitrile: water for negative ion mode ($[M - H]^- = 554.2615$), was employed at a flow rate of 60 μL/min via a lockspray interface. Systematic data was collected in centroid mode, the lockspray frequency was set at 5 s, and the lock-mass data was averaged over a 10 s scan for correction. The mass spectrometric data was collected in full scan mode, the m/z were from 100 to 900 in positive and negative ion.

3.2.3.3.2 Data Processing

By comprehensively searching databases such as Chemspider, PubMed, Pubchem compound, an in-house molecular formula database was established by use of Masslynx™ (V4.1) software MassFragment that includes the compound name, molecular formula, and chemical structure. For further confirmation of the structure and the source of the metabolites, the raw data were processed by a Metabolynx™ version 4.1, which employs a comprehensive list of potential biotransformation reactions with the elemental compositions of each possible metabolite and generates a sequence of extraction ion chromatograms (XICs). Comparing the XICs of the samples after administration and control samples allows for the identification of potential drug metabolites. Elemental composition for expected and unexpected metabolite peaks was created using an exhaustive list of potential formulas of the metabolites dependant on the ion mode.

3.2.4 Correlation Analysis between Marker Metabolites and Absorbed Constituents

A plotting of the correlation model between the marker metabolites and chemical composition (PCMS) was used to screen the constituents or metabolites absorbed into the blood after oral administration of YCHT. To determine which constituents contributed to the mainly therapeutic effect, the correlation coefficient (r) described the degree of correlation between two variables: all the markers and the chemical composition of the YCHT were *in vivo*. The r values were between -1 and $+1$; $r > 0$ indicates that two variables were positively correlated, whereas $r < 0$ indicates a negative correlation. The larger the absolute value, the higher the correlation; therefore, we defined $r = 0.90–1.00$ as extremely related and $r = 0.80 - 0.89$ as highly related. In this study, we extracted the relevant chemical component data when $r \geq 0.8$ and identified the individual components using the PCMS.

3.2.5 Cell Test to Verify the Efficacy of Chemical Components Correlated with Therapeutic Effect

3.2.5.1 *Chemicals and Reagents*

Dulbecco's modified Eagle's medium (DMEM) and HBSS was purchased from the HyClone Co., USA; fetal bovine serum (FBS) was sourced from Yuanheng-shengma Biotechnology Co., Ltd. (Beijing, China); trypsin was purchased from the Gibco Co., USA (Gibco company, USA); and collagenase I and thiazolyl blue was obtained from Sigma Chemical Co., USA. Hank solution, and 0.1 mol/L phosphate buffer solution (PBS) were purchased from the Cell Center of Peking Union Medical University. MTT was obtained from Sigma Chemical (Sigma, USA). 6,7-dimethylesculetin (D), geniposide (G), and rhein (R) were isolated within our laboratory, and identified by spectral analyses, primarily NMR and MS. After identification it was further purified by HPLC to yield an authorized compound with the purity of at least 99%. Freeze-dried YCHT powder was produced by our laboratory.

3.2.5.2 *Isolation of Hepatocytes*

Mice (8–10 weeks old) weighing 25–30 g, were anesthetized with 1% sodium pentobarbital and hepatocytes were isolated using a two-step collagenase perfusion method. The liver was perfused *in situ* with Ca^{2+}-free Hanks Balanced Salt Solution (HBSS) containing EGTA (0.5 mM), glucose, and penicillin/streptomycin (100 U/mL), pH 7.4 (12 mL/min, 37°C for 10 min), followed by perfusion with HBSS containing Ca^{2+} (1.5 mM), glucose (80 mM), penicillin/streptomycin (100 U/mL), and collagenase type I (0.05%) for 10 min. The liver was removed and gently minced in Dulbecco's Modification of Eagle's Medium (DMEM) containing fetal bovine serum (10%), penicillin/streptomycin (100 U/mL), dexamethasone (10^{-7} M), and insulin (10^{-7} M). The liver cell suspension was then filtered with 0.22 μm MILLEX GP Filter Unit and centrifuged at 50 g for 5 min at 4°C. The centrifugation was repeated twice. After washing by low-speed centrifugation at 50 g for 3 min at 4°C several times, cell viability and yield were determined by trypan blue exclusion. Cell viability was generally > 85%, and cell purity was > 95% hepatocytes. After 2–4 h, the medium was changed to remove unattached cells, and cultured at 37°C with 5% CO_2.

3.2.5.3 *Cell Culture and Treatment*

Cells from three independent biological replicates with viability of 85% were used and cultured on collagen gel-precoated 96-well plates at a density of 1×10^5 cells per well. Cells were allowed to attach for 2–4 h at 37°C in a humidified chamber with 95:5% air/CO_2 in Dulbecco's modified Eagle's medium supplemented with 10% fetal calf serum, 0.5 U/mL insulin, 7 ng/mL glucagon, and 2% penicillin/streptomycin. After attachment, medium containing unattached cells and debris was removed by washing, and the cultures were overlaid with a second collagen layer to form a collagen-collagen sandwich culture.

Primary cultures of mouse hepatocytes were cultured at 37°C in a humidified chamber with 95:5% air/CO_2 in serum-free culture medium.

Test compounds were dissolved in DMSO and the concentration of DMSO in the culture medium was 0.1%. One hour posttreatment, ethanol or HBSS was added, and the flasks were capped tightly to prevent ethanol evaporation. Medium and fresh chemicals were replaced every 24 h, and ethanol was added one hour after these treatments. The hepatocytes were harvested 24 h after the first ethanol treatment. Different concentrations of D, G, R and their combination, including control DMSO, were added to the relevant well and cells were incubated for 24 h at 37°C, 5% CO_2.

3.2.5.4 Flow Cytometric Analysis

Apoptosis cells were subjected to annexin V-PI dual staining assay according to the manufacturer's protocol. Stained cells were analyzed by Guava easyCyte (Millipore Co., CA, USA) and the apoptotic cell population was determined using GuavaSoft 2.6 software (Millipore Co., CA, USA).

3.2.5.5 Statistical Analyses

All statistical analyses were performed using the Student's t-test. Differences with a P-value of 0.05 or less were considered significant. Assays were performed in triplicate, and the results are expressed as mean ± SD.

3.3 RESULTS AND DISCUSSIONS

3.3.1 Metabolomics Analysis of Protective Function of YCHT

Using the optimal Waters ACQUITY UPLC-UV-Q-TOF/MS conditions, the representative base peak intensity (BPI) chromatograms of the urine or plasma samples that were collected from representative rats from each group are presented in Fig 3.1A and B. The low molecular mass metabolites were well separated at 11 min because of the minor particles (sub-1.7 μm) of UPLC. PCA, PLS, and OPLS-DA were used to classify the metabolic phenotypes and identify the differentiating metabolites. Analyzing the OPLS-DA score plots identified the control and HI rats based on differences in their metabolites, suggesting that their metabolic profiles significantly changed as a result of CCL_4 administration (Fig. 3.1C). YCHT helped prevent HI and maintained the animals in a normal state – there were no distinct clustering differences between the control group and YCHT treatment group. The PLS–DA loading plots displayed variables that were positively correlated with the score plots (Fig. 3.1D). S-plots (Fig. 3.1E) and VIP-value plots were combined for the structural identification of the biomarkers (Fig. 3.1F). Finally, potential biomarkers of significant contribution were identified, with 20 and 12 in the urine and plasma, respectively (Fig. 3.2).

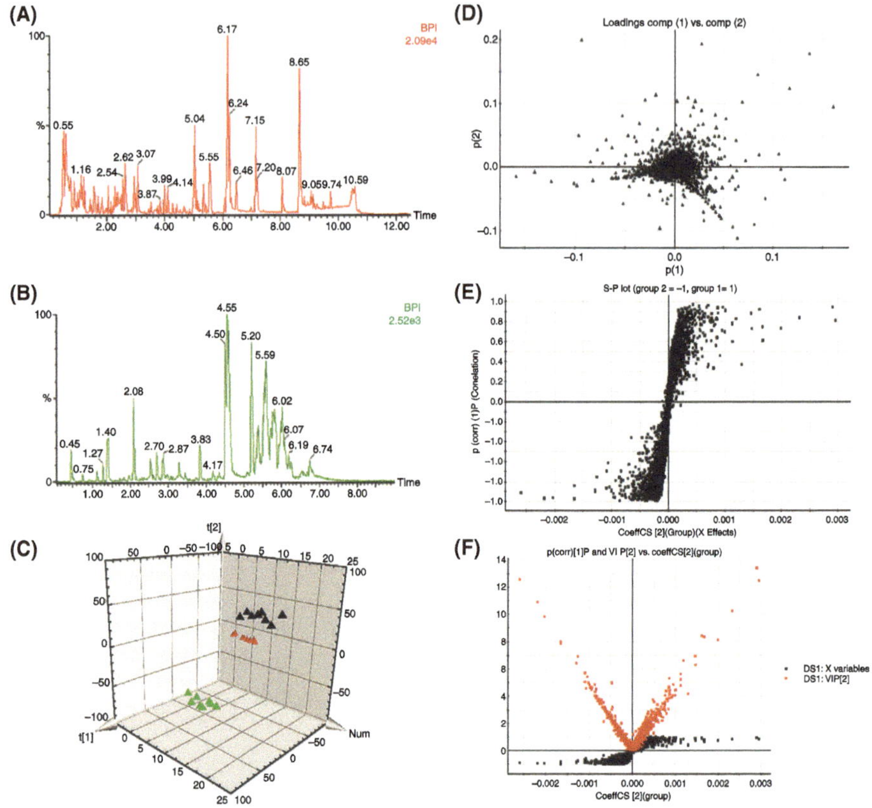

FIGURE 3.1 Metabolomic multivariate analysis.

(A) BPI chromatograms of urine samples; (B) BPI chromatograms of plasma samples; (C) 3D score plot of PCA for the control and model groups (▲, control group; ▲, model group; ▲, YCHT group); (D) the PLS-DA loading plot for the control and model groups; (E) S-plot of the OPLS-DA; (F) a combination plot of S-plot and VIP values. The red point graph is the VIP value plot, which represents the value of each metabolite. The farther from the origin, the higher the VIP value of the metabolites.

3.3.2 LC/MS Analysis for YCHT Samples *In Vitro* and *In Vivo*

3.3.2.1 Optimization of MS Conditions

To obtain better detection, the MS conditions should be optimized and a statistical orthogonal design was employed for optimization. The effects of formic acid concentration, desolvation temperature, source temperature, capillary voltage, and desolvation gas flow were examined. The experiments were arranged according to $L_{16}(4^5)$ orthogonal design table: formic acid concentration (0.005, 0.01, 0.05, 0.1%), desolvation temperature (250, 275, 300, 325 °C), source temperature (100, 105, 110, 115 °C), capillary voltage (positive ion ESI mode 2.6 kV, 2.8 kV, 3.0 kV, 3.2 kV; negative ion ESI mode 2.2 kV,

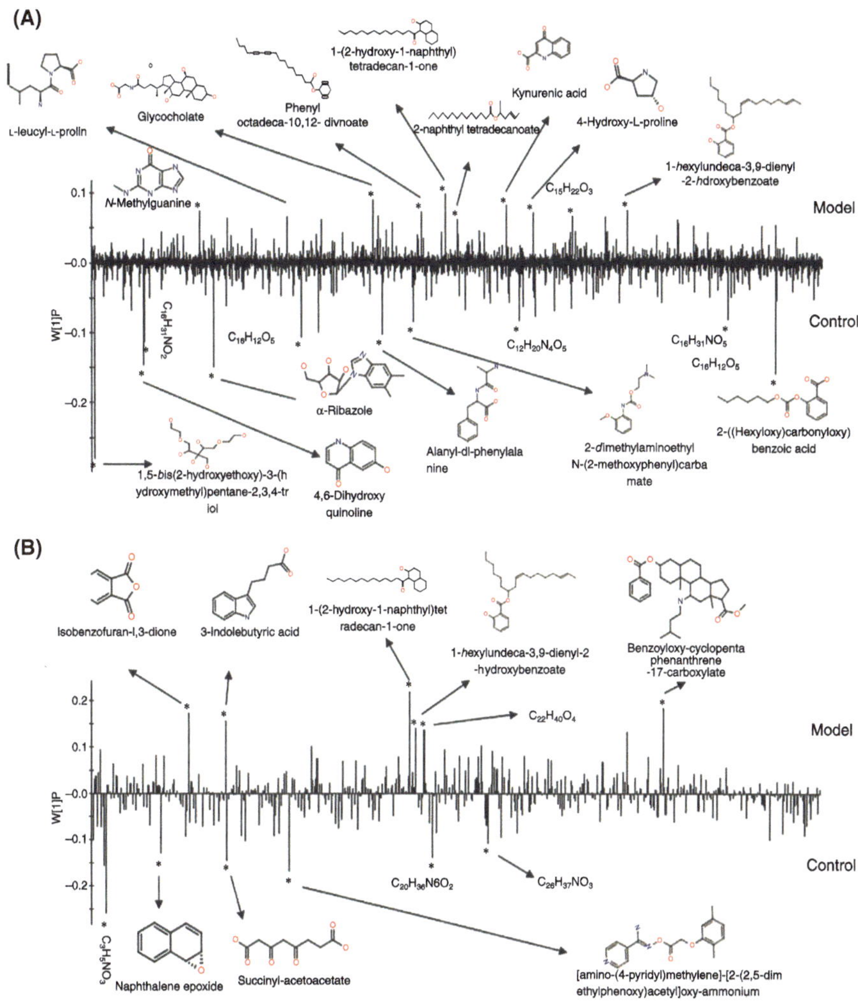

FIGURE 3.2 Loading plots of potential biomarkers identified in (A) urine and (B) plasma.

2.4 kV, 2.6 kV, 2.8 kV), desolvation gas flow (500, 550, 600, 650 L/h), and the peak area was taken as the criteria for optimization. The statistical results indicated that the influence of the factors on the ionization were as follows: desolvation temperature and desolvation gas flow were statistical significant at $p < 0.05$. Formic acid concentration, capillary voltage, and source temperature were insignificant at $p > 0.05$. The optimum conditions were shown to be as follows: 0.1% formic acid, desolvation temperature 300°C, source temperature 110°C, desolvation gas (N_2) 600 L/h, capillary voltage 3.0 kV for positive ion ESI mode, and 2.8 kV for negative ion ESI mode.

3.3.2.2 UPLC-UV-Q-TOF Analysis of YCHT

The chromatogram of the sample is shown in Fig. 3.3. The constituents in YCHT were well separated by using the UPLC-UV-Q-TOF method and 45 peaks were detected, the structures of the compounds are shown in Table 3.1. The MS data of (+) ESI-MS spectra and (−) ESI-MS spectra are shown in Table 3.2. For most of the constituents, $[M - H]^-$, $[M + HCOO]^-$, and $[M + Na]^+$ were observed. These results provided reliable information for confirming molecular weight and structure of the constituents. Peak 11, 16, 24, 36, 39, 44 were attributed

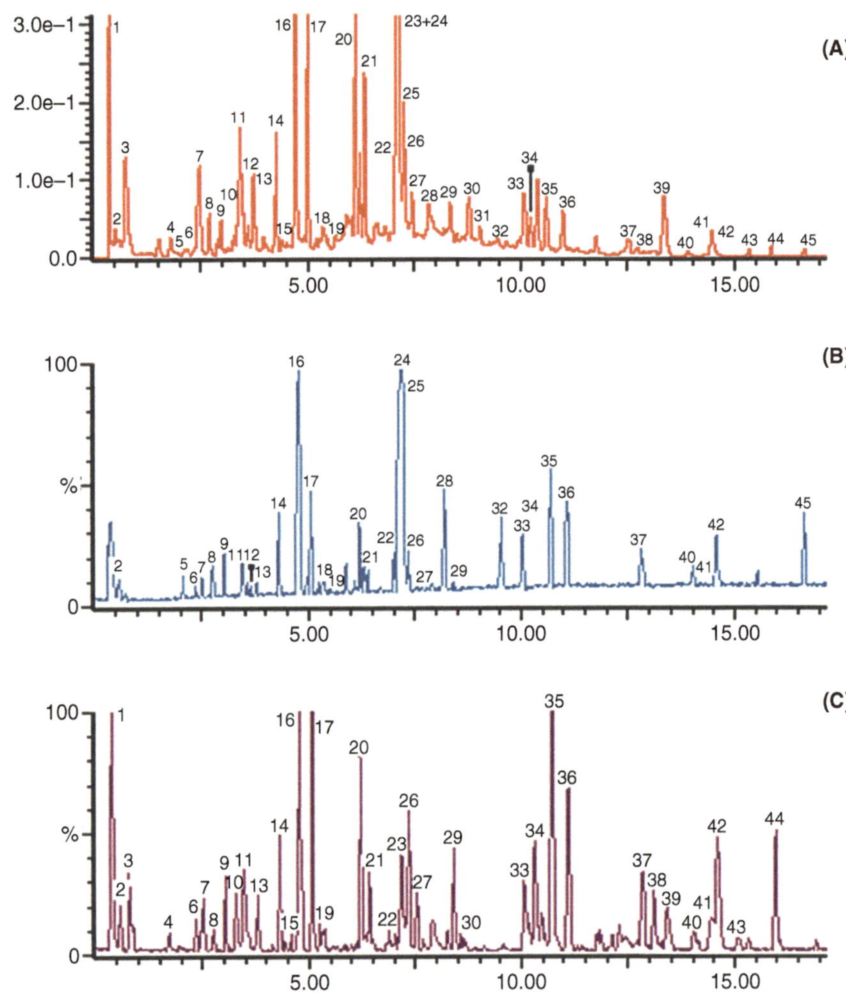

FIGURE 3.3 Chromatograms of YCHT by UPLC-UV-Q-TOF.

(A) UPLC-UV chromatogram at 254 nm. (B) TIC chromatogram in positive ESI mode. (C) TIC chromatogram in negative ESI mode.

Table 3.1 MS Data of (+) ESI-MS Spectra and (-) ESI-MS Spectra, and the Identification Results of the Constituents of YCHT

No.	Identified Compounds		T_R	Negative Ion (m/z)		Positive Ion (m/z)		MW (Da)	Elemental Composition
	In Vitro	*In Vivo*		Indicated	Actual	Indicated	Actual		
1	Gentiobiose	–	0.37	341.1083	341.1084	–	–	342.1162	$C_{12}H_{22}O_{11}$
2	Pyroglutamic acid	–	0.56	128.0345	128.0348	130.0500	130.0504	129.0426	$C_5H_7NO_3$
3	Gallic acid	–	0.77	169.0137	169.0137	–	–	170.0215	$C_7H_6O_5$
4	Protocatechuic acid-3-glucoside	–	1.72	315.0718	315.0716	–	–	316.0794	$C_{13}H_{16}O_9$
5	Not identificated	–	2.06	–	–	186.1291	186.1283	–	$C_{13}H_{15}N$
6	Villosolside	–	2.35	361.1495	361.1499	385.1538[M+Na]$^+$	385.1475	362.1577	$C_{16}H_{26}O_9$
7	Neochlorogenic acid/Isochlorogenic acid	–	2.53	353.0872	353.0873	377.0938[M+Na]$^+$	377.0849	354.0951	$C_{16}H_{18}O_9$
8	Gardenoside B	–	2.74	403.1237 449.1254[M-H+46]$^-$	403.1240	427.1252[M+Na]$^+$	427.1216	404.1319	$C_{17}H_{24}O_{11}$
9(1)*	Picrocrocininc acid	Picrocrocininc acid	3.04	345.1563	345.1549	369.158[M+Na]$^+$	369.1525	346.1628	$C_{16}H_{26}O_8$
10	Not identificated	–	3.28	289.0710	289.0712	–	–	–	$C_{15}H_{14}O_6$
11(2)	Chlorogenic acid	Chlorogenic acid	3.46	353.0876	353.0873	377.0927[M+Na]$^+$	377.0849	354.0951	$C_{16}H_{18}O_9$
12	Not identificated	–	3.63	–	–	369.1619	369.1549	–	$C_{18}H_{24}O_8$
13	Isochlorogenic acid/Neochlorogenic acid	–	3.78	353.0867	353.0873	377.0940[M+Na]$^+$	377.0849	354.0951	$C_{16}H_{18}O_9$
M1(3)	–	6-Methoxy coumarin-7-hydroxyl sulfate	3.88					270.9922	$C_{10}H_8O_7S$
M2(4)	–	7-Methoxy coumarin-6-hydroxyl sulfate	4.05					270.9922	$C_{10}H_8O_7S$
14(5)	Genipingentiobioside	Genipingentiobioside	4.31	549.1779 595.1921[M-H+46]$^-$	549.1819	573.1730[M+Na]$^+$	573.1795	550.1898	$C_{23}H_{34}O_{15}$

(Continued)

Table 3.1 MS Data of (+) ESI-MS Spectra and (-) ESI-MS Spectra, and the Identification Results of the Constituents of YCHT *(cont.)*

No.	Identified Compounds		T_R	Negative Ion (m/z)		Positive Ion (m/z)		MW (Da)	Elemental Composition
	In Vitro	*In Vivo*		Indicated	Actual	Indicated	Actual		
15(6)	Not identificated	Not identificated	4.53	353.0866	353.0873	377.0920[M+Na]$^+$	377.0849	354.0951	$C_{16}H_{18}O_9$
16(7)	Geniposide	Geniposide	4.77	387.1329 433.1343[M-H+46]$^-$	387.1291	411.1275[M+Na]$^+$	411.1267	388.1369	$C_{17}H_{24}O_{10}$
17(8)	Safflor yellow A	Safflor yellow A	5.08	593.1448	593.1506	595.1693	595.1663	594.1585	$C_{27}H_{30}O_{15}$
18(9)	Capillaridin A	Capillaridin A	5.48	–	–	353.1209	353.1178	352.1099	$C_{24}H_{16}O_3$
19(10)	5,6-Dimethoxy-7-hydroxy coumarin	5,6-Dimethoxy-7-hydroxy coumarin	5.99	221.0449	221.0450	223.0611	223.0606	222.0528	$C_{11}H_{10}O_5$
20(11)	Quercetin-3-0-glycoside	Quercetin-3-0-glycoside	6.20	463.0889	463.0877	487.0923[M+Na]$^+$	487.0852	464.0955	$C_{21}H_{20}O_{12}$
21	Isoquercitrin/ Quercetin-3-0-glycoside	–	6.40	463.0884	463.0877	487.0921	487.0852	464.0955	$C_{21}H_{20}O_{12}$
22(12)	6,8-Dimethoxy-7-hydroxycoumarin	6,8-Dimethoxy-7-hydroxycoumarin	7.00	221.0450	221.0450	223.0613	223.0606	222.0528	$C_{11}H_{10}O_5$
23	3,4-Di-o-caffeoylquinic acid	–	7.16	515.1200	515.1190	–	–	516.1268	$C_{25}H_{24}O_{12}$
24(13)	6,7-Dimethylesculetin	6,7-Dimethylesculetin	7.16	–	–	207.0655	207.0657	206.0579	$C_{11}H_{10}O_4$
25(14)	2,5-Dimethyl-7-hydroxy chromone	2,5-Dimethyl-7-hydroxy chromone	7.23	–	–	191.4302	191.4253	190.0630	$C_{11}H_{10}O_3$
26	Isorhamnetin-3-glucoside	–	7.34	477.1043	477.1033	501.1056[M+Na]$^+$	501.1009	478.1111	$C_{22}H_{22}O_{12}$
27	Cacticin	–	7.54	477.1041	477.1033	501.1047[M+Na]$^+$	501.1009	478.1111	$C_{22}H_{22}O_{12}$

28	Not identificated	–	8.19	–	–	349.1340	349.1499	–	$C_{15}H_{24}O_9$
29	6″-o-p-coumaroyl-genipingentiobi-oside	–	8.41	695.2194	695.2187	719.2181[M+Na]+	719.2163	696.2265	$C_{32}H_{40}O_{17}$
30(15)	Not identificated	Not identifi-cated	8.52	483.1032	483.0775	–	–	484.0853	$C_{20}H_{20}O_{14}$
31(16)	Not detected	Not detected	9.04	–	–	–	–	–	–
32(17)	Chimaphylin	Chimaphylin	9.52	–	–	187.0807	187.0759	186.0681	$C_{12}H_{10}O_2$
33	Naringenin	–	10.04	271.0600	271.0606	273.0870	273.0763	272.0685	$C_{15}H_{12}O_5$
34	Aloeemodin-w-o-β-D-gluco-pyranoside/ Emodin-1-o-β-D-glucopyranoside	–	10.32	431.0956	431.0978	–	–	432.1056	$C_{21}H_{20}O_{10}$
35(18)	6-Demethoxycapil-larisin	6-Demethoxy-capillarisin	10.72	285.0385	285.0399	287.0611	287.0556	286.0477	$C_{15}H_{10}O_6$
36(19)	Capillarisin	Capillarisin	11.11	315.0511	315.0505	317.0707	317.0661	316.0583	$C_{16}H_{12}O_7$
37	Cirsimaritin	–	12.83	313.0715	313.0712	315.0929	315.0869	314.0790	$C_{17}H_{14}O_6$
38	5,3′,4′-Trihy-droxy-6,7-dime-thoxyflavone	–	13.10	329.2329	329.2328	–	–	330.0740	$C_{17}H_{14}O_7$
39(20)	Rhein	Rhein	13.42	283.0242	283.0243	–	–	284.0321	$C_{15}H_8O_6$
40	5-Dehydroxy-6-de-methylcapillarisin	–	14.02	285.0754	285.0763	287.1006	287.0919	286.0477	$C_{15}H_{10}O_6$
41	Not identificated	–	14.46	207.1384	207.1385	209.1635	209.1542	–	$C_{13}H_{20}O_2$
42	Rhamnocitrin	–	14.58	299.0555	299.0556	301.0787	301.0712	300.0634	$C_{16}H_{12}O_6$
43	Not identificated	–	15.10	293.1760	293.1753	–	–	–	$C_{17}H_{26}O_4$
44(21)	Emodin	Emodin	15.96	269.0451	269.0450	–	–	270.0528	$C_{15}H_{10}O_5$
45	Not identificated	–	16.64	–	–	359.2462	359.2434	–	$C_{19}H_{34}O_6$

**Note: Number in ( ) is the number of Peak detected in serum postoral administration of YCHT*

Table 3.2 MS/MS Data of (+) ESI-MS Spectra and (-) ESI-MS Spectra, and the Identification Results of the Constituents of YCHT

No.	Identified Compounds	T_R (min)	Negative Ion (m/z)	Positive Ion (m/z)	MW (Da)	Elemental Composition
1	Gentiobiose	0.37	341[M-H]$^-$ 179[M-H-(Glc-H_2O)]$^-$ 161[(Glc-H_2O)-H]$^-$	–	342.1162	$C_{12}H_{22}O_{11}$
2	Pyroglutamic acid	0.56	128[M-H]$^-$	130[M+H]$^+$ 84[M-HCOO]$^+$	129.0426	$C_5H_7NO_3$
3	Gallic acid	0.77	169[M-H]$^-$ 125[M-HCOO]$^-$	–	170.0215	$C_7H_6O_5$
4	Protocatechuic acid-3-glucoside	1.72	315[M-H]$^-$ 153[M-H-(Glc-H_2O)]$^-$ 109[M-H-(Glc-H_2O)-COO]$^-$	–	316.0794	$C_{13}H_{16}O_9$
5	Not identificated	2.06	–	186[M+H]$^+$	–	$C_{13}H_{15}N$
6	Villosolside	2.35	361[M-H]$^-$ 181[M-H-Glc]$^-$ 137[M-H-Glc-COO]$^-$	385[M+Na]$^+$	362.1577	$C_{16}H_{26}O_9$
7	Neochlorogenic acid/Isochlorogenic acid	2.53	353[M-H]$^-$ 191[M-$C_9H_7O_3$]$^-$ 179[M-$C_7H_{11}O_5$]$^-$ 135[M-H-$C_7H_{10}O_5$-COO]$^-$	377[M+Na]$^+$ 215[M+Na+H-$C_9H_7O_3$]$^+$ 163[$C_9H_7O_3$]$^+$	354.0951	$C_{16}H_{18}O_9$
8	Gardenoside B	2.74	403[M-H]$^-$ 449[M+HCOO]$^-$ 241[M-H-(Glc-H_2O)]$^-$ 223[M-H-(Glc-H_2O)-H_2O]$^-$ 193[M-H-(Glc-H_2O)-H_2O-]$^-$	427[M+Na]$^+$ 265[M+Na-(Glc-H_2O)]$^+$ 233[M+Na-(Glc-H_2O)-CH_3OH]$^+$	404.1319	$C_{17}H_{24}O_{11}$
9	Picrocrocininc acid	3.04	345[M-H]$^-$ 165[M-H-Glc]$^-$ 179[Glc-H]$^-$ 119[M-H-Glc-HCOOH]$^-$ 89[M-H-Glc-HCOOH-2CH_3]$^-$	369[M+Na]$^+$	346.1628	$C_{16}H_{26}O_8$
10	Not identificated	3.28	289[M-H]$^-$	–	–	$C_{15}H_{14}O_6$
11	Chlorogenic acid	3.46	353[M-H]$^-$ 191[M-$C_9H_7O_3$]$^-$	377[M+Na]$^+$ 215[M+Na+H-$C_9H_7O_3$]$^+$ 163[$C_9H_7O_3$]$^+$	354.0951	$C_{16}H_{18}O_9$
12	Not identificated	3.63	–	369[M+H]$^+$	–	$C_{18}H_{24}O_8$
13	Isochlorogenic acid/Neochlorogenic acid	3.78	353[M-H]$^-$ 191[M-$C_9H_7O_3$]$^-$ 179[M-$C_7H_{11}O_5$]$^-$ 135[M-H-$C_7H_{10}O_5$-COO]$^-$	377[M+Na]$^+$ 215[M+Na+H-$C_9H_7O_3$]$^+$ 163[$C_9H_7O_3$]$^+$	354.0951	$C_{16}H_{18}O_9$

14	Genipingentiobioside	4.31	$549[M\text{-}H]^-$ $595[M\text{+}HCOO]^-$ $387[M\text{-}H\text{-}(Glc\text{-}H_2O)]^-$ $225[M\text{-}H\text{-}2(Glc\text{-}H_2O)]^-$	$573[M\text{+}Na]^+$ $248[M\text{+}Na\text{-}2(Glc\text{-}H_2O)]^+$	550.1898	$C_{23}H_{34}O_{15}$
15	Not identificated	4.53	$353[M\text{-}H]^-$ $191[M\text{-}C_9H_7O_3]^-$	$377[M\text{+}Na]^+$ $215[M\text{+}Na\text{+}H\text{-}C_9H_7O_3]^+$	354.0951	$C_{16}H_{18}O_9$
16	Jasminoidin	4.77	$387[M\text{-}H]^-$ $433[M\text{+}HCOO]^-$ $207[M\text{-}H\text{-}Glc]^-$ $163[M\text{-}H\text{-}Glc\text{-}COO]^-$ $119[M\text{-}H\text{-}Glc\text{-}2COO]^-$	$411[M\text{+}Na]^+$ $249[M\text{+}Na\text{-}(Glc\text{-}H_2O)]^+$ $217[M\text{+}Na\text{-}(Glc\text{-}H_2O)\text{-}CH_3OH]^+$	388.1369	$C_{17}H_{24}O_{10}$
17	Safflor yellow A	5.08	$593[M\text{-}H]^-$ $473[M\text{-}H\text{-}C_8H_8O]^-$	$595[M\text{+}H]^+$	594.1585	$C_{27}H_{30}O_{15}$
18	Capillaridin A	5.48	–	$353[M\text{+}H]^+$ $249[M\text{+}H\text{-}C_7H_5O]^+$	352.1099	$C_{24}H_{16}O_3$
19	5,6-Dimethoxy-7-hydroxy coumarin	5.25	$221[M\text{-}H]^-$ $206[M\text{-}H\text{-}CH_3]^-$ $191[M\text{-}H\text{-}2CH_3]^-$ $163[M\text{-}H\text{-}2CH_3\text{-}CO]^-$ $119[M\text{-}H\text{-}2CH_3\text{-}CO\text{-}COO]^-$	$223[M\text{+}H]^+$ $208[M\text{+}H\text{-}CH_3]^+$ $190[M\text{+}H\text{-}CH_3\text{-}H_2O]^+$ $162[M\text{+}H\text{-}CH_3\text{-}H_2O\text{-}CO]^+$ $134[M\text{+}H\text{-}CH_3\text{-}H_2O\text{-}2CO]^+$ $106[M\text{+}H\text{-}CH_3\text{-}H_2O\text{-}3CO]^+$ $78[M\text{+}H\text{-}CH_3\text{-}H_2O\text{-}4CO]^+$	222.0528	$C_{11}H_{10}O_5$
20	Quercetin-3-0-glycoside/Isoquercitrin	6.20	$463[M\text{-}H]^-$ $301[M\text{-}H\text{-}(Glc\text{-}H_2O)]^-$ $300[M\text{-}H\text{-}(Glc\text{-}H_2O)\text{-}H]^-$ $271[M\text{-}H\text{-}(Glc\text{-}H_2O)\text{-}H\text{-}HCO]^-$	$487[M\text{+}Na]^+$ $325[M\text{+}Na\text{-}(Glc\text{-}H_2O)]^+$ $185[Glc\text{-}H_2O\text{+}Na]^+$	464.0955	$C_{21}H_{20}O_{12}$
21	Isoquercitrin/Quercetin-3-0-glycoside	6.40	$463[M\text{-}H]^-$ $301[M\text{-}H\text{-}(Glc\text{-}H_2O)]^-$ $300[M\text{-}H\text{-}(Glc\text{-}H_2O)\text{-}H]^-$ $271[M\text{-}H\text{-}(Glc\text{-}H_2O)\text{-}H\text{-}HCO]^-$	$487[M\text{+}Na]^+$ $325[M\text{+}Na\text{-}(Glc\text{-}H_2O)]^+$ $185[Glc\text{-}H_2O\text{+}Na]^+$	464.0955	$C_{21}H_{20}O_{12}$
22	6,8-Dimethoxy-7-hydroxycoumarin	7.00	$221[M\text{-}H]^-$ $206[M\text{-}H\text{-}CH_3]^-$ $191[M\text{-}H\text{-}2CH_3]^-$ $163[M\text{-}H\text{-}2CH_3\text{-}CO]^-$ $119[M\text{-}H\text{-}2CH_3\text{-}CO\text{-}COO]^-$	$223[M\text{+}H]^+$ $208[M\text{+}H\text{-}CH_3]^+$ $190[M\text{+}H\text{-}CH_3\text{-}H_2O]^+$ $162[M\text{+}H\text{-}CH_3\text{-}H_2O\text{-}CO]^+$ $134[M\text{+}H\text{-}CH_3\text{-}H_2O\text{-}2CO]^+$ $106[M\text{+}H\text{-}CH_3\text{-}H_2O\text{-}3CO]^+$ $78[M\text{+}H\text{-}CH_3\text{-}H_2O\text{-}4CO]^+$	222.0528	$C_{11}H_{10}O_5$
23	3,4-Di-o-caffeoylquinic acid	7.16	$515[M\text{-}H]^-$ $353[M\text{-}H\text{-}C_9H_6O_3]^-$ $191[M\text{-}H\text{-}2C_9H_6O_3]^-$	–	516.1268	$C_{25}H_{24}O_{12}$

(Continued)

Table 3.2 MS/MS Data of (+) ESI-MS Spectra and (-) ESI-MS Spectra, and the Identification Results of the Constituents of YCHT *(cont.)*

No.	Identified Compounds	T_R (min)	Negative Ion (m/z)	Positive Ion (m/z)	MW (Da)	Elemental Composition
24	6,7-dimethoxy coumaric	7.16	–	207[M+H]$^+$ 192[M+H-CH_3]$^+$ 191[M+H-CH_3-H]$^+$ 163[M+H-CH_3-H-CO]$^+$	206.0579	$C_{11}H_{10}O_4$
25	2,5-Dimethyl-7-hydroxy chromone	7.23	–	191[M+H]$^+$ 151[M+H-C_3H_4]$^+$	190.0630	$C_{11}H_{10}O_3$
26	Isorhamnetin-3-glucoside	7.34	477[M-H]$^-$ 315[M-H-(Glc-H_2O)]$^-$ 314[M-H-(Glc-H_2O)-H]$^-$ 299[M-H-(Glc-H_2O)-H-CH_3]$^-$ 271[M-H-(Glc-H_2O)-H-CH_3-CO]$^-$ 243[M-H-(Glc-H_2O)-H-CH_3-2CO]$^-$	501[M+Na]$^+$ 339[M+Na-(Glc-H_2O)]$^+$ 338[M+Na-(Glc-H_2O)-H]$^+$ 185[Glc-H_2O+Na]$^+$	478.1111	$C_{22}H_{22}O_{12}$
27	Cacticin	7.54	477[M-H]$^-$ 315[M-H-(Glc-H_2O)]$^-$ 314[M-H-(Glc-H_2O)-H]$^-$ 299[M-H-(Glc-H_2O)-H-CH_3]$^-$ 271[M-H-(Glc-H_2O)-H-CH_3-CO]$^-$ 243[M-H-(Glc-H_2O)-H-CH_3-2CO]$^-$	501[M+Na]$^+$ 339[M+Na-(Glc-H_2O)]$^+$ 338[M+Na-(Glc-H_2O)-H]$^+$ 185[Glc-H_2O+Na]$^+$	478.1111	$C_{22}H_{22}O_{12}$
28	Not identificated	8.19	–	349[M+H]$^+$	–	$C_{15}H_{24}O_9$
29	6″-o-p-coumaroylgenipingentiobioside	8.41	695[M-H]$^-$ 469[M-H-$C_{11}H_{13}O_5$-H]$^-$ 225[$C_{11}H_{13}O_5$]$^-$ 163[$C_9H_7O_3$]$^-$	719[M+Na]$^+$ 556[M+Na-$C_9H_7O_3$]$^+$	696.2265	$C_{32}H_{40}O_{17}$
30	Not identificated	8.52	483[M-H]$^-$	–		$C_{20}H_{20}O_{14}$
31	Not detected	9.04	–	–	–	–
32	Chimaphylin	9.52	–	187[M+H]$^+$	186.0681	$C_{12}H_{10}O_2$
33	Naringenin	10.04	271[M-H]$^-$ 151[M-$C_7H_4O_4$]$^-$ 119[C_8H_7O]$^+$	273[M+H]$^+$ 153[M-C_8H_8O]$^+$	272.0685	$C_{15}H_{12}O_5$

34	Aloeemodin-w-o-β-D-glucopyranoside/Emodin-1-o-β-D-glucopyranoside	10.32	431[M-H]$^-$ 269[M-H-(Glc-H_2O)]$^-$ 225[M-H-(Glc-H_2O)-COO]$^-$	–	432.1056	$C_{21}H_{20}O_{10}$
35	6-Demethoxycapillarisin	10.72	285[M-H]$^-$ 193[M-C_6H_5O]$^-$ 192[M-C_6H_5O-H]$^-$ 177[M-C_6H_5O-H-CH_3]$^-$	287[M+H]$^+$ 195[M+H-C_6H_4O]$^+$	286.0477	$C_{15}H_{10}O_6$
36	Capillarisin	11.11	315[M-H]$^-$ 300[M-H-CH_3]$^-$ 207[M-H-CH_3-C_6H_5O]$^-$	317[M+H]$^+$ 302[M+H-CH_3]$^+$ 209[M+H-CH_3-C_6H_5O]$^+$	316.0583	$C_{16}H_{12}O_7$
37	Cirsimaritin	12.83	313[M-H]$^-$ 298[M-H-CH_3]$^-$ 283[M-H-2CH_3]$^-$ 255[M-H-2CH_3-CO]$^-$ 227[M-H-2CH_3-2CO]$^-$ 117[C_8H_5O]$^-$	315[M+H]$^+$ 300[M+H-CH_3]$^+$ 282[M+H-CH_3-H_2O]$^+$ 254[M+H-CH_3-H_2O-CO]$^+$ 226[M+H-CH_3-H_2O-2CO]$^+$	314.0790	$C_{17}H_{14}O_6$
38	5,3′,4′-Trihydroxy-6,7-dimethoxyflavone	13.10	329[M-H]$^-$ 299[M-H-2CH_3]$^-$	–	330.0740	$C_{17}H_{14}O_7$
39	Rhein	13.42	283[M-H]$^-$ 239[M-H-COO]$^-$ 211[M-H-COO-CO]$^-$ 183[M-H-COO-2CO]$^-$	–	284.0321	$C_{15}H_8O_6$
40	5-Dehydroxy-6-demethylcapillarisin	14.02	285[M-H]$^-$ 165[M-H-$C_7H_4O_2$]$^-$ 119[$C_7H_3O_2$]$^-$	287[M+H]$^+$ 167[M+H-$C_7H_4O_2$]$^+$	286.0477	$C_{15}H_{10}O_6$
41	Not identificated	14.46	207[M-H]$^-$	209[M+H]$^+$	–	$C_{13}H_{20}O_2$
42	Rhamnocitrin	14.58	299[M-H]$^-$ 284[M-H-CH_3]$^-$ 271[M-H-CO]$^-$ 255[M-H-CH_3-CO]$^-$ 165[$C_8H_5O_4$]$^-$	301[M+H]$^+$ 286[M+H-CH_3]$^+$ 258[M+H-CH_3-CO]$^+$	300.0634	$C_{16}H_{12}O_6$
43	Not identificated	15.10	293[M-H]$^-$	–	–	$C_{17}H_{26}O_4$
44	Emodin	15.96	269[M-H]$^-$ 241[M-H-CO]$^-$ 240[M-H-CO-H]$^-$ 225[M-H-CO-H-CH_3]$^-$ 197[M-H-2CO-H-CH_3]$^-$ 182[M-H-2CO-H-2CH_3]$^-$	–	270.0528	$C_{15}H_{10}O_5$
45	Not identificated	16.64	–	359[M+H]$^+$	–	$C_{19}H_{34}O_6$

to chlorogenic acid, geniposide, 6,7-dimethylesculetin, capillarisin, rhein, and emodin respectively by comparing the retention time and mass data with those of the reference compounds. Other peaks were identified from several aspects, utilizing Elemental Composition of MassLynx software to carefully study their MS and MS–MS spectra. For example, a series of fragment ions of peak 14 were given in negative and positive ion ESI mode, 549 at $[M - H]^-$ and 573 at $[M + Na]^+$, fragments, their molecular weight could be confirmed to be 550, $C_{23}H_{34}O_{15}$, $C_{30}H_{30}O_{10}$, and $C_{36}H_{70}O_3$ were possible molecular composition being deduced by elemental composition. Accordingly, the literature data peak 14 was deduced as genipingentiobioside, 6,6′-diacetyl-2,2′,4,4′,5,5′-hexamethoxy; 1,1′-binaphthalene-7,7′-diol; or stearic anhydride. Among three compounds, peak 14 was perhaps genipingentiobioside, which have similar fragment ion spectra, 549 $[M - H]^-$, 225 $[M\text{-}H\text{-}2Glc]^-$, 207 $[M\text{-}H\text{-}2Glc\text{-}H_2O]^-$, and 179 $[M\text{-}H\text{-}2Glc\text{-}H_2O\text{-}CO]^-$ were determined in the (−) ESI-MS-MS spectra. Some peaks had same m/z value, such as peak 7, 13, and 15, peak 20 and 21. They had the same m/z value of quasimolecular ion in the MS spectra and a similar fragment ion in MS–MS spectra, respectively but the retention times were different, and the structural identification of these compounds are in progress. Peak 31 was detected at UV spectrum, however, there is no signal in the MS spectrum, and further identification is in progress.

3.3.2.3 UPLC-UV-Q-TOF Analysis of Plasma Sample after Oral YCHT Administration

Study on the adaptability of 6,7-dimethylesculetin, capillarisin, geniposide, emodin, and rhein in rat blood postoral administration was carried out. The limit of detection of them was determined to be 41.44 ng/L, 102.20 ng/L, 301.80 ng/L, 4.70 ng/L, 31.56 ng/L, respectively ($S/N \geq 3$). Kaempferol was as an internal standard. The typical chromatograms of the samples were shown in Figs 3.4–3.6. The constituents in rat plasma postoral administration of YCHT were well separated and identified by using their retention time and mass spectra. By comparing the chromatograms of plasma containing drug and control plasma, 21 peaks were identified in plasma. As a result, peak 2, 8, 9, 10, 11, 12, 13, 15, 17, 18, and 19 were original form compounds existing in Flos Artemisiae. Peak 1, 2, 5, 7 came from Gardeniae Jasminoidis, Fructus; Peak 9, 14, 16, 20, 21 resulted from Radix et Rhizoma Rhei; and Peaks 3 and 4 were metabolites of 6,7-dimethylesculetin. The MS data of (+) ESI-MS spectra and (−) ESI-MS spectra, and the identification results are shown in Table 3.1. Two (Peak 3 and Peak 4) of the 21 compounds were not an original form of the compounds that existed in the YCHT, they were identified as 6-methoxy coumarin-7-hydroxyl sulfate and 7-methoxy coumarin-6-hydroxyl sulfate by analyzing the mass data, and they were the metabolites of 6,7-dimethylesculetin.

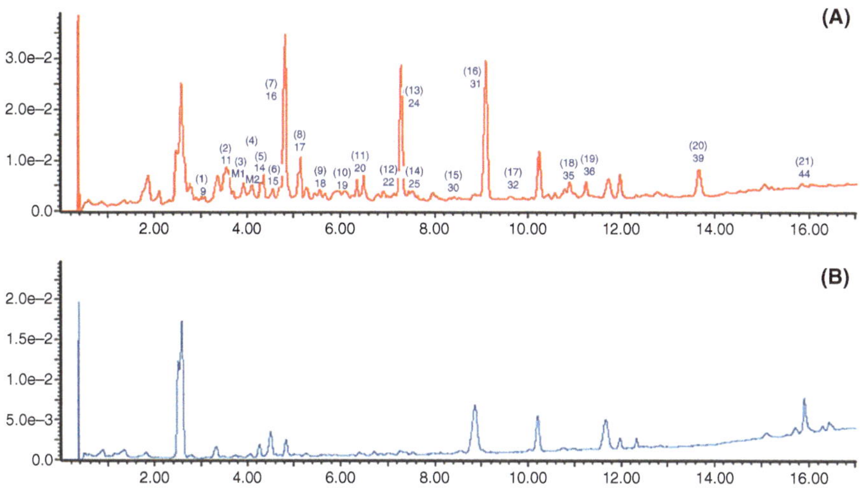

FIGURE 3.4 **Chromatograms of rat plasma by UPLC-UV detected at 254 nm.**
(A) Plasma postoral administration of YCHT; (B) control serum.

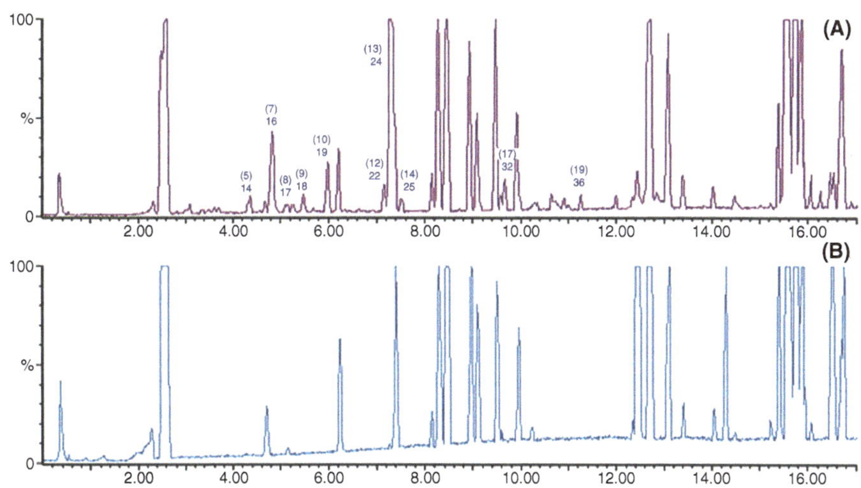

FIGURE 3.5 **TIC chromatograms of rat plasma by UPLC-Q-TOF in positive ESI mode.**
(A) Plasma postoral administration of YCHT; (B) control serum.

3.3.3 Correlation Analysis between Marker Metabolites and Absorbed Chemical Components

Under the premise of therapeutic efficacy, a correlation model between the marker metabolites and their chemical composition was established to discover and screen the constituents or metabolites absorbed into blood postoral administration of YCHT. The potential active constituents were

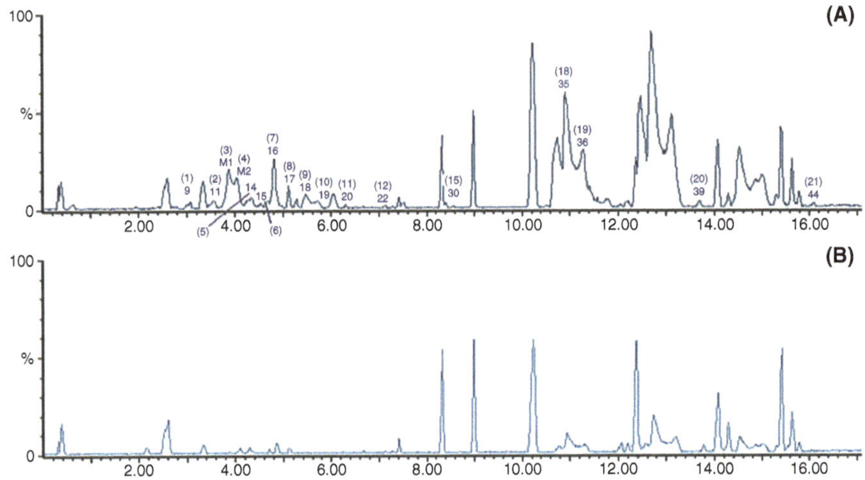

FIGURE 3.6 TIC chromatograms of rat plasma by UPLC-Q-TOF in negative ESI mode.
(A) Plasma postoral administration of YCHT; (B) control plasma.

screened using the plotting of correlation between marker metabolites and serum constituents (PCMS) in YCHT (graphed in Fig. 3.7). To determine which constituents contributed to the mainly therapeutic effect, we extracted the relevant chemical component data when $r \geq 0.8$ and identified the individual components using the PCMS and chemical composition databases. PCMS analysis showed that, the content of three potential active constituents changed significantly and had a trend toward a high correlation with effect. According to the correlation coefficient, nine components were highly positively correlated ($r \geq 0.8$) to therapeutic effect. However, eight of these could be identified and all came from YCHT. Interestingly, PCMS's correlation coefficient identified 16 components that were highly negatively correlated ($r \leq -0.8$) with therapeutic effect. However, 14 of these could be identified and all came from YCHT. In addition, top three compositions (6,7-dimethylesculetin (D), geniposide (G), and rhein (R)) had an extremely positive and negative relationship with therapeutic effect, and these constituents contributed most of the therapeutic effect. Therefore, we hypothesized that these components might play important roles in the therapeutic effect of YCHT. Currently, the activity of these compounds is under ongoing verification by cell biological experiments. Our results significantly narrow the range of potentially active compounds in YCHT, and build a solid foundation for future research on its action mechanism. The works could provide key information for the identification and elucidation of bioactive constituents, and providing a greater understanding of the metabolism as well as pharmacological actions for YCHT.

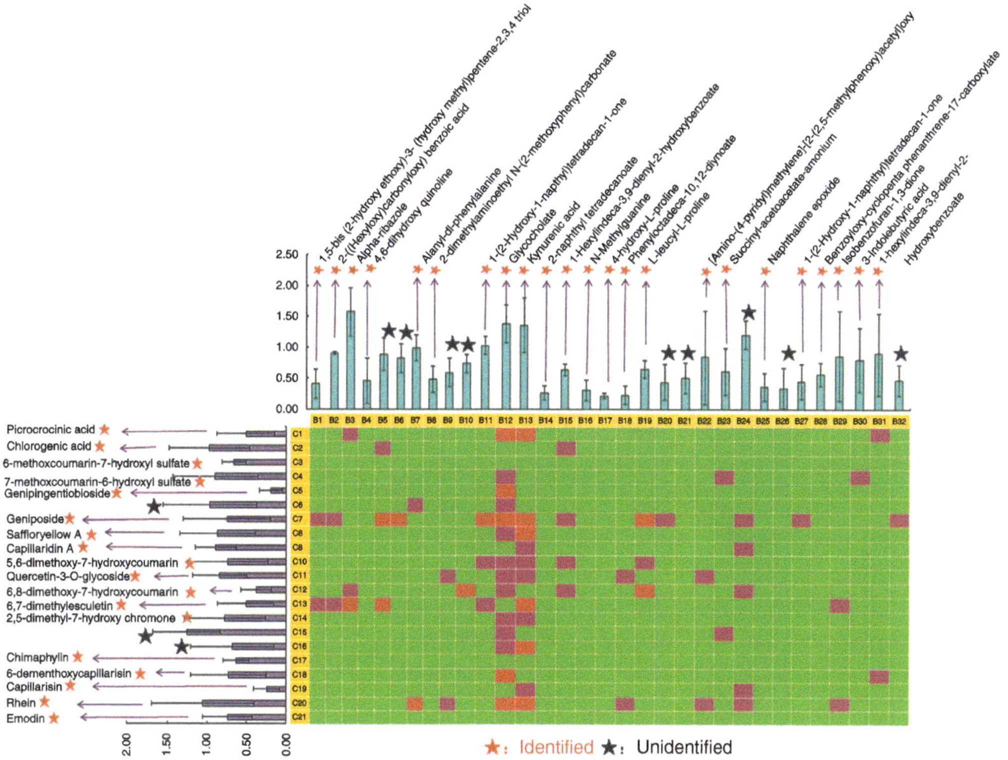

FIGURE 3.7 Plotting of the correlation between marker metabolites and serum constituents in YCHT.

Note: [orange], highly positively correlated; [purple], highly negatively correlated; [green], low correlation; C, chemical components; B, marker metabolites.

3.3.4 Cell Test to Verify the Therapeutic Efficacy of the Correlated Chemical Components

In order to maximize the experimental conditions, primary hepatocytes were treated with different concentrations of ethanol (50–200 mM) for 24–48 h. Hepatocytes tend to lose their phenotype after long-term culture. Therefore, a time course was performed to determine the earliest time point at which ethanol inhibited cell viability. However, a dose dependent decrease in cell count was observed after 24 h of ethanol treatment, and 50, 100, and 200 mM ethanol decreased cell viability by 17, 32, and 74.6%, respectively. Two hundred millimole of ethanol significantly decreased cell viability compared with control. Longer treatment (24 h) did not seem to further reduce the cell count. Based upon this finding, a 24 h time point was used for the remaining experiments.

To test whether compounds D, G, R and their combination had an effect equal to that of YCHT, the combination of these individual components was administered to primary hepatocytes alcohol-induced apoptosis. As shown in Fig. 3.8, after 24 h of treatment, the apoptotic index was significantly increased in the model group compared to that in the control group, whereas the apoptotic index was significantly decreased by treatment with D, G, R and their combination DGR, compared with that of the model group. Interestingly, the antiapoptotic effect of DGR combination was markedly stronger than the efficacy of monotherapy with each of the three components individually of its constituents, and the flow cytometric analysis showed that D played a stronger role in inhibiting alcohol-induced apoptosis in primary hepatocytes, than did G and R. It also shows that the DGR combination statistically intensified the

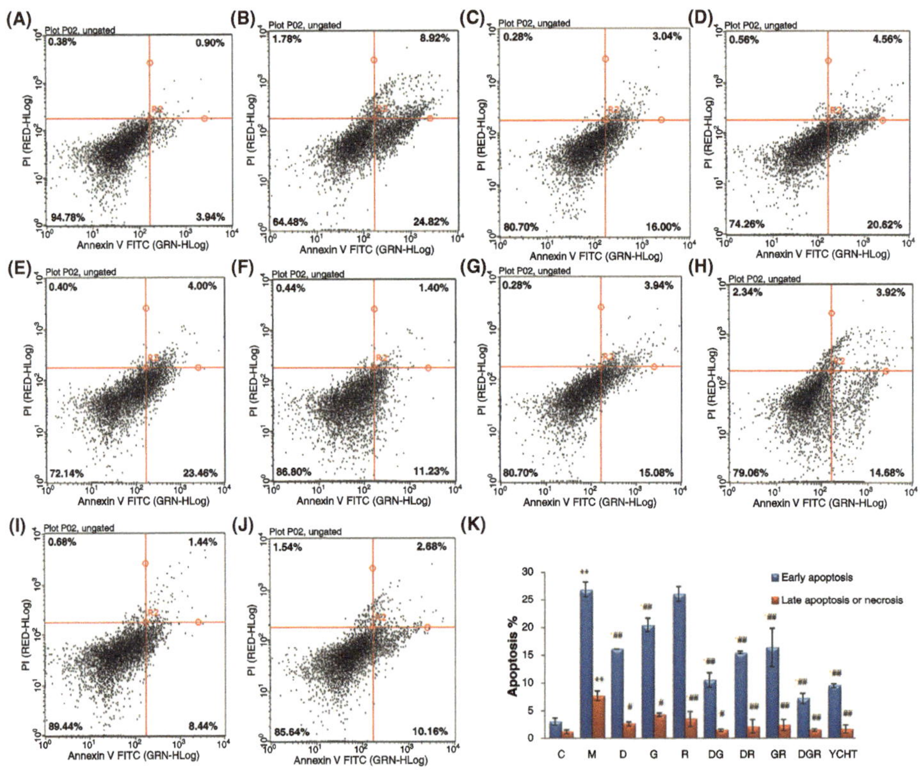

FIGURE 3.8 The antiapoptotic effect of 6,7-dimethylesculetin, geniposide, and rhein and their combination.

The flow cytometry data showed one representative dual staining result. Early- (%) and late-stage (%) apoptotic cell deaths were detected by staining cells with Annexin V-FITC and PI and analyzing by flow cytometry. (Note: (A), control; (B) model group; (C) 6,7-dimethylesculetin group; (D) geniposide group; (E) rhein group; (F) 6,7-dimethylesculetin + geniposide group; (G) 6,7-dimethylesculetin + rhein group; (H) geniposide + rhein; (I) 6,7-dimethylesculetin + geniposide + rhein group; (J) YCHT group; (K) the quantification of flow cytometry analyses.)

therapeutic efficacy relative to the control condition or monotherapy using D, G, R, and was better than the original YCHT. The mechanisms of combination treatment warrant further investigation and penetrating research is still in progress.

3.4 CONCLUSIONS

In this study, the establishment and implementation of chinmedomics was to explore the intrinsic link between the essence of syndrome and formulae efficacy, as well as potentially active constituents. YCHT has a hepatoprotective effect through regulating the perturbed metabolic pathways to the normal state, which are helpful to further understand the therapeutic mechanism of YCHT. Under the premise of therapeutic efficacy, a total of 21 peaks were tentatively characterized *in vivo*, and a correlation model between the marker metabolites and their chemical composition was used to screen the potential active constituents. PCMS analysis showed that top three compositions (6,7-dimethylesculetin, geniposide, and rhein) had an extreme relationship, with therapeutic effects, and might play important roles in the therapeutic effect of YCHT. Activity of these compounds has also been verified by the cell biological assay. It suggests that the chinmedomics strategy is a powerful approach used for elucidation of the therapeutic mechanisms of TCM, and exploiting a bright future for drug discovery from TCM, as well as providing assistance in the exploration of the effective substances for further research of TCM.

CHAPTER 4

Chinmedomics Approach Dissects Therapeutic Properties of ShenQiWan Acting on ShenYangXu Syndrome

Xijun Wang, Aihua Zhang, Hui Sun

CONTENTS

4.1 INTRODUCTION

Traditional Chinese Medicine (TCM) prescriptions have been widely used, as these TCM prescriptions possess antibacterial, antifungal, anticancer, antiviral, anti-inflammatory activity, and other pharmacological activities of benefit to humankind. The pharmacological effects of TCM prescription are demonstrated to express their effects through multiactive chemical constituents that have active function on some diseases. For the screening and analysis of effective compounds in TCM prescription, a conventional phytochemical approach that isolates and identifies individual components one by one is both time-consuming, and labor intensive. In view of this, it is essential to develop the available methods that could overcome these limitations, to reveal the complex compounds in TCM prescription. Recently, the chinmedomics strategy based on the serum pharmacochemistry of TCM combined with metabolomics has been introduced; for the biological ingredients analysis of herbal medicine.

With the development of Chinese medicine, many Chinese medicine preparations such as granules, capsules, pills, and injections are now produced in order to cater to the demands of customers. TCM has played an indispensable role in the prevention and treatment of diseases in most Asian countries. TCM consider the kidney is the most important system in the body. The TCM view as "Kidney yang" motivates the power of human vitality, and controls the storage of vital essence, and is the pilot light for our energy system. Since the TCM kidney affects so many vital functions, ShenYangXu (SYX, Kidney-yang deficiency syndrome) is a complex kidney disorder involving many diseases, such as, nephrotic syndrome and chronic renal failure. ShenQiWan (abbreviated as SQW) is a typical TCM prescription for invigorating SYX, used for thousands of years in Asia, and was first recorded in Synopsis of Golden Chamber, consisting of *Radix Rehmanniae Preparata, Fructus Macrocarpii, Rhizoma Dioscoreae Oppositae, Poria, Rhizoma Alismatis, Cortex Moutan Radicis, Achyranthes bidentata,* and *Plantago asiatica*. It may yield multiactive chemical constituents that may affect various pathways and maintain the

Chinmedomics. http://dx.doi.org/10.1016/B978-0-12-803117-9.00004-4

balance of metabolic pathways in diesease treatment. Nevertheless, due to its complex constituents, it is not easy to explore the action mechanism of SQW using traditional ways, and little is known about the change of the potential biological active constituents, and metabolites in SYX treated with SQW. In this study, we have selected SQW as the case study for biological ingredient analysis via chinmedomics strategy.

4.2 EXPERIMENTAL DETAIL

4.2.1 Chemicals and Materials

Acetonitrile, HPLC grade, was obtained from Merck (Darmstadt, Germany); methanol (HPLC grade) was purchased from Fisher Scientific Corporation (Loughborough, UK); leucine enkephalin was purchased from Sigma–Aldrich (St. Louis, MO, USA). The deionized water (18.2 MΩ) was further purified using a Milli-Q system (Millipore, Billerica, USA); formic acid and phosphoric acid were of an analytical grade purchased from Beijing Reagent Company (Beijing, P.R.C.). *Radix Rehmanniae Preparata, Fructus Macrocarpii, Rhizoma Dioscoreae Oppositae, Poria, Rhizoma Alismatis, Cortex Moutan Radicis, Achyranthes bidentata,* and *Plantago asiatica* were obtained from Harbin Shiyitang Drug Store (Harbin, China), and authenticated by Prof. Xijun Wang, Department of Pharmacognosy of Heilongjiang University of Chinese Medicine. Voucher specimens were deposited at the authors' laboratory. Commercial SQW were purchased from Shiyitang Co., Ltd (Harbin, China).

4.2.2 Metabolomics Study

4.2.2.1 Animal Handling

Male Wistar rats (weighing 200–240 g) were supplied by the GLP center of Heilongjiang University of Chinese Medicine (Harbin, China). The room temperature was regulated at 21 ± 5°C with 60 ± 5% humidity. A 12 h light–dark cycle was set, and there was free access to standard diet and water. The animals were allowed to acclimatize in metabolism cages for 7 days prior to dosing and were put in metabolism cages during the urine collection periods. After acclimatization, the animals were randomly divided into 5 groups with 15 rats in each group: the control, model, and SQW group. The rats in the control group were firstly induced by i.p. injection of olive oil for 21 consecutive days. The rats in model group were induced by i.p. injection of corticosterone at the dose of 10 mg/kg per day for 21 consecutive days. The SQW group was administrated by the same method as the model group during the first 21 days. After that the SQW group was orally administrated: an accurate volume of solution of SQW (3.389 g/kg) for 15 consecutive days. All animal care and experimental procedures were performed in compliance with the Ethical Committee of Heilongjiang University of Chinese Medicine.

4.2.2.2 Sample Collection and Preparation

Blood was collected from the abdominal aorta, and plasma and serum were separated via centrifugation at 4000 rpm for 10 min at 4°C after standing for 30 min. The supernatants were collected in microtubes and stored at −80°C before analysis. The levels of biochemical assay according to the manufacturer's instructions of commercial kits (data not shown). All procedures completely complied with the manufacturer's guidelines. The serum samples were thawed before analysis. Proteins were precipitated from the defrosted serum samples (200 μL) by adding four volumes of methanol in 1.5 mL micro tubes at room temperature. After vortexing for 10 s, supernatants (800 μL) were collected after centrifugation at 13,000 r/min for 10 min. Supernatants were dried under nitrogen and redissolved with 80% methanol to volumes of 200 μL. The mixture solution was filtered through a PTFE membrane of 0.22 mm, and transferred to vials for UPLC-Q-TOF-HDMS analysis.

4.2.2.3 UPLC-Q-TOF-HDMS Analysis

Chromatography was performed on an ACQUITY UPLC system (Waters Corp., Milford, MA/USA) with autosampler at 4°C. The separation was carried out on an ACQUITY UPLC HSS T_3 column (100 × 2.1 mm i.d., 1.8 μm, Waters Corporation, Milford, USA). The column temperature was maintained at 40°C. The analysis was performed with gradient elution using (A) acetonitrile with 0.1% formic acid and (B) water, with 0.1% formic acid as the mobile phase, and the flow rate was set at 0.4 mL/min. The gradient eluting condition was: 0–2 min, linear increasing from 5–50% A; 2–4 min, 50–60% A; 4–7 min, 60–80% A; 7–10 min, 80–100% A; 10–12 min, maintain 100% A; 12–12.5 min, linear decreasing from 100–5% A; held at 5% A for 2 min for equilibration of the column. The sample injection volume was 3 μL. The QC samples were used to optimize the condition of UPLC/MS, as it contained the most information of whole plasma samples.

The Waters ACQUITY™ Mass Spectrometer (Waters Corp., Milford, USA) was connected to the UPLC system via an electrospray ionization (ESI) interface. The ESI source was operated in positive ionization mode with the capillary voltage at 3.0 kV, sample cone voltage at 30 V, and extraction cone voltage at 4 V. The temperature of the source and desolvation was set at 110 and 450°C, respectively. The cone and desolvation gas flow rates were maintained at 50 and 500 L/h, respectively. In the negative ion mode sample cone voltage at 40 V, and desolvation temperature was 300°C. All data collected in centroid mode were acquired using Masslynx™ NT4.1 software (Waters Corp., Milford, MA/USA) and digitally stored. The data acquisition rate was set to 0.3 s/scan, with a 0.1 s (0.02 interscan time) interscan delay. Data were collected in centroid mode from 50 Da–1000 Da. For accurate mass acquisition, a lock-mass of leucine enkephalin at a concentration of 0.2 ng/mL was used via a lockspray interface at flow rate of 100 mL/min, monitoring in positive ion mode

($[M + H]^+ = 556.2771$) and in negative ion mode ($[M - H]^- = 554.2615$) to ensure accuracy during the MS analysis.

4.2.2.4 Metabolic Profiling and Metabolite Analysis

The UPLC-HDMS data acquired were performed to Markerlynx within Masslynx software (version 4.1) for peak detection and alignment. The intensity of each ion was normalized with regard to the total ion count to generate a data matrix that consisted of the retention time, m/z value, and the normalized peak area. The multivariate data matrix was induced to EZinfo software 2.0 (Waters Corp., Milford, USA) for principal component analysis (PCA), partial least-squares-discriminant analysis (PLS–DA) and orthogonal projection to latent structures (OPLS) analysis. The molecular ion was found from the mass spectrometry and subsequently, accurate molecular weight was calculated. The MassFragment™ application manager (Waters corp., Milford, USA) was used to facilitate the MS/MS fragment ion analysis process by way of chemically intelligent peak-matching algorithms. Furthermore, MassFragment conducts rapid putative ion identification and putative biologically relevant analysis via incorporation of ChemSpider, HMDB, KEGG, and Lipid Maps. The pathway analysis of potential biomarkers was performed with MetaboAnalyst based on database source.

4.2.3 Constituents Analysis *In Vitro* and *In Vivo*

4.2.3.1 Animals

Male Wistar rats (220 ± 20 g) were obtained from the Laboratory Animal Center of Heilongjiang University of Chinese Medicine (Harbin, China). Animals were bred in a breeding room with temperature of 24 ± 2°C, humidity of 60 ± 5%, and 12 h dark–light cycle. They were given tap water and fed normal food *ad libitum*. All the experiment animals were housed under the above conditions for 3 days acclimation, and were fasted overnight before the experiments.

4.2.3.2 Preparation of SQW Samples for Analysis

According to the original composition and preparation method of SQW recorded in "Synopsis of Golden Chamber", the SQW was prepared in the following procedure: *Radix Rehmanniae Preparata, Fructus Macrocarpii, Rhizoma Dioscoreae Oppositae, Poria, Rhizoma Alismatis, Cortex Moutan Radicis, Achyranthes bidentata,* and *Plantago asiatica* in proportion 24:6:6:6:3:6:6:1:6:6. Half grams was ground into crude powders, mixed, and then ultrasonic extraction with 100 mL volume of 100% ethanol for 40 min twice. The methanol extraction was centrifuged at 13,000 rpm for 15 min at 4°C, and the supernatant was filtered through a 0.20 μm filter, the 5 μL filtrate was used as UPLC sample.

The raw powder of SQW was dissolved with distilled water as stock solution (0.3389 g/mL). The above solution was orally administrated to male Wistar rats (1 mL/100 g body weight). Sixty minutes after administration, the animals were anesthetized by intraperitoneal injection of 3% pentobarbital sodium

(0.3 mL/100 g body weight). The blood was collected from the hepatic portal vein and then centrifuged at 4000 rpm for 15 min at 4 °C. The supernatant obtained were frozen immediately and stored at −80 °C, and thawed before analysis. Phosphoric acid (40 μL) was added to 2 mL of the above supernatant, and ultrasonicated for 1 min, and vortexed for 30 s. The mixed solution was applied to a preactivited OASIS HLB solid phase extraction C_{18} column (30 μm, 60 mg, Waters Corporation, USA). The column was washed with 1 mL of water and 2 mL of 100% methanol. The 100% methanol eluates were collected and dried under nitrogen gas at 45 °C. The residues were redissolved in 150 μL of methanol and at 13,000 rpm for 15 min at 4 °C, filtered through a 0.20 μm filter, the 5 μL filtrate was used as UPLC sample.

4.2.3.3 *Instrumentation and Conditions*

4.2.3.3.1 Ultra-High Performance Liquid Chromatography

Chromatographic separation for samples was performed using a Waters Acquity™ ultra performance LC system controlled with Masslynx (V4.1). Separation was performed with an acquity UPLC BEH C_{18} column (100 × 2.1 mm i.d., 1.7 μm particle size, Waters Corporation, Milford, MA, USA) maintained at 35 °C. A gradient with eluent A (HCOOH: H_2O = 0.1: 100, v/v) and B (HCOOH: CH_3CN= 0.1:100, v/v) was used at a flow rate of 0.5 mL/min. The linear elution gradient program was used as follows: 0–2.0 min, 90–65 % A; 2.0–4.0 min, 65–40% A; 4.0–8.0 min, 40–20% A; 8.0–8.5 min, 20–1% A; 8.5–10.5 min, 1% A; 10.5–11.5 min, 1–90 % A; 11.5–13.0 min, 90 % A. The sample-tray temperature was kept at 4 °C.

4.2.3.3.2 Mass Spectrometry Condition

MS instrument consisted of a Waters Synapt™ QTOF/ HDMS (Waters Corp., Milford, MA, USA). Ionization was performed in both positive and negative ion modes. The MS source temperature was set at 110 °C, and the desolvation temperature was set at 300 °C with desolvation gas flow at 500 L/h. The capillary voltage was 4 kV. The mass spectrometric data was collected in full scan mode across the range of m/z 100–1000 Da, with accurate mass measurement of all mass peaks. Ar gas was used as collision gas at a pressure of 0.2 Mpa. The collision energy was set as 10–30 eV for low-energy scans, and 30–50 eV for high-energy scans. Leucine-enkephalin ($[M + H]^+$ = 556.2771, $[M - H]^-$ = 554.2615) was used as the lock-mass generating a reference ion for accurate mass acquisition. The instrument was controlled by Masslynx 4.1 (Waters Corp.), and the m/z was from 50–1000 in positive and negative ion.

4.2.3.3.3 Data Processing

LC–MS data of all determined samples were further processed by Markerlynx V4.1 software (Waters Corporation, Milford, USA) for peak detection and peak alignment. The method parameters for data processing were set as follows:

retention time range 0.1–10.0 min, mass range 100–1000 Da, retention time tolerance 0.2 min, mass tolerance 0.05 Da, noise elimination level 6; peak intensity threshold 50. The peak width at 5% height and the peak-to-peak baseline noise were automatically determined. Pareto scaling of data was performed prior to analysis that generated the less noise level than other scaling method.

4.2.3.3.4 Metabolynx Processing Approach

Expected and unexpected metabolites were found out using Metabolynx™ software (Waters Corporation, Milford, USA) by automatically comparing MS(E) data from the sample data. Metabolynx in combination with the elemental composition of the substrate molecules generated a series of extracted ion chromatograms. The extracted ion chromatograms which present in the analyte were evaluated relative to the controls sample by comparing their retention time and peak area. The method parameters for data processing were set as follows: retention time range 0.1–20.0 min, mass range 50–2000 Da, retention time tolerance 0.1 min, mass tolerance 0.05 Da, noise elimination level 5, peak intensity threshold 50. If the different peaks were found in the analyte by comparing with control sample, only those peaks with the area ratio of analyte to control sample to be above three were considered to be the compound-related metabolites.

4.2.4 Correlation Analysis of Biomarkers and Absorbed Constituents

A plotting of a correlation model between the marker metabolites and chemical composition (PCMS) was used to screen the constituents or metabolites absorbed into blood postoral administration of SQW. To determine which constituents contributed to the mainly therapeutic effect, PCMS were performed using Pearson's correlation, and the correlation coefficient (r) described the degree of correlation between two variables: all the markers and the chemical composition of the SQW *in vivo*. The r values were between -1 and $+1$; $r > 0$ indicates that two variables were positively correlated, whereas $r < 0$ indicates a negative correlation. The larger the absolute value, the higher the correlation; therefore, we defined $r = 0.90$–1.00 as extremely related, and $r = 0.80$–0.89 as highly related.

4.3 RESULTS AND DISCUSSIONS

4.3.1 Metabolomics Analysis

4.3.1.1 Multivariate Statistical Analysis of Metabolite Profiling

Using optimal conditions, the representative base peak intensity (BPI) chromatograms of the plasma samples that were collected from representative rats from each group are presented in Fig. 4.1. Raw data from UPLC/MS were analyzed by the MarkerLynx software. RT, m/z and peak height intensities were

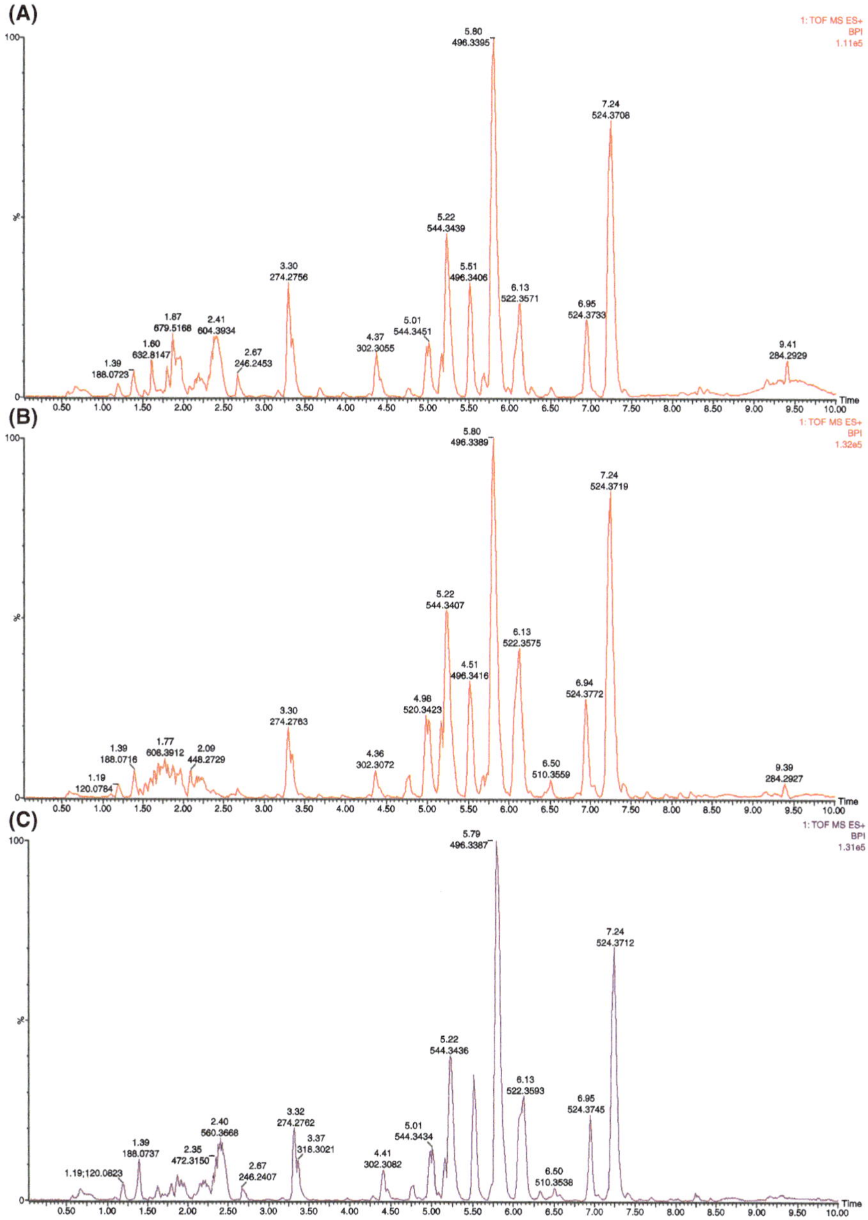

FIGURE 4.1 BPI chromatograms of plasma samples.

(A) Control group; (B) model group; and (C) SQW group.

imported into EZinfo 2.0 software for data analysis. Multivariate data analysis was performed using PCA method, and there is an obvious classification between the clustering of the model and control groups (Fig. 4.2A). The obvious separation between the control and model group suggest that biochemical perturbation significantly happened in the model group. S-plot of OPLS was carried out to find biomarkers of SYX in the study (Fig. 4.2B). The furthest metabolite ions from the origin exhibiting a higher value of S-plot were potential biomarkers, which were responsible for the difference between control and model groups.

4.3.1.2 Identification of Serum Biomarkers Related to SYX

The S-plot suggests the ions furthest away from the origin having a higher value contribute significantly to the clustering of the two groups, and they may be regarded as the potential biomarkers in SYX rats. Metabolite identification was conducted with high-resolution MS and MS/MS fragments, as well as database analyses. In this study, a total of 23 ions ($p < 0.05$) contributed to the classification of the control and model groups. Among them, 5 of the 23 biomarkers identified were upregulated and 18 of them were downregulated in the SYX group (shown in Fig. 4.3). To identify these metabolites, these variables were predicted by comparing the accurate MS and MS/MS fragments with the metabolites searching in online databases.

4.3.1.3 Effects of SQW on SYX Based on Metabolite Profile

Using the presented UPLC–HDMS method, the serum metabolic profiles of SQW treated group were obtained. As the 23 potential biomarkers from UPLC–HDMS have been found, it is reasonable to take them as the potential drug targets for further investigating the therapeutic effects of SQW. From the 3D plot of score plot of the PCA, a clear separation among the control, model, and SQW group was easily seen in Fig. 4.2C. The score plot of the PCA showed that the control, model, and SQW treatment groups were separated clearly, and the SQW treatment group was much closer to the control group than the model group, which suggested that SQW could restore the pathological process of SYX. To determine whether SQW was possible to influence the metabolic pattern induced by SYX, the intensities of the metabolites were compared among the control, SYX, and SQW treated groups. Of note, 17 of the identified biomarkers are completely reversed by SQW (graphed in Fig. 4.3).

4.3.2 Constituents Analysis of SQW

4.3.2.1 Optimization of Chromatographic and Mass Spectrometry Conditions

UPLC and MS conditions were optimized to obtain better detection. UPLC chromatograms with good separation, different mobile-phase compositions

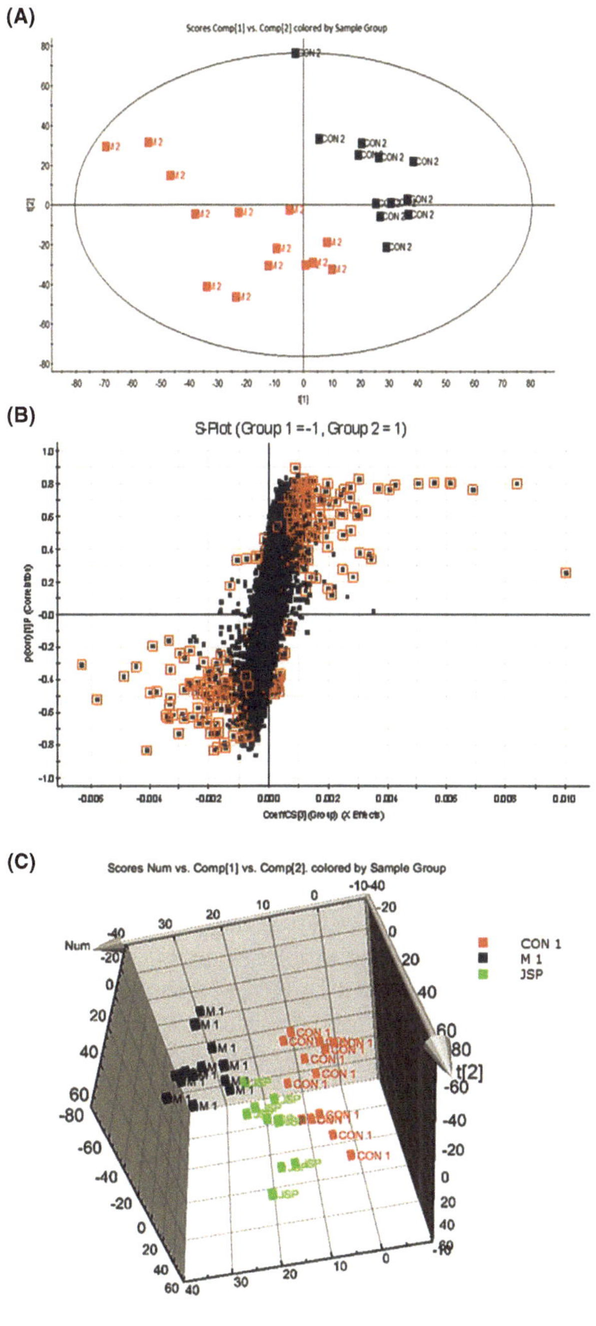

FIGURE 4.2 Multivariate data analyses of serum metabolite profiling.

(A) Score plots of PCA of serum metabolite profiling (■, control group; ■, model group); (B) S-plot of PLS-DA for the control and model groups; (C) 3D plot of PCA of SQW treatment on SYX (■, control group; ■, model group; ■, SQW group).

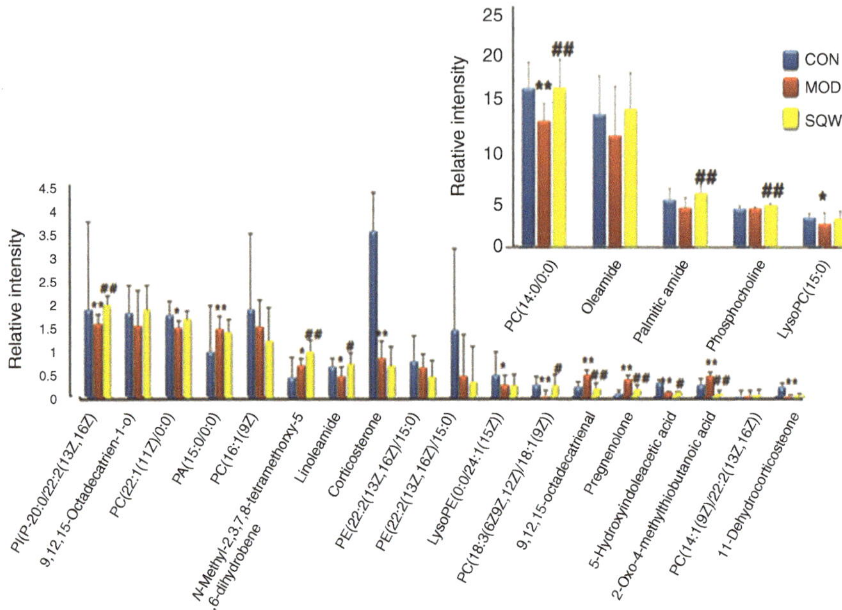

FIGURE 4.3 **Relative signal intensities of serum metabolites identified by UPLC/MS.** Bar plots represent the mean relative metabolite concentration and standard deviations.

were screened (data not shown) and it was found that acetonitrile and formic acid was the most suitable eluting solvent system. The mobile phase played an important role in achieving good chromatographic behavior and appropriate ionization. A mixture of 0.1% v/v aqueous formic acid and acetonitrile was finally chosen as the preferred mobile phase because it produced the desired separation and acceptable tailing factors within the 13 min run time. In the course of optimizing separation conditions, mobile phase, gradient program, column temperature, and detection wavelength were investigated. The final results showed that best resolution, shortest analysis time and lowest pressure variations were achieved when a gradient elution mode composed of acetonitrile (containing 0.1% formic acid, eluent A) and water (containing 0.1% formic acid, eluent B) was programmed as follows: 0–1 min, 1% A; 1–2.5 min, 1–13% A; 2.5–6.5 min, 13–40% A; 6.5–8.0 min, 40–99% A; 8.0–10.5 min, 99% A; 10.5–11.0 min, 99–1% A; 11–13 min, 1 A. The most efficient way to obtain rapid analysis time, while generating high column efficiency and resolution, is by utilizing a small particle size column. Therefore, chromatographic analysis was executed through a 1.7 μm particle size BEH C_{18} column. Considering sensitivity and resolution, the flow rate was optimized at 0.5 mL/min.

Furthermore, in order to reduce the column pressure resulting from a higher flow rate, the liquid chromatography column temperature was set at 40°C. TOF/MS enables peak deconvolution of analytes of differing m/z after ionization was utilized in order to achieve fast scanning characteristics for rapid chromatographic separations. For the MS conditions, a series of parameters including the ion spray voltage, turbo spray temperature, declustering potential, collision energy, nebulizer gas, heater gas, and curtain gas were all optimized. Both positive and negative ion modes were employed to screen as many potential compound signals as possible. The BPI chromatogram in rat plasma postoral administration of SQW analyzed by UPLC/Q-TOF-MS/MS in positive ion mode and negative ion mode was shown in Fig. 4.4.

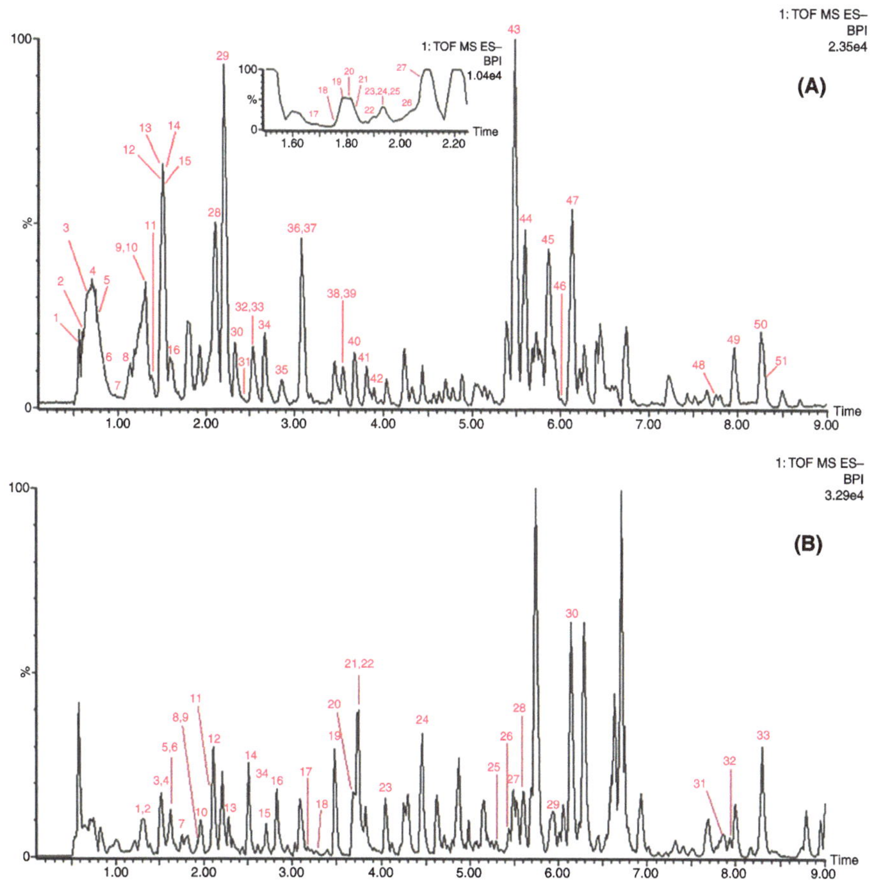

FIGURE 4.4 ESI base peak ion (BPI) chromatogram of the SQW analyzed by UPLC/QTOF MS in negative ion mode (A) and positive ion mode (B).

4.3.2.2 Rapid Discovery and Global Characterization of Multiple Constituents From SQW

UPLC/QTOF MS system with automated MS^E (E represents collision energy) data analysis software (MetaboLynx™) was used to analyze and identify the chemical components of SQW. Mass spectrometry was performed in a full-spectrum (m/z 100–2000) mode using an MS^E acquisition of both molecular and fragment ion data at low (10–30 eV) and ramped (30–50 eV) collision energies. We developed a broad-spectrum screening method on an UPLC/QTOF MS with collision energy function (MS^E mode and accurate mass) that provides a unique way to incorporate fragment ion information without the need of precursor ion selection. Identification was based on the presence of one characteristic m/z ion obtained with the low collision energy function, and at least one fragment ion obtained with the high collision energy function, both with mass errors of less than 5 ppm permitted. The molecular formulas with relative mass error between the measured, and theoretical mass within 5 ppm were chosen, and then applied to MassFragment analysis for further confirmation of their structures. High- and low-energy MS^E tool generates product ion spectra that were successfully applied to structural elucidation of detected SQW.

Detailed clarification of the chemical constituents in SQW is essential for its holistic quality control. In order to complete the comprehensive characterization, multiple approaches, including database searching, reference standard comparison, and QTOF–MS and MS/MS data analysis, were employed for structural characterization of the SQW constituents. The chemical formula of an unknown structure was deduced based on the high-accuracy $[M-H]^-$/ $[M+HCOO]^-$ (in the negative ion mode) or $[M+H]^+$/$[M+Na]^+$ (in positive ion mode) precursor ion. The compound(s) matching the determined chemical formula and considered to undergo the observed MS/MS fragmentation was selected as the final identity. As a result, a total of 84 compounds, were identified or tentatively characterized from the constituents of SQW. The ESI BPI chromatogram of the SQW by UPLC/QTOF MS is shown in Fig. 4.4. By UPLC/QTOF MS analysis, a total of 84 compounds including 51 compounds in the negative ion mode and 33 compounds in the positive ion mode were identified from SQW.

4.3.2.3 SQW In Vivo Analysis Using Metabolynx Tool

The bioactive compounds in SQW could be discovered by comparative analysis of the chemical profiles of control plasma and dosed plasma, and then identified based on their MS and MS/MS spectra (Figs 4.5 and 4.6). It indicated that 24 components in the SQW were absorbed into the rat's body. In the present study, metabolites *in vivo* rat's system were identified and elucidated by UPLC/Q-TOF-MS/MS with the Metabolynx software tool. Using the

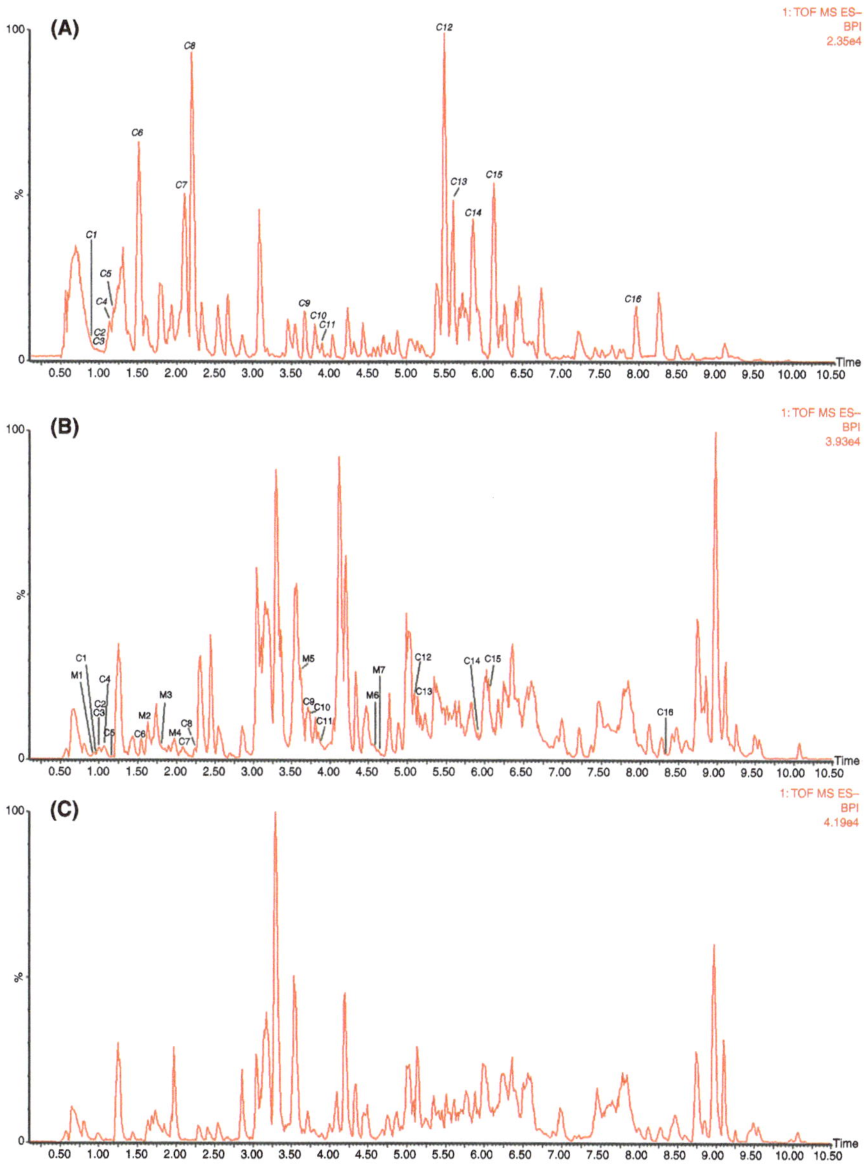

FIGURE 4.5 **The UPLC–HDMS chromatograms of SQW in negative ion mode.**
(A) *In vitro*; (B) dosed serum; (C) control serum.

Metabolynx™ software, peaks present in the analyte were evaluated relative to the control samples by comparing their retention times with postacquisition analyses, MS spectra and the peak area. When a peak was included as a metabolite, the peak area in the analyte had to be at least three times greater than that of the control. After postacquisition data processing of plasma samples (using

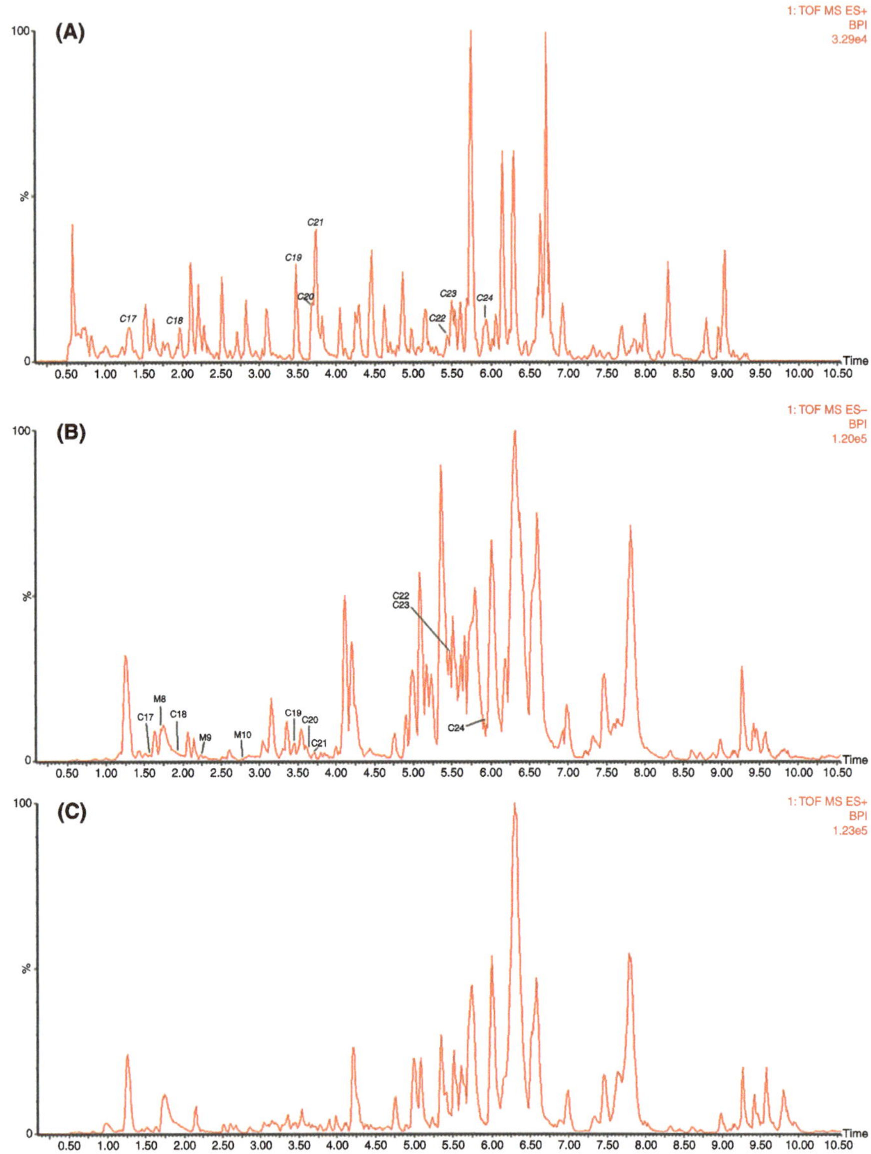

FIGURE 4.6 The UPLC–HDMS chromatograms of SQW in positive ion mode.
(A) *In vitro*; (B) dosed serum; (C) control serum.

the MetaboLynx program) the results demonstrated that the vast majority were metabolites. The extracted ion chromatogram of plasma samples with the Metabolynx tool are presented in Fig. 4.7, and the products were well separated using the developed UPLC method. Through a comprehensive analysis of the peaks, we found 10 metabolites that indicated that glucuronide conjugation,

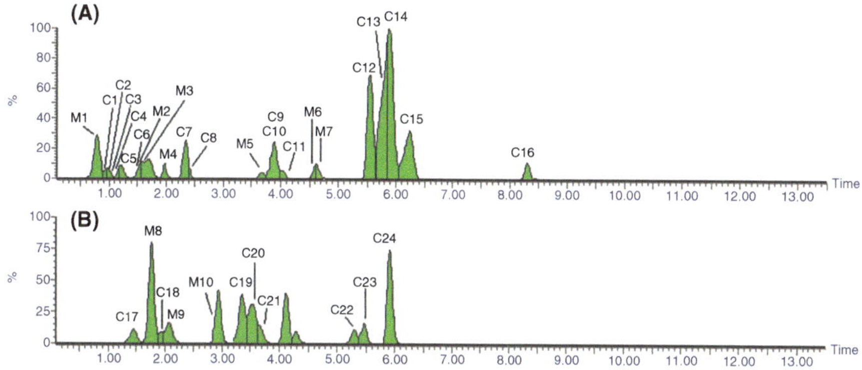

FIGURE 4.7 The extracted ions chromatogram of rat serum postoral administration of SQW with Metabolynx in the negative ion mode (A) and positive ion mode (B).

decarbonylation, and de-ethylation were the major metabolic pathways of constituents *in vivo*. The automated data analysis (MetaboLynx) was developed and validated for simultaneous identification, and determination of metabolites produced by a body *in vitro*. Expected and unexpected metabolites were detected by the Metabolynx software, which could automatically compare MS data from the sample and control. From these results, it could be concluded that the proposed method could be used to simultaneously analyze and screen the multiple absorbed bioactive constituents and metabolites in TCM. The present method provides a quick and reliable analytical tool to investigate the prototype components absorbed and metabolite metabolism of TCM. In order to illustrate the dynamic metabolism profiles of absorbed constituents and metabolites *in vivo* further research would be required on pharmacokinetics to understand the absorption, distribution, and excretion of these components.

4.3.3 Correlation Analysis Between Marker Metabolites and Absorbed Constituents

A correlation model between the marker metabolites and their chemical composition was established, under the premise of therapeutic efficacy, to discover the constituents or metabolites absorbed into blood postoral administration of SQW. To determine which constituents contributed to the mainly therapeutic effect of SQW against SYX syndrome, we extracted the relevant data when $r \geq 0.8$ using the PCMS software. The absorbed constituents that were highly correlated with therapeutic effect of SQW are shown in Fig. 4.8. According to the correlation coefficient, 23 absorbed components such as gallic acid-O-glucuronide, gallic acid, geniposidic acid, etc. were highly positively and negatively correlated ($r \leq -0.8$) with the therapeutic effect of SQW acting on SYX syndrome. It also proves that TCM has features of multicomponent and

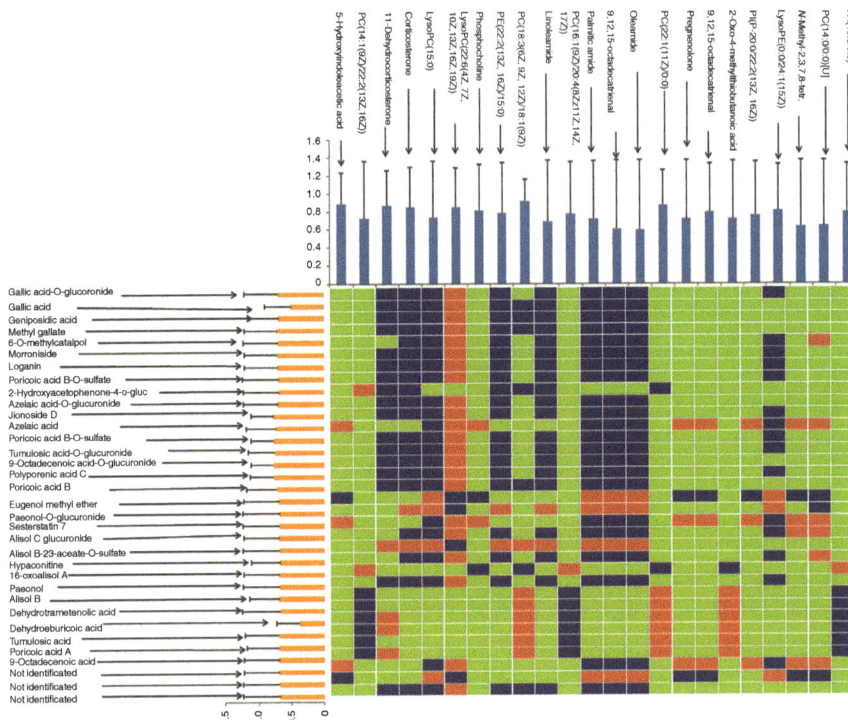

FIGURE 4.8 **Plotting of the correlation between marker metabolites and serum constituents in SQW.**

(Note: , highly positively correlated; , highly negatively correlated; , low correlation; left column, chemical components; top column, marker metabolites.)

multitarget effects. Therefore, we hypothesized that these components might play important roles in the therapeutic effect of SQW. Interestingly, top three compositions (azelaic acid, eugenol methyl ether, and sesterstatin) had an extremely positive and negative relationship with therapeutic effect of SQW against SYX syndrome. Additional experiments are ongoing to further reveal the activity of these compounds.

4.4 CONCLUSIONS

In this study, the chinmedomics approach has been successfully established for biomarker exploration in SYX and the potentially active components of SQW. As a result, a total of 84 compounds, were identified or tentatively characterized from the SQW. Under the premise of therapeutic effects, a total of 34 peaks were tentatively characterized *in vivo*, 24 of which were parent components, and 10 metabolites were detected. PCMS correlation analysis model showed that 23 absorbed components, such as, gallic acid-O-glucuronide, gallic acid, geniposidic acid, etc. were highly correlated ($r \leq -0.8$) with the therapeutic effect of SQW acting on SYX syndrome. In this work, we conclude that the multicomponents of SQW caused therapeutic effect on the SYX syndrome. This approach yielded a series of potential bioactive compounds in a high-throughput manner, which is useful for the drug discovery of compounds from TCM prescriptions.

CHAPTER 5

Metabolic Profiling and Biomarkers Analysis of Jaundice Syndrome

Xijun Wang, Aihua Zhang, Hui Sun

CONTENTS

5.1 INTRODUCTION

Traditional markers of the conventional clinical chemistry and histopathology method are not region-specific, and only increase significantly after substantial disease injury. Therefore, more sensitive biomarkers of disease are needed. Compared with traditional diagnostic methods, even small changes of metabolites can help to detect early pathologic changes more sensitively. Particularly, for the early detection of disease, highly sensitive, and specific biomarkers as primary indicators in biofluids are relatively more useful because these can be used for nonbiopsy tests. Monitoring certain metabolite levels in urine fluid has become an important way to detect early stages in disease (Arakaki et al., 2008). Analyzing and verifying the early biomarkers of a disease −metabolomics enables us to better understand pathological processes, substance metabolic pathways (Zhang et al., 2014). Metabolite changes that are observed in diseased individuals as a primary indicator have become possible, and hence the measurement of metabolites has become an important part of clinical practice. Urinary metabolomic approaches have been used to screen for potentially earlier diagnostic and prognostic biomarkers of disease (Kim et al., 2009).

Traditional Chinese Medicine (TCM) treats syndrome that is an outcome after a careful analysis of all symptoms and signs as the core and the essence of human life systems (Zhang et al., 2013a). Understanding syndromes is a core research to develop more efficient therapeutic strategies, classification, and diagnostic criteria for patients (Li et al., 2013). However recently, TCM syndrome has been found to suffer from several obstacles that are not able to capture real characteristics. With the increasing pace of modern life, many people have suffered from liver diseases that are common diseases in China. Clinical practice has shown that patients with liver diseases such as cholestasis, hepatitis C, primary biliary cirrhosis, and liver fibrosis, are commonly complicated by jaundice syndrome (JS; Huangdan Zheng in Chinese) (Selmi et al., 2011; Arias, 1998). JS is a common and fatal disease requiring early diagnosis and

Chinmedomics. http://dx.doi.org/10.1016/B978-0-12-803117-9.00005-6

effective treatment, and is mainly focused on the two main subtypes of YangHuang (YAH) and YinHuang (YIH) Syndrome. Current diagnostic methods, measuring levels of alanine aminotransferase and aspartate aminotransferase, etc., in blood samples, have been the main diagnostic markers for JS and can be considered as the current "gold standard" for initial diagnosis and surveillance. However, sensitivity of these markers is relatively low, it is difficult to observe outcomes immediately, and is not particularly effective in separating cases of JS from other non-JS disorders. Fortunately, the rapid development of metabolomics technology has been used to explore the particular metabolites, potentially diagnostic and prognostic biomarkers, and pathways of syndrome and provides a methodological basis for deeply understanding the essence of the syndrome from the aspect of systems biology (Zhang et al., 2012). Consequently, this work was designed to investigate a comprehensive metabolic profiling of JS, and its subtypes by ultraperformance liquid-chromatography/quadrupole-time-of-flight coupled with high-definition mass spectrometry (UPLC-Q-TOF-HDMS) combined with pattern recognition methods and pathways analysis, in order to define the new potential biomarkers, and generate a better understanding of the pathophysiology.

5.2 MATERIALS AND METHODS

5.2.1 Subjects

Written informed consents were obtained from all subjects. The outcomes of the Health Survey Questionnaire in patients with JS, YAH, YIH, and the normal controls were assessed, and the related clinical information including gender, age, body mass index (BMI), basic syndromes of disease, and the main parameters of liver makers were collected (data not shown). Exclusion criteria are that patients (nonsmokers) had cancer, cardiac insufficiency, hepatosis, renal inadequacy, respiratory failure, alimentary tract hemorrhage, or other diseases that will affect the clinical observations and biological indicators.

5.2.2 Chemicals and Reagents

Acetonitrile, HPLC grade, was obtained from Merck (Darmstadt, Germany); methanol (HPLC grade) was purchased from Fisher Scientific Corporation (Loughborough, UK); water was produced by a Milli-Q Ultra-pure water system (Millipore, Billerica, USA); formic acid was of HPLC grade, and obtained from Honeywell Company (Morristown, New Jersey, USA); leucine enkephalin was purchased from Sigma-Aldrich (St. Louis, MO, USA). All other reagents were HPLC grade.

5.2.3 Sample Preparation

The subjects were given insulated ice packs in which they were asked to store the urine samples immediately until they were received by the study investigator.

On arrival at the laboratory, the samples were centrifuged at 10,000 rpm for 10 min at 4°C to remove any solid debris. Fractions (5 mL) of the urine supernatants were then stored at −80°C until UPLC-Q-TOF-HDMS analysis. Thawed urine samples were collected after centrifugation at 13,000 rpm for 10 min at 4°C, and the supernatant was transferred to a 1.5 mL polypropylene tube, and then filtered through a syringe filter (0.22 μm), 5 μL of the supernatant were injected into the UPLC-Q-TOF-HDMS.

5.2.4 Metabolic Profiling Platform

5.2.4.1 Chromatographic Condition

The UPLC/MS analysis was carried out using a Waters ACQUITY ultra performance liquid chromatography (UPLC) system (Waters Corp., Milford, USA) coupled with Time-of-Flight Mass Spectrometry (TOF–MS) (from Waters Corp., Milford, MA). Chromatography was carried out with an ACQUITY BEH C_{18} chromatography column (2.1 mm × 100 mm, 1.7μm). The column temperature was maintained at 45°C, and then the gradient mobile phase conditions were composed of phase A (water with 0.1% formic acid) and phase B (acetonitrile containing 0.1% formic acid). The gradient for the urine sample was as follows: 0–5 min, 1–25% B; 5–9 min, 25–50% B; 9–9.1 min, 50–99% B; 9.1–11 min, 99% B; 11–11.1 min, 99–1% B; 11.1–13 min, 1% B. The flow rate was 0.40 mL/min and 5 μL aliquot of each sample was injected onto the column. The eluent was introduced to the mass spectrometry directly without a split. To ensure the stability and repeatability of the system, pooled quality control (QC) samples were prepared from 10 μL of each sample. The QC samples were also inserted and analysed in every 10 samples.

5.2.4.2 TOF–MS Condition

Mass spectrometry and accurate mass acquisition was performed with a Waters QTOF Premier (Milford, MA) operating at positive-ion (ESI^+) and negative-ion (ESI^-) electrospray ionization mode. The optimal capillary voltage was set at 3200 V, and cone voltage at 35 V. Nitrogen was used as the dry gas, the desolvation gas flow rate was set at 500 L/h, and cone gas flow was maintained at 50 L/h. The desolvation temperature was set at 350°C, and the source temperature at 110°C. The scan time and interscan delay were set to 0.4 s and 0.1 s, respectively. MS data were collected in the full scan mode from m/z 50–1000. All the data were acquired using an independent reference lock-mass via the LockSpray™ interface to ensure accuracy, and reproducibility during the MS analysis. Leucine enkaphalin was used as the reference compound (positive ion mode ($[M + H]^+ = 556.2771$) and $[M - H]^- = 554.2615$) at a concentration of 0.2 ng/mL under a flow rate of 100 μl/min. The data were collected in the centroid mode, and the LockSpray frequency set at 10 s and averaged over 10 scans for correction.

5.2.4.3 Data Processing

All data were processed using the MarkerLynx application manager for MassLynx 4.1 software (Waters Corp., Milford, USA). The UPLC/MS data are detected and noise reduced in both the UPLC and MS domains such that only true analytical peaks are further processed. A list of intensities of the peaks detected is then generated for the first chromatogram, using the Rt-m/z data pairs as identifiers. The resulting normalized peak intensities form a single matrix with Rt-m/z pairs for each file in the dataset. All processed data of each chromatogram were normalized and pareto scaled, prior to multivariate statistical analysis. SPSS 13.0 for Windows was used for the statistical analysis. The data were analysed using the Wilcoxon Mann–Whitney Test, with $p < 0.05$ set as the level of statistical significance.

5.2.4.4 Multivariate Data Analysis

The multivariate data matrix was analyzed by EZinfo software (Waters Corp., Milford, USA). Potential markers of interest were extracted from the combining S- and VIP-plots that were constructed from the OPLS–DA, and markers were chosen based on their contribution to the variation, and the correlation within the data set. With the completion of the OPLS–DA analysis, we can try computational systems analysis with MetaboAnalyst data annotation approach (Xia et al., 2012) to distinguish control and JS subjects. Heatmap, commonly used for unsupervised clustering, was constructed based on the potential candidates of importance, which were extracted with OPLS–DA analysis.

5.2.4.5 Biomarkers Identification

Exact molecular mass data from redundant *m/z* peaks corresponding to the formation of different parent and product ions were first used to help confirm the metabolite molecular mass. The MassFragment™ application manager (Waters MassLynx v4.1, Waters Corp., Milford, USA) was used to facilitate the MS/MS fragment ion analysis process by way of chemically intelligent peak-matching algorithms. The identities of the specific metabolites were confirmed by comparison of their mass spectra and chromatographic retention times with those obtained using commercially available reference standards. This information was then submitted for database searching, either inhouse or using the online ChemSpider database, and MassBank data source. Pathway analysis and visualization using the KEGG (www.genome.jp/kegg/) pathway database was carried out using Metaboanalyst.

5.3 RESULTS

5.3.1 Urine Metabolic Profiling

Urine metabolic profiling was established to explore important biomarkers and metabolic pathways related to JS, and developed a prediction model for

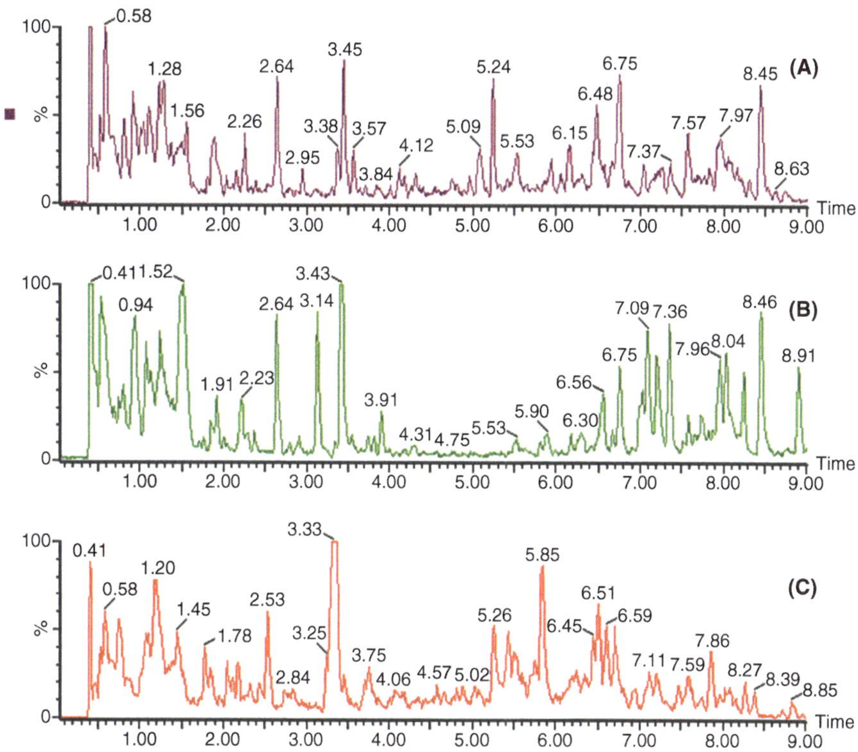

FIGURE 5.1 TIC chromatograms of Yanghuang (A), Yinhuang (B), and healthy persons (C): in positive ion mode.

aided diagnosis of the disease. The representative UPLC-Q-TOF-HDMS total ion chromatograms (TIC) from positive ion modes of human urine from the patients with YAH and YIH, and controls were compared visually (Fig. 5.1). The stable TIC reflected the stability of UPLC-Q-TOF-HDMS analysis and reliability of the metabolomic data. Using the optimized UPLC-Q-TOF-HDMS analysis protocol and subsequent processes, such as baseline correction, peak deconvolution, alignment and normalization, we obtained a three-dimensional matrix, including data filename, retention time exact mass pair, and normalized peak areas. Overall 9986 retention time-exact mass pairs were determined in each sample profile. Although, some differences could be visually noted among the three sets of the detail illustrated in the chromatogram, more subtle changes could be found using a pattern recognition approach.

5.3.2 Pattern Recognition Approach Analysis

Using our metabolomics platform, the statistically important metabolites were studied. PCA was used firstly to investigate general interrelation between groups, including clustering and outliers among the samples. And then, PLS–DA was used to maximize the difference of metabolic profiles between control and JS groups, and to facilitate the detection of metabolites consistently present in the biological samples. Data were standardized using the pareto-scaling technique. The supervised OPLS–DA can improve biomarker discovery efforts, and separate samples into two blocks were applied to obtain enhanced discrimination between the control and JS groups. Score plots from the supervised OPLS–DA showed obvious separation between the JS, YAH, YIH groups, and healthy group in both positive ion mode (Figs 5.2A, 5.3A, and 5.4A) and negative ion mode (data not shown) – this suggests that urinary biochemical perturbation significantly happened in JS, and its subtype groups. Trajectory analysis of score plots (3D) for the control, and JS groups showed clear segregation (Fig. 5.2B). Additionally, the similar analysis based on subtype groups and control group did reveal clearer separation (Figs 5.3B and 5.4B). From the corresponding loading plots, the ions furthest away from the origin contribute significantly to be responsible for the separation between control and JS groups, and may be therefore regarded as the differentiating metabolites for the JS, YAH, YIH groups (Figs 5.2C, 5.3C, and 5.4C). Combining the results of S- and VIP-plots from the OPLS analysis (Figs 5.2D, 5.3D, and 5.4D), the UPLC-Q-TOF-HDMS platform could provide the retention time, precise molecular mass, and MS/MS data for the structural identification of biomarkers. Finally, according to the variable importance in the projection (VIP > 2), the variables (ions) were identified based on the metabolite identification strategy. The parallel analysis of the samples with MetaboAnalyst allows for the ability to verify that ions, which are identified through both ways (i.e., Heatmaps, OPLS–DA) are highly significant, as depicted through two completely different algorithms. Analysis of the control and JS samples utilizing MetaboAnalyst's data annotation tools revealed differences between the two groups. Figures 5.2E, 5.3E, and 5.4E showed the top 15 significant features of the metabolite markers based on the VIP projection. The heatmaps were constructed based on the potential candidates of importance that were extracted with an OPLS–DA analysis. The parallel heatmap visualization (Figs 5.2F, 5.3F and 5.4F) using Ward's method in computational systems analysis for the JS, YAH, YIH, and controls showed distinct segregation. To test the usefulness of the marker metabolites for diagnosis, a second set of JS, YAH, and YIH patients, and control subjects were blindly selected. Using our metabolomics platform, interestingly, the heatmap visualization (Fig. 5.5) used in computational systems analysis could be achieved for the discrimination of all of the patients.

5.3.3 Metabolite Identification

The robust UPLC-Q-TOF-HDMS analysis platform provides the retention time, precise molecular mass, and MS/MS data for the structural identification of biomarkers. Structure identification was performed according to their

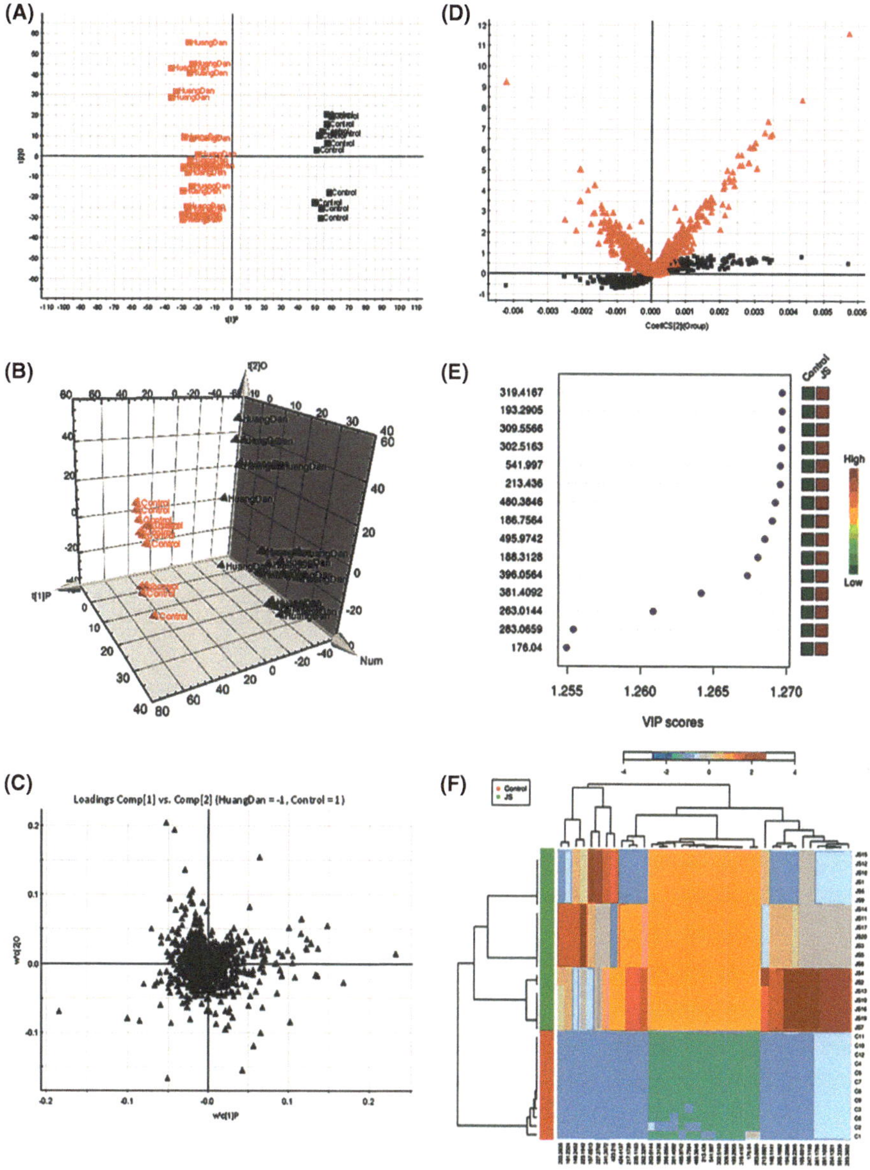

FIGURE 5.2 Urine metabolomic profiling of jaundice syndrome (JS).
(A) OPLS–DA model results for JS group in positive mode. (B) 3D of OPLS–DA model for JS group. (C) Loading plot of OPLS–DA of JS in positive mode. (D) Combination of S- and VIP-score plots constructed from the supervised OPLS analysis of urine (ESI^{+} mode). (E) Top 15 significant features of the metabolite markers based on the VIP projection. (F) Heat map visualization for the urine samples.

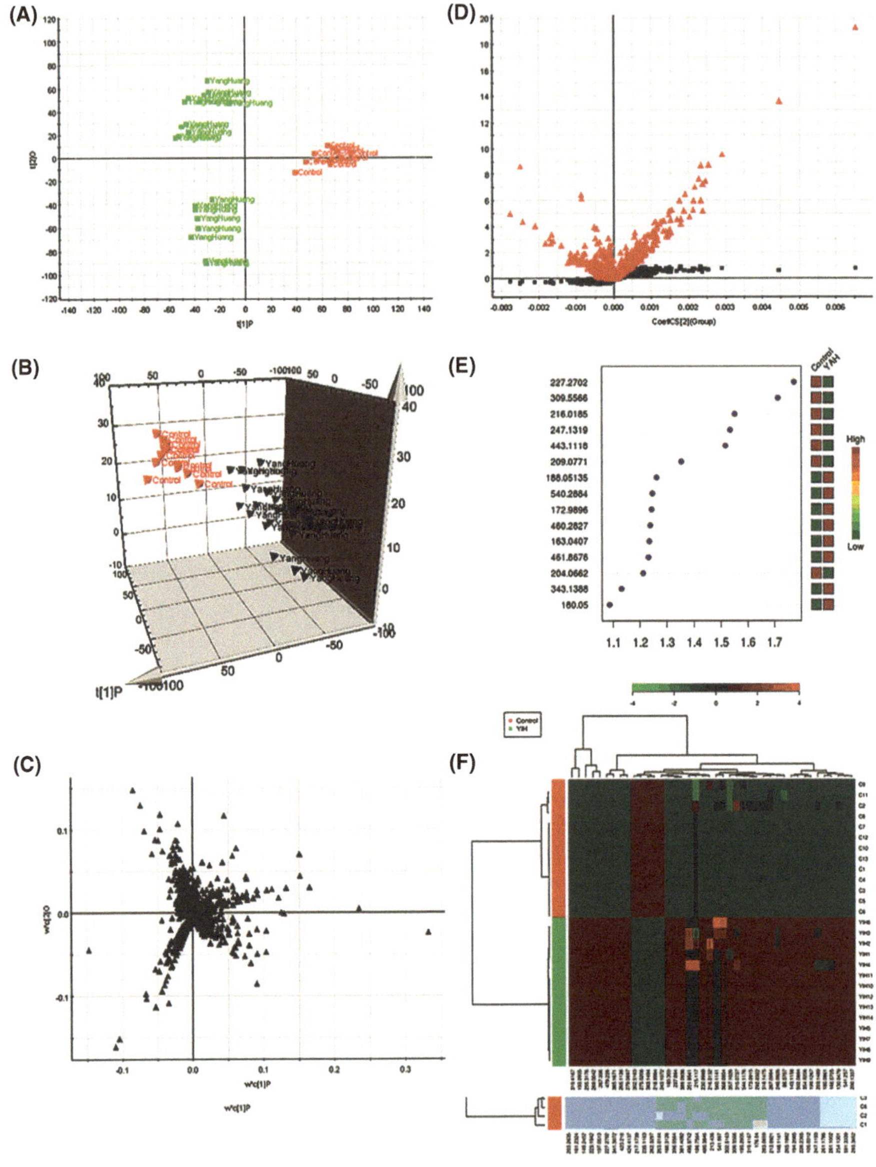

FIGURE 5.3 Metabolomic profiling of YAH.

(A) OPLS–DA model results for YAH group in positive mode. (B) 3D of OPLS–DA model for YAH group. (C) Loading plot of OPLS–DA of YAH in positive mode. (D) Combination of S- and VIP-score plots constructed from the supervised OPLS analysis of urine (ESI$^+$ mode). (E) Top 15 significant features of the metabolite markers based on the VIP scores of OPLS–DA. (F) Heat map visualization constructed based on the differential metabolites of importance for the urine of YAH.

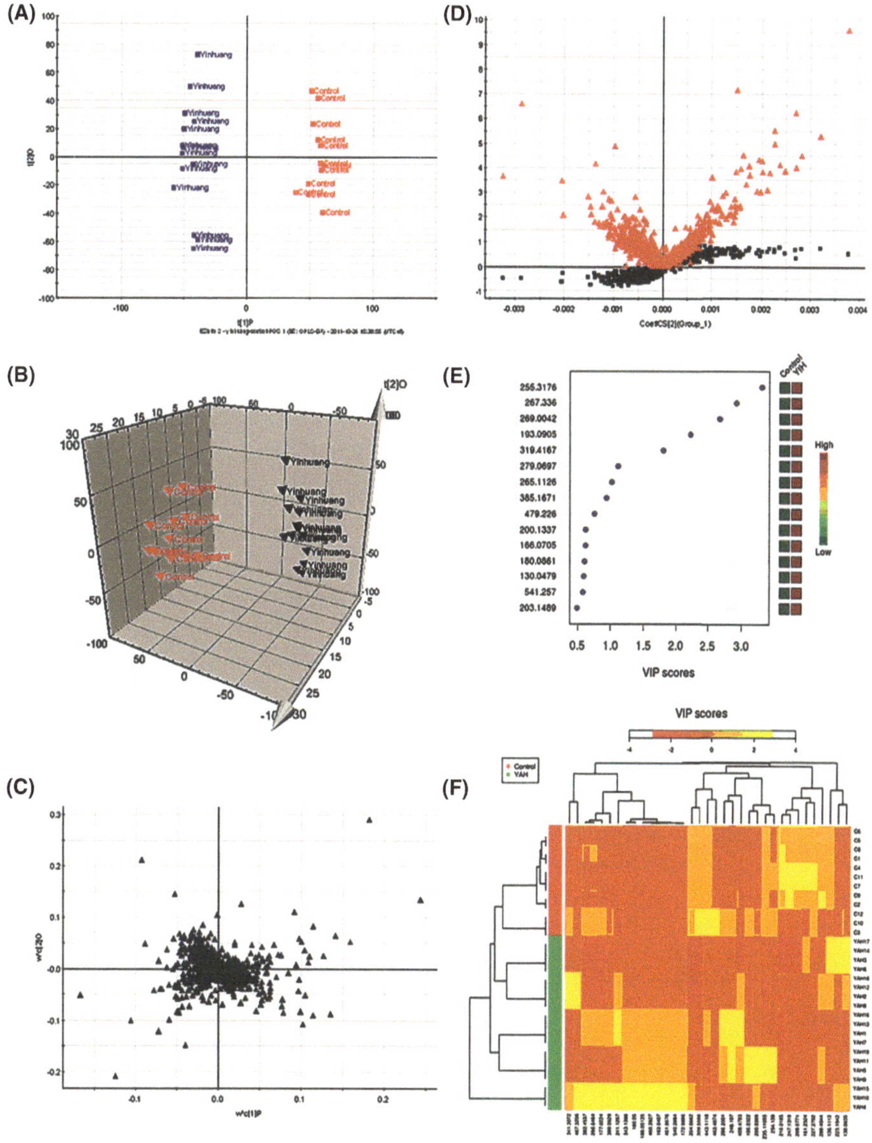

FIGURE 5.4 Metabolomic profiling of YIH.

(A) OPLS–DA model results for YIH group in positive mode. (B) 3D of OPLS–DA model for YIH group. (C) Loading plot of OPLS–DA of YIH in positive mode. (D) Combination of S- and VIP-score plots constructed from the supervised OPLS analysis of urine (ESI$^+$ mode). (E) Top 15 significant features of the metabolite markers based on the VIP scores of OPLS–DA. (F) Heat map visualization constructed based on the differential metabolites of importance for the urine of YIH.

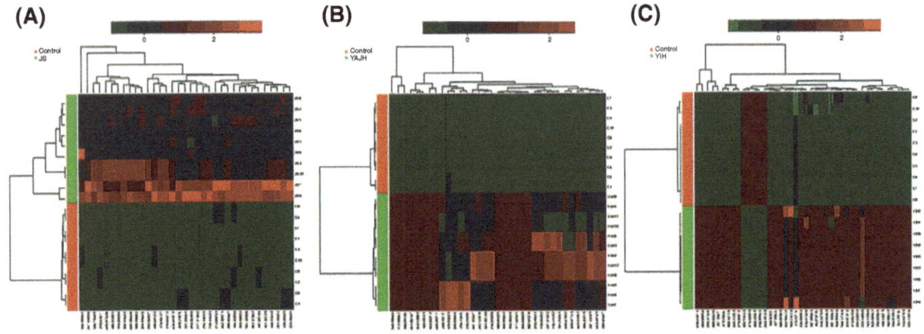

FIGURE 5.5 Hierarchical clustering and diagnostic potential of JS metabolites.
(A) Unsupervised hierarchical clustering of JS metabolite profiles. (B) Profiles of the signatures that distinguish YAH from control samples. (C) Dendrogam obtained from unsupervised hierarchical clustering of metabolite profiles for YIH and control samples.

molecular ion masses, and MS/MS production analysis was compared with authentic standards or database resources. Some parameters, such as deviation from calculated mass (mDa or ppm), double bond equivalent (DBE), and i-fit value (the isotopic pattern of the selected ion) were used to evaluate the accuracy of possible formula. Based on the relative intensities of the metabolites from the normalized spectrum, the combination of S- and VIP-score plots constructed from the supervised OPLS analysis of urine was used to reveal the significant differences of identified metabolites between the JS group and the control group, as well as between the subtype groups, and the control group. The presumed molecular formula was searched in Chemspider, Human Metabolome Database, and other databases to identify the possible chemical constitutions, and MS/MS data were screened to determine the potential structures of the ions.

Here, the biomarker with Rt-m/z of 6.76–591.3339 in positive ion mode was detailed as an example to illustrate the identification process. Using a mass tolerance of 5mDa, $C_{33}H_{42}N_4O_6$ was located as the candidate because of its high mass accuracy and low i-fit value among the possible chemical formulas. And then, $C_{33}H_{42}N_4O_6$ was input in the KEGG ligand for possible compound, and D-urobilinogen was finally emerged, which was further confirmed by comparing it to its authentic standards. In this study, we performed UPLC-Q-TOF-HDMS based metabolomic profiling in urine, to detect the potential biomarkers associated with JS patients. Finally, 44 marker metabolites contributing to the complete separation of JS from matched healthy controls were identified, with epinephrine, deoxycytidine, methylglutarylcarnitine, etc. exhibiting the best combined classification performance. Forty endogenous metabolites contributing to a good separation between YAH, and that matched healthy controls were identified, with kinetin, porphobilinogen, indoleacrylic acid, etc. showing the maximum combined classification performance (data not shown). Forty-nine metabolites contributing to the complete separation between YIH and matched healthy controls were identified, with 2-octenedioic acid, pyroglutamic acid, alpha-N-phenylacetyl-L-glutamine, etc. showing the best combined classification performance (data not shown).

5.3.4 Metabolic Pathway and Function Analysis

With pattern recognition analysis of metabolites, a clear separation of JS groups and healthy groups was achieved. Metabolite profiling focuses on the analysis of a group of metabolites related to a metabolic pathway in biological states. More detailed analysis of the most relevant pathways and networks of JS were performed by Metaboanalyst (a free, web-based tool) that combines the results from powerful pathway enrichment analysis involved in the conditions under study. Metaboanalyst, directed graph, uses the high quality the KEGG pathway database as the backend knowledgebase (Xia et al., 2012). Consequently, potential targets of metabolic pathway analysis (Impact-value $\geq$ 0.10) with Metaboanalyst revealed that metabolites which were identified together are important for the host response to JS, and are responsible for D-glutamine and D-glutamate metabolism, synthesis and degradation of ketone bodies, alanine, aspartate and glutamatemetabolism (Fig. 5.6A). The top three metabolic pathways of importance including vitamin B6 metabolism, tryptophan metabolism, arginine, and proline metabolism were found to be disturbed in YAH patients (Fig. 5.6B). Distinct metabolic pathway analysis (Impact-value $\geq$ 0.10) that were all related to YIH, are responsible for steroid hormone biosynthesis, primary bile acid biosynthesis, cysteine, and methionine metabolism (Fig. 5.6C). The detailed construction of the D-glutamine and D-glutamate metabolism (Fig. 5.7A), vitamin B6 metabolism (Fig. 5.7B), primary bile acid biosynthesis (Fig. 5.7C) – pathways with higher scores were generated using the reference map by searching KEGG.

5.4 DISCUSSION

Identification of molecular markers and metabolic pathways has the potential to improve diagnostic, prognostication, and therapy of interest (Zhang et al., 2013b). The determination of metabolic profiles may probably provide a biomarker for the discrimination of diseases as well as differentiation of healthy volunteers from patients. Current diagnostic methods, measuring levels of alanine aminotransferase and aspartate aminotransferase, etc. in blood samples, have been the main diagnostic markers for JS, and can be considered as the current "gold standard" for initial diagnosis and surveillance. However, sensitivity of these markers is relatively low, and it is difficult to obtain an outcome immediately. Hence, novel approaches for the detection of JS are

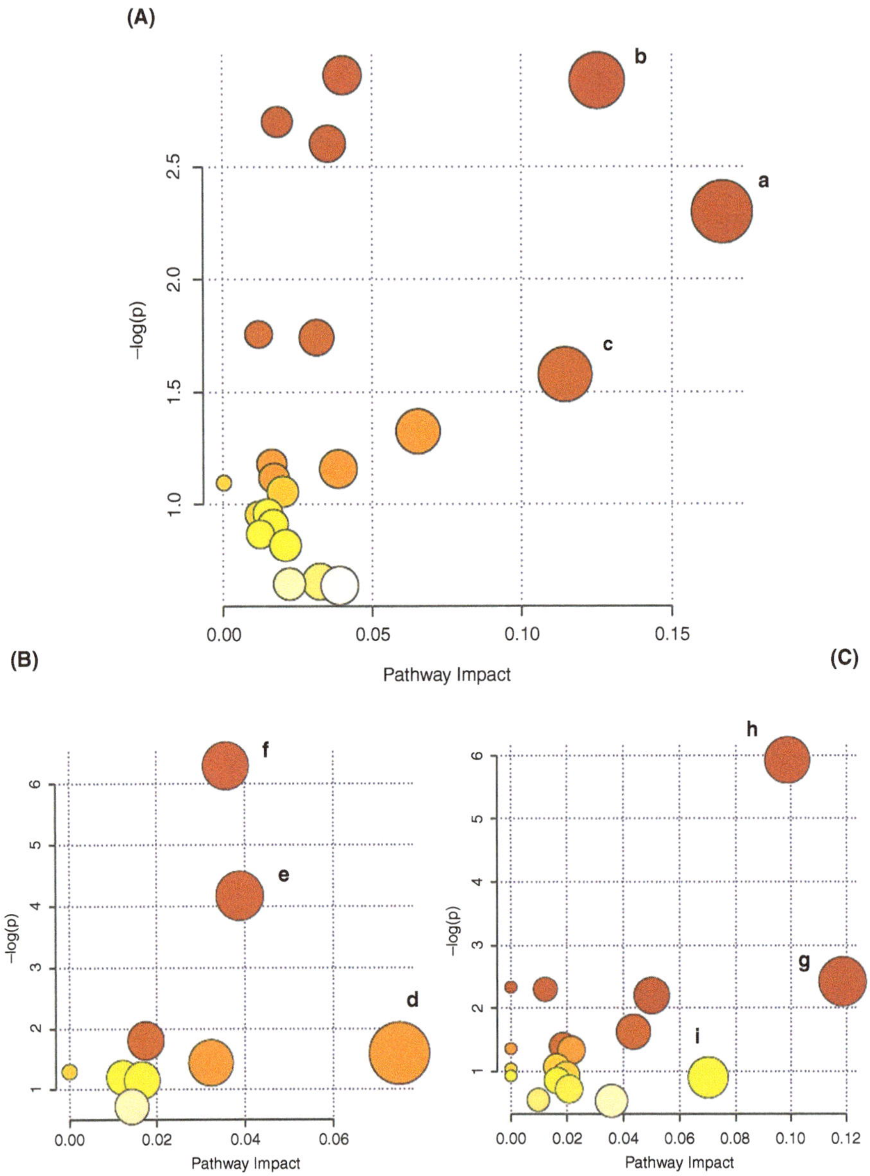

FIGURE 5.6 **Summary of pathway analysis with Metaboanalyst.**
(A) D-Glutamine and D-glutamate metabolism. (B) Synthesis and degradation of ketone bodies. (C) Alanine, aspartate, and glutamatemetabolism. (D) Vitamin B6 metabolism; e, tryptophan metabolism; f, arginine and proline metabolism; g, steroid hormone biosynthesis; h, primary bile acid biosynthesis; i, cysteine and methionine metabolism;

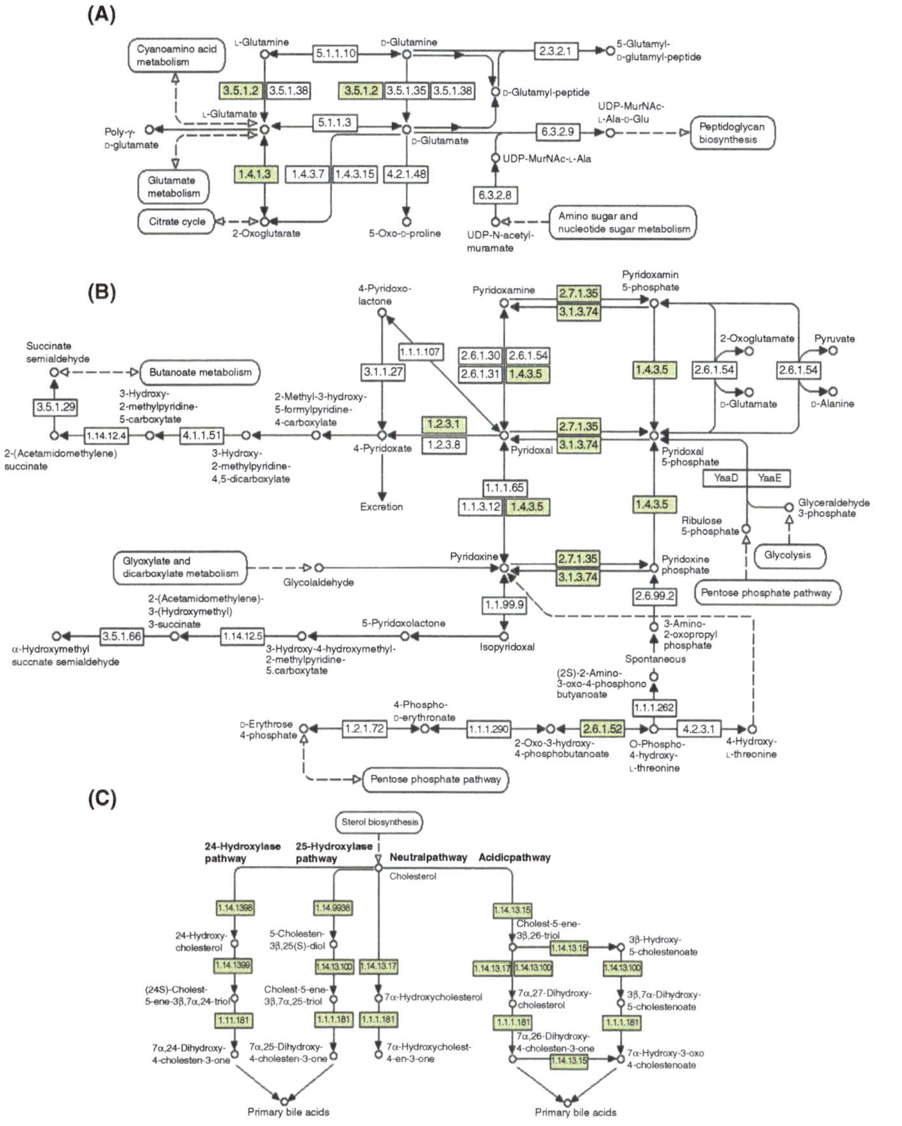

FIGURE 5.7 Construction of the D-glutamine and D-glutamate metabolism.
(A) Vitamin B6 metabolism. (B) Primary bile acid biosynthesis. (C) Pathways in the human. The map was generated using the reference map by KEGG (http://www.genome.jp/kegg/). The green boxes: enzymatic activities with putative cases of analogy in the human. Simplified altered metabolism pathways for the most relevant metabolic differences between patients with JS (A), YAH (B), YIH (C), and control subjects.

urgently required. The nontargeted metabolomics provides a global view of the organism, and can be used to monitor the dynamic metabolic alterations that occur in different pathological processes.

Efficient diagnosis methods in clinical practice are needed for the metabolite biomarkers. Metabolomics is a powerful new technology that allows for the assessment of global metabolic profiles in easily accessible biofluids and biomarker discovery in order to distinguish between diseased and nondiseased status information. Offering a physiologically holistic, noninvasive platform, the emergent field of metabolomics has the potential to provide new diagnostic tools. Recent technological breakthroughs have provided researchers with the capacity to measure hundreds or even thousands of small-molecule metabolites in as little as a few minutes per sample, paving the way for hypothesis generation studies ideally suited to complex diseases such as JS. Early or clinical diagnosis of JS would be helpful for both prevention and treatment. Metabolomics is particularly suited to liver injury assessment applications, where urine is the most common samples made available for laboratory tests. Although gene and protein expression have been extensively profiled in human disease, little is known about the global metabolomic alterations that characterize the disease. Multiple, complex molecular events characterize disease development and progression. Deciphering the molecular networks that distinguish disease from liver disease may lead to the identification of critical biomarkers for disease aggressiveness.

We utilized the metabolomics approach in a pilot study in urine samples of JS patients from control subjects (Wang et al., 2012). It was demonstrated that based on UPLC-Q-TOF-HDMS measurements of urine samples with statistically significant multivariate models, it could be constructed to distinguish between diseased and nondiseased subjects. The global metabolic profiling and subsequent multivariate analysis clearly distinguished JS patients from matched controls. Pattern recognition approach revealed an evident and statistically significant separation between the JS and control samples. Results indicate that JS-related metabolites play an essential role in glutamate metabolism, synthesis and the degradation of ketone bodies, alanine and aspartate metabolism that are tightly correlated with the genes of neurotransmitters, hormones, and cytokines in the metabolites interaction network. By using our metabolomics platform, 44 statistically important variables with VIP > 2 were defined. Interestingly, 44 distinct metabolites identified from these pathways, many are in various stages of progress at the JS. Further study of these metabolites may facilitate the development of noninvasive biomarkers and more efficient therapeutic strategies for JS. Furthermore, vitamin B6 metabolism, tryptophan metabolism, arginine, and proline metabolism were also the top functions as listed by Metaboanalyst for the YAH patient. Significant changes associated with YAH disease, defined as metabolite changes in YAH vs. controls were identified for 40 metabolites that are potential candidates for biomarkers. Additionally, steroid hormone biosynthesis, primary bile acid biosynthesis, cysteine and methionine metabolism were all related with YIH. It is noteworthy that 49 metabolites together are important for the host response to JS through target metabolism pathways.

We specifically introduced a new way to examine metabolome expression information of typical patients vs. controls. A significant separation using Heatmap analysis was obtained for the difference between JS vs. controls as well as between subtype patients vs. controls. Computational systems analysis with the MetaboAnalyst tool (Heatmap) provides a powerful approach to the metabolic profiling of urine to differentiate patients from control subjects. The patients with JS were easily distinguished from control subjects by the Heatmap approach when using autoscaling methods. To determine whether metabolic markers were identified using our metabolomics platform, we performed a preliminary validation using a second set of patients and control subjects that were blindly selected. Our results indicated that the model had high sensitivity and specificity for the clinical JS diagnosis. Targeted metabolite analysis studies have already shown that alterations of critical JS metabolic pathways, such as glutamate metabolism, synthesis, and the degradation of ketone bodies, alanine, and aspartate metabolism, are strongly associated with JS development. Such changes are expected to be reflected in wider coverage metabolic profiles, which may in turn be explored as potential biomarkers for JS assessment and treatment. Recently, metabolic pathways incorporate complex networks of protein-based interactions and modifications, and are usually considered to provide information on the mechanisms of disease, and have become a common and probably the most popular form of representing biochemical information for hypothesis generation and validation. These maps store wide knowledge of complex molecular interactions and regulations occurring in the living organism in a simple and obvious way, often using intuitive graphical notation. Based on the KEGG, a detailed construction of the D-glutamine and the D-glutamate metabolism, vitamin B6 metabolism, and primary bile acid biosynthesis; pathways map with higher scores (as shown in Fig. 5.7). Our results indicated that these target pathways show the marked perturbations over the entire time-course of JS and could therefore contribute to the development of JS. Furthermore, the inclusion of more patients and the detection of metabolomic biomarkers make further evaluative studies necessary, before the significance of our results can be assured. Further investigations are also underway to clarify the precise pathogenesis: concerning why JS induced these results. Therefore, metabolomics suggests that there is a great potential for candidate metabolite biomarker discovery, metabolite signatures may also have the potential to be used as diagnostic biomarkers. It opens out the mechanism of the classical TCM theory on syndrome, and also enriches current research on complex diseases as well as systems biology. Future research will focus on the discovery of additional biomarkers using metabolomics platforms, and the validation of the explorative biomarkers. In addition, more effort will be directed to the biological interpretation: it will be investigated which pathways were involved in the biochemical changes associated with the onset, development and progression of JS, and whether these changes are the same during onset and progression, or if different changes of biochemistry occur at the different stages of JS.

5.5 CONCLUSIONS

In summary, metabolomics provides a powerful approach to discover diagnostic and therapeutic biomarkers by analyzing global changes by metabolic profile. We here, for the first time, report a comprehensive analysis of metabolic patterns of JS and subtypes. We have identified some significantly changed metabolites associated with JS, and specifically noted 44 metabolic signatures in JS. Glutamate metabolism, synthesis, and degradation of ketone bodies, alanine and aspartate metabolism was found to exhibit the most altered functional pathway associated with JS according to ingenuity pathway analysis. The results not only indicated that the metabolomic methods had sufficient sensitivity and specificity to distinguish JS from healthy controls, but also have the potential to be developed into a clinically useful diagnostic tool, and could also contribute to a further understanding of disease mechanisms. In conclusion, our findings suggested that the metabolomics method would be helpful in establishing a suitable model for reasonably evaluating disease syndrome, exploring pathological mechanism of the syndrome, and clarifying the relationships between the syndrome and related diseases.

References

Arakaki, A.K., Skolnick, J., McDonald, J.F., 2008. Marker metabolites can be therapeutic targets as well. Nature 456 (7221), 443.

Arias, I.M., 1998. New genetics of inheritable jaundice and cholestatic liver disease. Lancet 352 (9122), 82–83.

Kim, K., Aronov, P., Zakharkin, S.O., Anderson, D., Perroud, B., Thompson, I.M., Weiss, R.H., 2009. Urine metabolomics analysis for kidney cancer detection and biomarker discovery. Mol. Cell Proteomics 8 (3), 558–570.

Li, R., Ma, T., Gu, J., Liang, X., Li, S., 2013. Imbalanced network biomarkers for traditional Chinese medicine syndrome in gastritis patients. Sci. Rep. 3, 1543.

Selmi, C., Bowlus, C.L., Gershwin, M.E., Coppel, R.L., 2011. Primary biliary cirrhosis. Lancet 377 (9777), 1600–1609.

Wang, X., Zhang, A., Han, Y., Wang, P., Sun, H., Song, G., Dong, T., Yuan, Y., Yuan, X., Zhang, M., Xie, N., Zhang, H., Dong, H., Dong, W., 2012. Urine metabolomics analysis for biomarker discovery and detection of jaundice syndrome in patients with liver disease. Mol. Cell. Proteomics 11 (8), 370–380.

Xia, J., Mandal, R., Sinelnikov, I.V., Broadhurst, D., Wishart, D.S., 2012. MetaboAnalyst 2.0 – a comprehensive server for metabolomic data analysis. Nucleic Acids Res. 40, W127–W133.

Zhang, A., Sun, H., Wang, X., 2012. Recent highlights of metabolomics for traditional Chinese medicine. Pharmazie 67 (8), 667–675.

Zhang, A.H., Sun, H., Qiu, S., Wang, X.J., 2013a. Recent highlights of metabolomics in Chinese medicine syndrome research. Evid. Based Complement. Alternat. Med. 2013, 402159.

Zhang, A., Sun, H., Han, Y., Yan, G., Wang, X., 2013b. Urinary metabolic biomarker and pathway study of hepatitis B virus infected patients based on UPLC-MS system. PLoS One 8 (5), e64381.

Zhang, A., Sun, H., Yan, G., Wang, P., Han, Y., Wang, X., 2014. Metabolomics in diagnosis and biomarker discovery of colorectal cancer. Cancer Lett. 345 (1), 17–20.

CHAPTER 6

Metabolic Profiling and Biomarkers Analysis of the GanYu PiXu Syndrome

Ying Han, Hui Sun, Aihua Zhang, Xijun Wang

CONTENTS

6.1 INTRODUCTION

Metabolomics is the comprehensive assessment of endogenous metabolites and attempts to systematically identify, and to quantify metabolites from a biological sample (Sreekumar et al., 2009). The small-molecule metabolites as primary indicators have an important role in biological systems, and represent attractive candidates to understand disease phenotypes. Comprehensive study of metabolites is a desirable tool for diagnosing disease, identifying new therapeutic targets, and enabling appropriate treatments. Metabolomics attempts to capture global changes, and the overall physiological status in biochemical networks, and pathways in order to elucidate sites of perturbations, and has shown great promise as a means to identify biomarkers of diseases. A key task of medicine is to detect the disease as early as possible, and urine is without a doubt the preferred sample for metabolome analysis.

Understanding syndromes is a core research in developing more efficient therapeutic strategies and diagnostic criteria for patients. Modern clinical evidence has shown that patients with liver diseases complicated by GanYu PiXu (GYPX, liver stagnation, and spleen deficiency) Syndrome that is characterized by the metabolic disorder of body fluid; causing liver failure and weakness, and pathological changes (Guo et al., 2012; Zeng et al., 2011). The incidence of GYPX is rising at a rapid rate, and the currently available biomarkers have an unacceptably low sensitivity, and specificity for diagnostic use. Therefore, the development of biomarkers with higher sensitivity, and specificity will surely emerge. The lack of sensitive and efficient platforms have limited the ability to simultaneously study the dozens of different pathways that may be affected in GYPX in relation to clinical phenotype. Recent advances in metabolomic technology have made it possible to identify the metabolites in the clinical samples, and thus extensive efforts are now attempted to search for the biomarkers (Zhang et al., 2012a; Wang et al., 2011; Chen et al., 2011). Consequently, this work was designed to investigate a comprehensive metabolome of

Chinmedomics. http://dx.doi.org/10.1016/B978-0-12-803117-9.00006-8

GYPX by UPLC-Q-TOF-HDMS combined with pattern recognition methods and pathways analysis, in order to establish specific metabolites phenotype and explore the mechanisms.

6.2 MATERIALS AND METHODS

6.2.1 Subjects

Patients were collected from the Hospital of Heilongjiang University of Chinese Medicine, China for the period of June 1, 2010 to December 30, 2010. GYPX patients and control subjects were recruited in this study. The clinical information with GYPX and the normal controls were collected. A second set of GYPX patients and control subjects were blindly selected and tested using our approach. This research was conducted according to the principles expressed in the Declaration of Helsinki.

6.2.2 Sample Preparation

The urine samples were centrifuged at 10,000 rpm for 10 min at 4°C to remove any solid debris. The supernatants were then stored at −80°C until analysis. Thawed urine samples were collected after centrifugation at 13,000 rpm for 10 min at 4°C, and the supernatant was transferred to a 1.5mL polypropylene tube, and then filtered through a syringe filter (0.22 μm), 5 μL of the supernatant were injected into the UPLC-Q-TOF-HDMS.

6.2.3 Metabolic Profiling Platform

6.2.3.1 Chromatographic and MS Condition

Chromatography was carried out with an ACQUITY BEH C_{18} chromatography column (2.1 × 100 mm, 1.7μm) maintained at 45°C. A Waters UPLC BEH C_{18} (2.1 mm i.d. × 100 mm ACQUITY) column packed with 1.7 mm beads was used to separate the molecules in the biofluid. The injection volume was 5 μL and the flow rate of the LC system was 0.5 mL/min. The eluent was introduced to the mass spectrometry directly, that is, without a split. After every 10 sample injections, a pooled sample was injected in order to ensure the stability and repeatability of the LC–MS systems. The MS system was operated using the positive ion (ESI^+) and negative ion (ESI^-) mode so as to monitor as many ions as possible, and the mass range was set at 100–1000 m/z in the full scan mode. All the data were acquired using an independent reference lock-mass via the LockSpray™ interface to ensure accuracy and reproducibility during the MS analysis.

6.2.3.2 Multivariate Data Analysis

The multivariate data matrix was analyzed by EZinfo software (Waters Corp., Milford, USA). The unsupervised segregation was checked by PCA. Potential markers of interest were extracted from the combining S- and VIP-plots from the OPLS-DA, and markers were chosen based on their contribution to the variation, and correlation within the data set. With the completion of OPLS-DA, we can try computational systems analysis with the MetaboAnalyst data annotation approach: to distinguish between control and GYPX subjects. PLS-plot and Heatmap have long been used in prediction pattern recognition, in exploratory data analysis, and in data mining problems.

6.2.3.3 Biomarkers Identification

Exact molecular mass data from redundant *m/z* peaks corresponding to the formation of different parent and product ions were first used to help confirm the metabolite molecular mass. The MassFragment™ application manager was used to facilitate the MS/MS fragment ion analysis process by way of chemically intelligent peak-matching algorithms. The identities of the specific metabolites were confirmed by comparison of their mass spectra and chromatographic retention times with those obtained using commercially available reference standards.

6.2.3.4 Construction of Metabolic Pathway

The construction, interaction, and pathway analysis of potential biomarkers of GYPX patients was performed with MetPA based on database source including the KEGG and HMDB to identify the top altered pathways analysis and visualization. The possible biological roles were also evaluated by the enrichment analysis using MetPA.

6.2.4 Data Analysis

SPSS 17.0 software for Windows was used for the statistical analysis. The data were analyzed using the Wilcoxon Mann–Whitney Test, with $p < 0.05$ set as the level of statistical significance. Receiver operating characteristic curve (ROC) was made by using the SPSS 17.0. The optimized cut-off values in this study were those corresponding with the highest accuracy. The MetaboAnalyst approach was used for the hierarchical clustering analysis (HCA) and significance analysis for microarrays (SAM).

6.3 RESULTS

6.3.1 Analysis of Metabolic Pattern

To evaluate the capability of the UPLC–MS-based metabolomic approach, it is useful to differentiate GYPX patients from controls, PLS–DA, OPLS–DA, and Heatmap – as carried out in the study. Score plots showed obvious separation between the GYPX and healthy group in both positive ion mode (Fig. 6.1A) and negative ion mode (Fig. 6.2A) – this suggests that urinary biochemical perturbation had significantly occurred in GYPX. Trajectory analysis of score plots

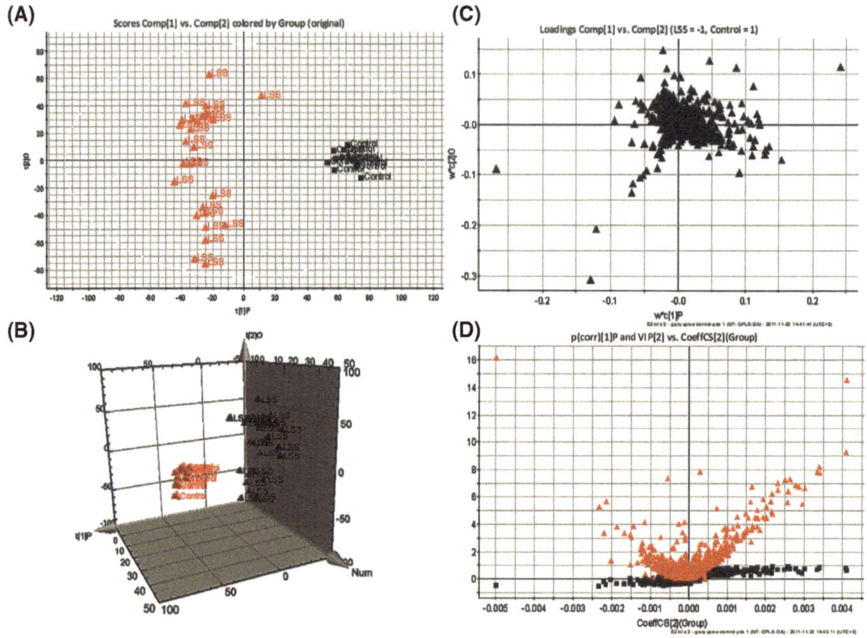

FIGURE 6.1 **Metabolomic profiling of GYPX in positive ion mode.**
(A) OPLS–DA model results for GYPX group. (B) 3D of OPLS–DA model for GYPX group. (C) Loading plot of OPLS–DA of GYPX in positive mode. (D) Shows the combination of S- and VIP-score plots constructed from the supervised OPLS analysis of urine (ESI+ mode). *Note*: LSS, GYPX.

(3D) for the control and GYPX groups showed clear segregation (Figs 6.1B and 6.2B). From the corresponding loading plots, the ions furthest away from the origin contribute significantly to be responsible for the separation between control and GYPX groups and may, therefore, be regarded as the differentiating metabolites for GYPX groups (Figs 6.1C and 6.2C). Combining the results of S- and VIP-plots from the OPLS analysis (Figs 6.1D and 6.2D), the UPLC-MS platform provided the retention time, precise molecular mass, and MS/MS data for the structural identification of biomarkers. As an ideal biomarker, it should be related to only the disease state, and have very few, or no interference, from the nondisease factors. A Wilcoxon Mann–Whitney test was performed and it was found that 12 ions (VIP > 8) significantly changed ($p < 0.05$) between the disease and the control groups. Finally, potential biomarkers of significant contribution were listed in Table 6.1. The MetaboAnalyst's data annotation tool revealed differences between the two groups (Fig. 6.3A). Correlation analysis of the 12 differential metabolites were marked on the hierarchical clustering plot, which was performed to understand the potential relationships among the metabolites that were shown on the plot as different colors (Fig. 6.3B). Figure 6.3C showed top significant features of the metabolite markers based

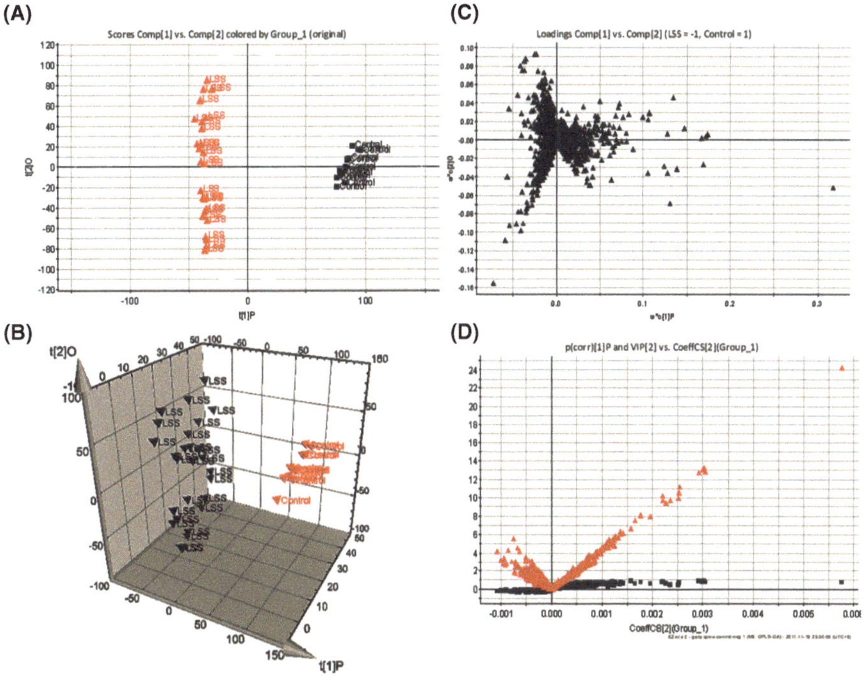

FIGURE 6.2 Metabolomic profiling of GYPX in negative ion mode.

(A) OPLS–DA model results for GYPX group. (B) 3D of OPLS–DA model for GYPX group. (C) Loading plot of OPLS–DA of GYPX in positive mode. (D) Shows the combination of S- and VIP-score plots constructed from the supervised OPLS analysis of urine (ESI$^-$ mode). *Note*: LSS, GYPX.

Table 6.1 Identification of Urinary Biomarkers in GYPX Cases

No.	VIP	Rt	$[M+H]^+$	$[M-H]^-$	Formula	Metabolites	*P*-value
1	24.2906	0.66		211.6081	$C_{10}H_{15}N_2O_4$	Prolylhydroxyproline	0.01
2	13.2855	6.39		310.2018	$C_{21}H_{26}NO$	Methadone	0.03
3	13.1444	2.29		263.1882	$C_{13}H_{16}N_2O_4$	Alpha-*N*-phenylacetyl-L-glutamine	0.00
4	12.8996	5.81		284.2001	$C_{15}H_{27}NO_4$	2-Octenoylcarnitine	0.01
5	12.7669	3.93		186.655	$C_9H_{13}NO_3$	Epinephrine	0.02
6	11.2109	5.45		381.4089	$C_{14}H_{17}N_5O_8$	Succinyladenosine	0.03
7	10.2331	3.02		207.0773	$C_{11}H_{10}O_4$	Benzylsuccinic Acid	0.04
8	16.1878	0.92	166.7153		$C_5H_{12}NO_3S$	Methionine sulfoxide	0.03
9	14.5521	0.89	195.0905		$C_6H_{10}O_7$	D-Glucuronic acid	0.03
10	9.2834	3.4	269.3362		$C_8H_{16}N_2O_4S_2$	L-Homocystine	0.00
11	8.1888	6.19	345.1384		$C_{16}H_{24}O_8$	Iridotrial glucoside	0.02
12	8.0438	3.93	188.3165		$C_{10}H_7NO_3$	Kynurenic acid	0.02

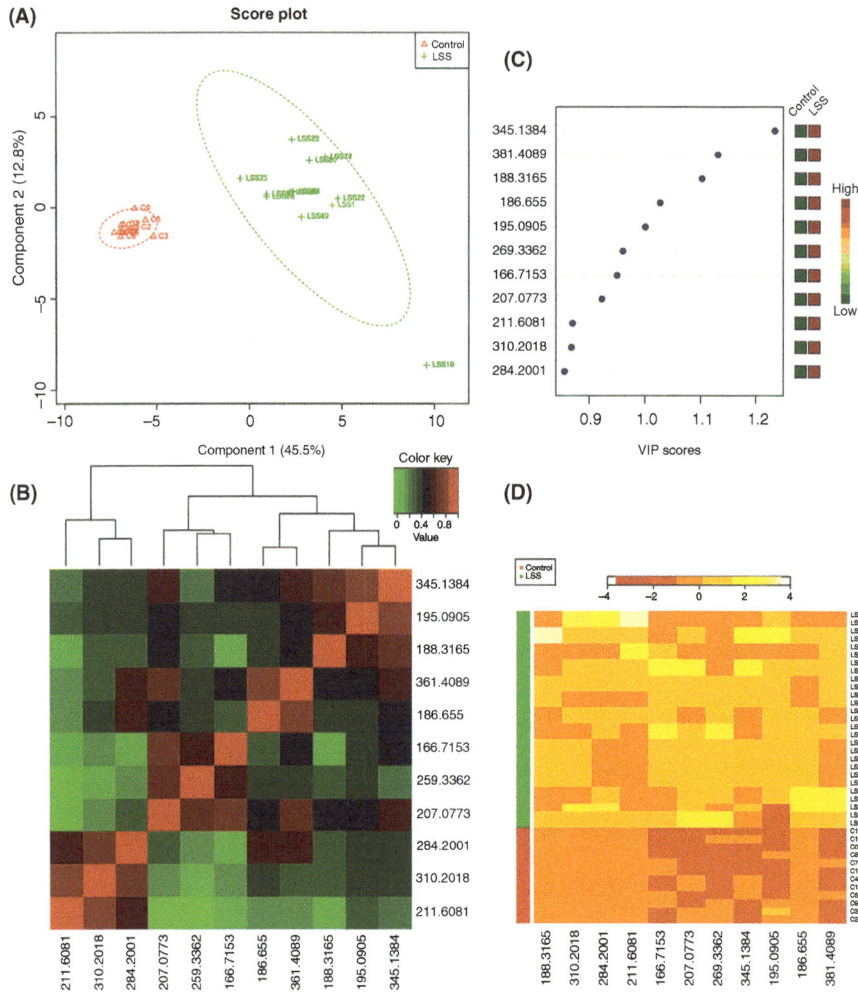

FIGURE 6.3 Systems analysis of metabolomic alterations of the GYPX and control samples with MetaboAnalyst's data annotation tools. PLS-plot revealed differences between the two groups.

Note: LSS, GYPX. (A) Hierarchical clustering analysis of the differential metabolites. (B) Correlation analysis of the 12 differential metabolites marked on the plot. Three clusters were identified representing the different groups of the metabolites. The top significant features of the metabolite markers based on the VIP projection. (C) Heat map visualization constructed based on the differential metabolites of importance for the urine of GYPX; (D) The heatmaps were constructed based on the potential candidates of importance that were extracted with OPLS–DA analysis. Rows, samples; columns, metabolites; color key indicates metabolite expression value: blue, lowest; red, highest.

on the VIP projection. The parallel heatmap visualization (Fig. 6.3D) analysis for the GYPX and controls showed distinct segregation.

6.3.2 Selection of the Differential Metabolites

At the defined operational conditions, all of the metabolites in a total ion chromatogram were extracted and aligned using the EZinfo 2.0 software; 12 differential metabolites for distinguishing the GYPX from matched healthy controls were identified (Table 6.1). Finally, these metabolites were selected as candidate markers for further validation. The SAM method was used to select the most differential oriented and interesting biomarkers. The results indicated that prolylhydroxyproline, alpha-*N*-phenylacetyl-L-glutamine, 2-octenoylcarnitine, and L-homocystine were the most significant differential metabolites for the classification of the GYPX, and the control (Fig. 6.4A).

6.3.3 Validation of Metabolite Markers

To test the usefulness of the four "metabolite markers" for human GYPX diagnosis (especially discriminating GYPX from controls) a second set of GYPX patients and control subjects was blindly selected and tested using our metabolomics platform. The four metabolite markers reached a sensitivity of 93.0%, and a specificity of 81.2% (the cut-off value: 0.63, Fig. 6.4B) whereas the results of ALT and AST (the cutoff value: 0.55) for these patients were 56.5 and 43.3%, respectively. Furthermore, the ROC curve analysis of the three metabolite markers yielded an AUC of 0.807, which was greater than that of ALT (0.599) and AST (0.594), respectively. The use of the four metabolite markers could significantly increase the specificity and the diagnostic performance of the metabolic markers (AUC: 0.807).

6.3.4 Metabolic Pathway Analysis

With pattern recognition analysis of metabolites, a clear separation of GYPX groups and a healthy group was achieved. The potential targets metabolic pathway analysis (Impact-value ≥ 0.10) with MetPA revealed that metabolites, which were identified together, are important for the host response to GYPX. The top three metabolic pathways of importance included are ascorbate and aldarate metabolism, pentose and glucuronate interconversions, and tyrosine metabolism that were found to be disturbed in GYPX patients (Fig. 6.4C). The detailed construction of the altered tyrosine metabolism pathways (Fig. 6.4D) with higher scores in humans was generated using the reference map by searching KEGG.

6.4 DISCUSSION

GYPX, one of the common syndromes, has a prevalence, and is often seen in the course of diseases such as cholestasis, hepatitis C, primary biliary cirrhosis, and liver fibrosis. Since the development and prognosis of GYPX liver disease

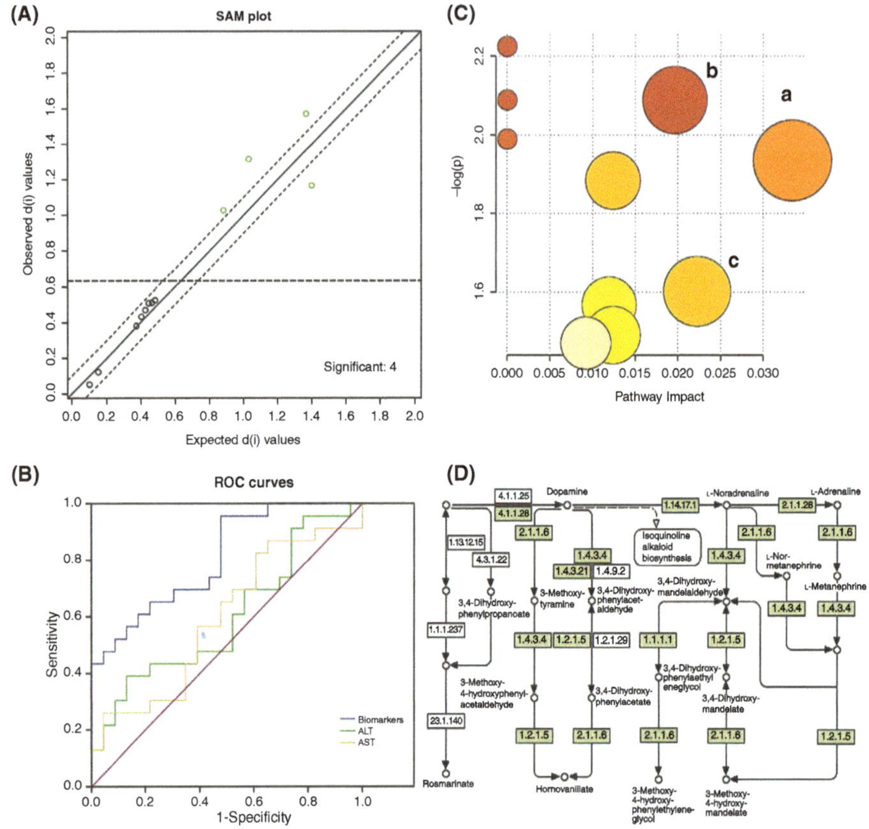

FIGURE 6.4 Selection and diagnostic potential of the significant metabolites.
(A) The significance analysis for the microarrays method used to select four marker metabolites. (B) ROC curves for the differential diagnosis by conventional clinical chemistry markers and the metabolic markers using data from controls and GYPX patients. (C) Summary of pathway analysis with MetaboAnalyst. (a) Pentose and glucuronate interconversions; (b) ascorbate and aldarate metabolism; (c) tyrosine metabolism. (D) Construction of the altered tyrosine metabolism pathways in human GYPX disease. The map was generated using the reference map by KEGG database. The green boxes: enzymatic activities with putative cases of analogy in humans.

vary significantly with genetic background, noninvasive GYPX biomarkers would significantly improve screening and diagnosis. The UPLC/MS-based metabolomics approach coupled with multivariate statistical methods provide a powerful approach to clearly differentiate patients with GYPX from matched control subjects, and identify the potential biomarkers. (Zhang et al., 2012b). Results indicate that OPLS–DA revealed an evident and statistically significant separation between the GYPX and the control samples. Interestingly, 12 distinct metabolites (such as prolylhydroxyproline, methadone, alpha-*N*-phenylacetyl-L-glutamine, epinephrine, benzylsuccinic acid, succinyladenosine, 2-octenoylcarnitine, methionine sulfoxide, D-glucuronic acid, L-homocystine, kynurenic acid, iridotrial glucoside) were identified in the urine sample, and many were in various stages of progress at the GYPX. GYPX-related metabolites were tightly correlated with the ascorbate and aldarate metabolism, pentose and glucuronate interconversions, cysteine and methionine metabolism, tyrosine metabolism, tryptophan metabolism, cytochrome P450, amino sugar and nucleotide sugar metabolism interaction network. Furthermore, ascorbate and aldarate metabolism, pentose and glucuronate interconversions, tyrosine metabolism were also the top function for GYPX patients. Further study of these metabolites may facilitate the development of noninvasive biomarkers and more efficient therapeutic strategies for GYPX. The patients with GYPX were easily distinguished from control subjects by the Heatmap approach when using autoscaling methods. With the HCA analysis 12 metabolite markers were clustered into three types, and finally defined for their potential applications. Among these metabolic alterations, the three clusters of metabolites covered most of these differential metabolites, and represented the major metabolic disturbance.

The SAM method, a well-established statistical method for identification of differentially expressed metabolites on high-dimensional data, was used to select the most discriminant and interesting biomarkers. The results indicated that prolylhydroxyproline, alpha-*N*-phenylacetyl-L-glutamine, 2-octenoylcarnitine, and L-homocystine were the most significant differential metabolites for the classification of the GYPX and the control (Fig. 6.4A). To determine whether the four metabolic markers that were identified could be used for clinical GYPX diagnosis, we performed a preliminary validation using the sera from GYPX patients. Compared with the traditional GYPX biomarker ALT and AST, the combinational metabolic markers showed a better sensitivity. Similarly, data from all nine patients showed that four metabolite markers have a better AUC value than the ALT and AST (Fig. 6.4B). Moreover, to test their diagnosis potential, in urine from patients with LSS it was found that four metabolite markers were effective for the discrimination of LSS patients, achieving a high sensitivity. These results indicate the metabolomics method has the potential of finding biomarkers for the early diagnosis of GYPX. It should be pointed out that metabolites are regulated by diverse intrinsic or extrinsic factors, the specificity of the metabolic markers is still to be improved, one of ways is the combinational use of the ALT, AST, and the metabolic markers. Based on the KEGG, a detailed construction of the altered tyrosine metabolism pathways map with a higher score is shown in Fig. 6.4D. These results suggest that these target pathways show the marked perturbations over the entire time-course of GYPX, and could contribute to the development of GYPX.

Metabolomics provides a powerful approach to discover diagnostic and therapeutic biomarkers by analyzing global changes in an individual's metabolic profile (Sun et al., 2013). We here, for the first time, report a comprehensive analysis of metabolic patterns of GYPX. Metabolite profiling analysis was shown to be an effective approach to differentiating GYPX patients and healthy subjects. We have identified 12 differential metabolites associated with GYPX. More importantly, four metabolite markers provided the effective diagnosis for human GYPX. Interestingly, ascorbate and aldarate metabolism, pentose and glucuronate interconversions, tyrosine metabolism were found to be the most altered functional pathways associated with GYPX, according to the ingenuity pathway analysis. Based on our findings, this makes these pathways possible therapeutic targets for advanced GYPX. A prediction model was developed to indicate GYPX, and an accuracy of 93.0% was obtained. Furthermore, studies with more patients included and the detection of metabolomic biomarkers makes evaluation of such further studies necessary, before the significance of our results can be assured. In conclusion, it indicated that metabolomics has sufficient sensitivity and specificity to distinguish GYPX from healthy controls, and could also contribute to a further understanding of disease mechanisms.

References

Chen, T., Xie, G., Wang, X., Fan, J., Qiu, Y., Zheng, X., Qi, X., Cao, Y., Su, M., Wang, X., Xu, L.X., Yen, Y., Liu, P., Jia, W., 2011. Serum and urine metabolite profiling reveals potential biomarkers of human hepatocellular carcinoma. Mol. Cell. Proteomics 10 (7), M110.004945.

Guo, Z., Yu, S., Guan, Y., Li, Y.Y., Lu, Y.Y., Zhang, H., Su, S.B., 2012. Molecular mechanisms of same TCM Syndrome for different diseases and different TCM Syndrome for same disease in chronic hepatitis B and liver cirrhosis. Evid. Based Complement. Alternat. Med. 2012, 120350.

Sreekumar, A., Poisson, L.M., Rajendiran, T.M., Khan, A.P., Cao, Q., Yu, J., Laxman, B., Mehra, R., Lonigro, R.J., Li, Y., Nyati, M.K., Ahsan, A., Kalyana-Sundaram, S., Han, B., Cao, X., Byun, J., Omenn, G.S., Ghosh, D., Pennathur, S., Alexander, D.C., Berger, A., Shuster, J.R., Wei, J.T., Varambally, S., Beecher, C., Chinnaiyan, A.M., 2009. Metabolomic profiles delineate potential role for sarcosine in prostate cancer progression. Nature 457 (7231), 910–914.

Sun, H., Zhang, A., Yan, G., Piao, C., Li, W., Sun, C., Wu, X., Li, X., Chen, Y., Wang, X., 2013. Metabolomic analysis of key regulatory metabolites in hepatitis C virus-infected tree shrews. Mol. Cell. Proteomics 12 (3), 710–719.

Wang, T.J., Larson, M.G., Vasan, R.S., Cheng, S., Rhee, E.P., McCabe, E., Lewis, G.D., Fox, C.S., Jacques, P.F., Fernandez, C., O'Donnell, C.J., Carr, S.A., Mootha, V.K., Florez, J.C., Souza, A., Melander, O., Clish, C.B., Gerszten, R.E., 2011. Metabolite profiles and the risk of developing diabetes. Nat. Med. 17 (4), 448–453.

Zeng, X.X., Bian, Z.X., Wu, T.X., Fu, S.F., Ziea, E., Woon, W.T., 2011. Traditional Chinese medicine syndrome distribution in chronic hepatitis B populations: a systematic review. Am. J. Chin. Med. 39 (6), 1061–1074.

Zhang, A., Sun, H., Wang, X., 2012a. Serum metabolomics as a novel diagnostic approach for disease: a systematic review. Anal. Bioanal. Chem. 404 (4), 1239–1245.

Zhang, A., Sun, H., Han, Y., Yuan, Y., Wang, P., Song, G., Yuan, X., Zhang, M., Xie, N., Wang, X., 2012b. Exploratory urinary metabolic biomarkers and pathways using UPLC-Q-TOF-HDMS coupled with pattern recognition approach. Analyst. 137 (18), 4200–4208.

CHAPTER 7

Metabolic Profiling and Biomarkers of Yinhuang Syndrome and Evaluation of Yinchensini Tang

Xijun Wang, Xin Tong, Hui Sun, Aihua Zhang

CONTENTS

7.1 INTRODUCTION

Jaundice syndrome (JS) is a common symptom caused by some chronic liver diseases such as hepatitis, moderate and severe fatty liver, alcoholic liver disease, and hepaticcarcinoma (Wang et al., 2012). Jaundice syndrome, caused by various hepatopathies, is an often-encountered disease in clinical practice. In TCM, JS is sorted into two syndromes, Yanghuang syndrome and Yinhuang syndrome (YIS). YIS is known to be caused by Dan-channel obstruction with various factors of yin-nature, such as Pi-Wei asthenia cold constitution, overstrain, protracted illness, Pi-yang deficiency due to excessive taking of drugs in cold nature, mistreatment or error treatment of Yanghuang syndrome, yin-qi injury, bibulosity, qi stagnancy due to cold/dampness evil-pathogen invasion, and blood-stasis blocking collaterals (Sun et al., 2006). YIS is always accompanied by yellow-colored sclera, skin, and urine. The color of yellow is tarnished. Viewing from TCM pathogenesis and pathologic mechanism, an experimental animal model of YIS was established with triplet factors, rhubarb, ethanol, and α-nephthylisothiolyanate (abbreviated as R, E, and A) and evaluated with the metabolomic method (Tong et al., 2011).

Improving jaundice in the treatment of chronic liver injury is critical, but there is a lack of effective drugs for the relieving of severe jaundice to treat the underlying diseases. Compared with Western medicine treatment, traditional Chinese herbal formulae can regulate pathological states of diseases via multitarget pathways (Sun et al., 2013; Wang et al., 2013; Zhang et al., 2014). This is because they can act on multiple targets and have synergistic effects. It is important to take discoveries from clinical application to produce innovations in drug development in translational medicine, which is thought to be a suitable route for the development of TCM. Yinchensini Tang (YCSNT) is a TCM recorded originally in *Shanghan Weizhi Lun* during the Song Dynasty (Yan et al., 2010). The ingredients of the formulae include *Herba Artemisiae Scopariae, Radix Aconiti Lateralis Preparata, Rhizoma Zingiberis,* and *Radix et Rhizoma Glycyrrhizae Preparata Cum Melle*. Traditionally, YCSNT is used for

Chinmedomics. http://dx.doi.org/10.1016/B978-0-12-803117-9.00007-X

the treatment of liver disorders and YIS. However, due to its complex constituents, it is not easy to explore the action mechanism of YCSNT. In light of this, the purpose of the study is to investigate the use of a metabonomic method to characterize metabolic profiling of YIS and intervention role of YCSNT in rats.

7.2 METHODS

7.2.1 Drugs, Reagents, and Instrument

Raw material of rhubarb was purchased from Harbin Stores of Beijing Tongrentang, *Herba Artemisiae Scopariae* was purchased from the China Branch Office of Xinneitian Pharmaceutical Company of Japan (Tianjin, China), *Radix Aconiti Lateralis Preparata* from Zhongba Fuzi Company (Jiangyou, Sichuan, China); *Rhizoma Zingiberis*, and *Rhizoma Glycyrrhizae Preparata Cum Melle* from Harbin Tongrentang Drug Store (Harbin, China). All crude drugs were authenticated by Prof. Xijun Wang, Department of Pharmacognosy of Heilongjiang University of Chinese Medicine. α-Nephthylisothiolyanate was purchased by Sigma Co. USA; ethanol (analytic pure) was purchased from Yongda Developing Center of Chemical Reagents (Tianjin City, China); methanol (chromatographic grade) was purchased from Merck Co., USA; the distilled water was purchased from Watsons Co., China; leucine enkephalin was purchased from Sigma Co., USA; isopropanol (analytic grade) was purchased from Bidi Chemical Engineering Co., Ltd., Tianjin City, China; and sodium hydroxide (analytic grade) was purchased from Sixth Tianjin Municipal Chemical Reagent Co., China.

7.2.2 Solution Preparation

Raw rhubarb (2500 g) was immersed in cold water for 60 min and boiled for 20 min. After filtration with eight-layer gauze, the filtrate was concentrated to 1 g crude drug per milliliter by decompression at 50 °C and then freeze-dried. The percentage of yielding was 16.42%. The freeze-dried rhubarb powder was screened (pharmacopoeial sieve No. 5) poured in mortar, hot water was added, and then milled along the same direction to ensure the powder dissolves completely for use. Ethanol (50 mL) was mixed with equal volume of pure water by ultrasonic mixer to prepare its 50% (v/v) solution. α-Nephthylisothiolyanate powder (1.6 g) was measured precisely, dissolved in 40 mL olive oil, and mixed well in an ultrasonic mixer to prepare the 4% solution.

7.2.3 YCSNT Preparation

YCSNT was prepared in our laboratory according to the method recorded in *Shanghan Weizhi Lun'*, and the decoction was transformed into freeze-dried powder. With the temporary use of YCSNT, its freeze-dried powder was added to hot water to ensure the powder was fully dissolved. The low dose was equivalent to 0.2 g crude drug/mL; medium dose was equivalent to 0.6 g crude drug/mL; high dose was equivalent to 1.2 g crude drug/mL; low, medium and high dose were equivalent to adult daily dose of l, 3, 6 times, respectively.

7.2.4 Collection and Preparation of Samples

Male Wistar rats, weighing 200 ± 20 g, were supplied by the GLP of Heilongjiang University of Chinese Medicine. After being fed in groups separately with free access to food and water for 3 days, the eighty Wistar rats were divided at random into eight groups; the blank group (GA) rats were infused through gastrogavage with 0.9% normal saline (N) everyday for 14 successive days; the model group (GB) rats were infused with rhubarb solution (R) for 14 successive days; 50% ethanol solution (E) everyday on days 10–14 and 4% α-nephthylisothiolyanate solution (A) on days 12 and 13 were administered by gastrogavage. The dosages of N, R, and E given to each rat were all 1 mL/100 g body weight per day; for A, it was 15 mg/100 g on day 12, and 5 mg/100 g on day 13. Imitatively, the solution administered to GC was R on days 1–14; to GD were N on days 1–9, E on days 10–14; to GE were N on days 1–11 and A on days 12 and 13; to GF were R on days 1–14 and E on days 10–14; to GG were R on days 1–14, and A on days 12 and 13; and to GH were N on days 1–9, E on days 10–14, and A on days 12 and 13. The dosages administered were the same as mentioned above. Rats' urine was collected everyday from days 1 to 14 successively. The urine sample was centrifuged at 13,000 r/min for 5 min to obtain the supernatant, which was screened with 0.22 μm filter membrane for UPLC/TOF-MS analysis.

7.2.5 Chromatography Analysis

Acquity UPLC™ BEH C_{18} column: 50 × 2.1 mm i.d., 1.7 μm (Waters Corp., Milford, USA); flow speed: 0.40 mL/min; temperature: 45°C; mobile phase A: 0.1% water solution of formic acid; mobile phase B: 0.1% methylcyanide solution of formic acid; gradient elution program is not shown. Full wave screen was performed with diode array detector. Ultraviolet detector effluent was admitted directly into the mass spectrogram system for detection.

7.2.6 Mass Spectrometry

Mode of detection: electrospray positive ion source; precipitation agent, gas feed rate: 600 L/h, temperature: 300°C; atomizing gas feed rate: 100 L/h, temperature of ion source: 110°C, capillary voltage: 3200 V, counterbore voltage: 35 V, microchannel plate voltage: 2300 V, impact energy: 3 V. The mass spectrogram was picked up every 0.14 s; leucineenkephalin solution was used for accurate mass detection ($[M + H]^+ = 556.2771$). The screen scope was m/z100–900, data collected with Centriod mode.

7.2.7 Data Management and Statistical Analysis

Metabolomic data were managed with MarkerLynx software to make chromatographic peak differentiation and matching, and the principal component analysis was adopted to analyze the metabolites of rats' urine in different groups. The results were expressed as the mean ± standard deviation and the treatment differences were considered to be significant when the P value was less than 0.05.

7.3 RESULT AND DISCUSSION

7.3.1 Change of the UPLC/TOF-MS BPI Chromatogram

New type packing material (1.7 μm) was used in this experiment for chromatographic column at a flow speed of 0.40 mL/min. Favorable result of analysis were obtained under the optimized condition (Fig. 7.1). As shown in Fig. 7.2, the BPI chromatogram in GB was markedly changed after a 14-day treatment, significantly different from that in GA. This difference reached its peak on day 14, illustrating that quantitative or qualitative different low molecular metabolites were produced in GB, which is evidential for the successful modeling of YIS by the R-E-A combination.

7.3.2 Principal Component Analysis

Adopting nonsupervision of the pattern recognition method, the primary PCA disposal on fingerprint data of urine sample showed that the score plots in different groups were well assorted, indicating that the three factors (R, E, and A) either used individually or in different combinations, could intervene in the normal metabolic network of rats to induce apparent changes in urine metabolic profiling (Fig. 7.3).

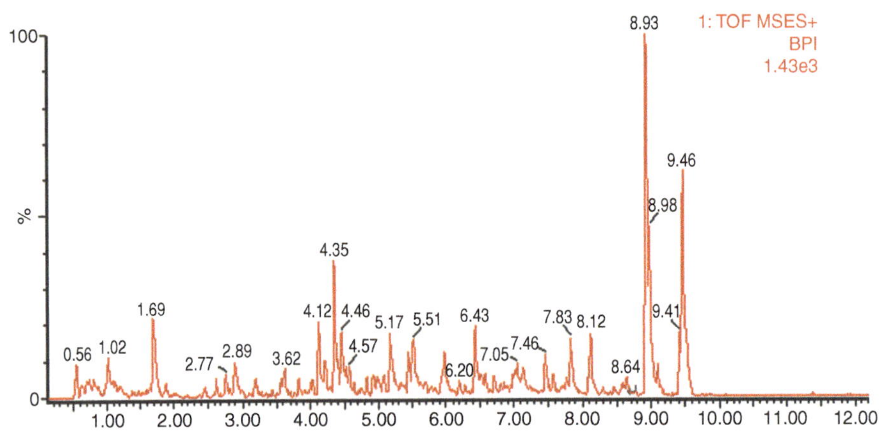

FIGURE 7.1 BPI chromatogram of urine samples of rats in positive ion mode.

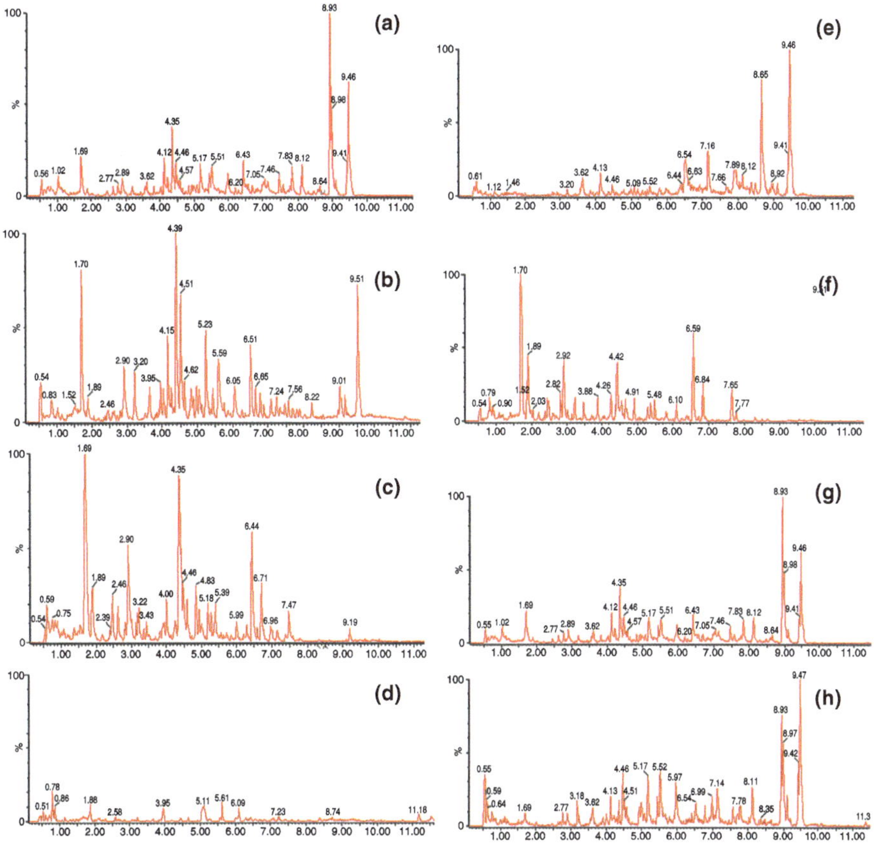

FIGURE 7.2 UPLC/TOF-MS BPI chromatogram of urine samples of rats induded by rhubarb, ethanol, and α-nephthylisothiolyanate (abbreviated as R, E, and A).

(A) Control group; (B) R + E + A group; (C) R group; (D) E group; (E) A group; (F) R + E group; (G) R + A group; (H) E + A group.

7.3.3 Orthogond Projection to Latent Structure Analysis

Adopting supervision pattern recognition method, OPLS analysis on outcomes of GA and GB showed that the scores were well differentiated in both groups, especially in P1 principal component orientation, while the P2 principal component reflected the intra group individual difference, which would worsen the convergence in the group (Fig. 7.4).

7.3.4 Biomarkers Confirmation

Biomarkers for the cooperation of R-E-A internal intervening on rats' physiologic metabolism were sought out by analysis of the loading plot (Fig. 7.5), which represented the principal components, and the contribution of intrinsic

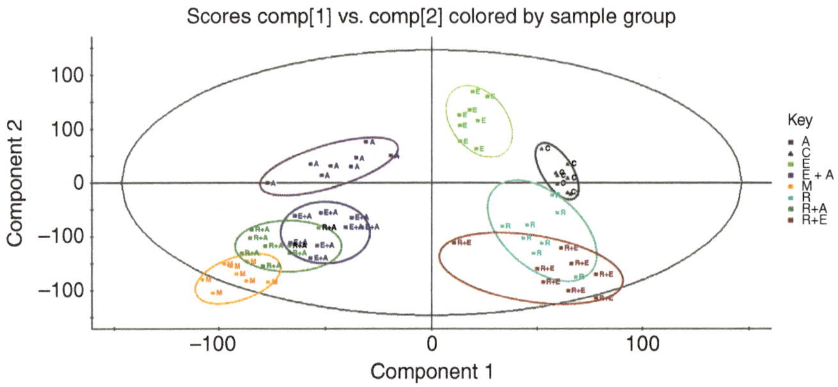

FIGURE 7.3 **PCA score plot of urine metabolic profiling of rats treated with rhubarb (R), ethanol (E), and α-nephthylisothiolyanate.**
(A). ▲ C (control group); ■ A group; ■ E group; ■ E + A group; ■ M group; ■ R group; ■ R + A group; ■ R + E group.

metabolites' ions on dispersion, and illustrated the ions maximally contributed to the assorting. The mass of the biomarkers was estimated accurately using TOF/MS, and the element structure and unsaturation of corresponding compounds within error range were obtained. The possible structure of compounds was then obtained through searching database of molecular formulas or molecular weights to screen one or more compounds by combined consideration on the MS/MS data. Finally, nineteen metabolites in rats' urine were confirmed through the retention behavior on the chromatogram contrast to the control, and the MS/MS data (data not shown).

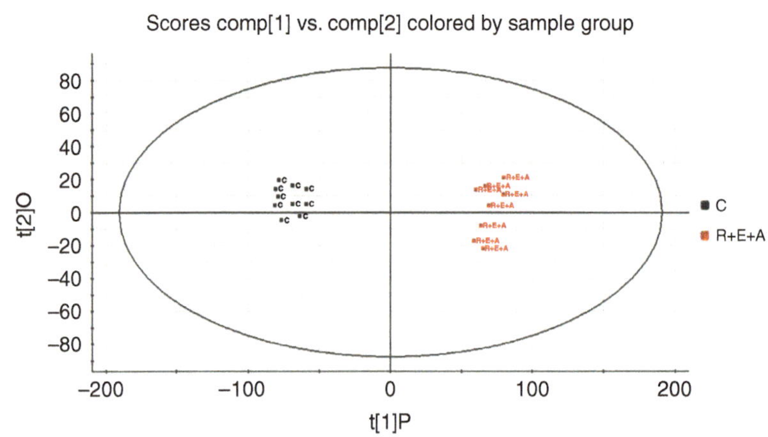

FIGURE 7.4 **OPLS score plot of urine metabolites of Yinhuang syndrome induced by combination of rhubarb, ethanol, and α-nephthylisothiolyanate.**

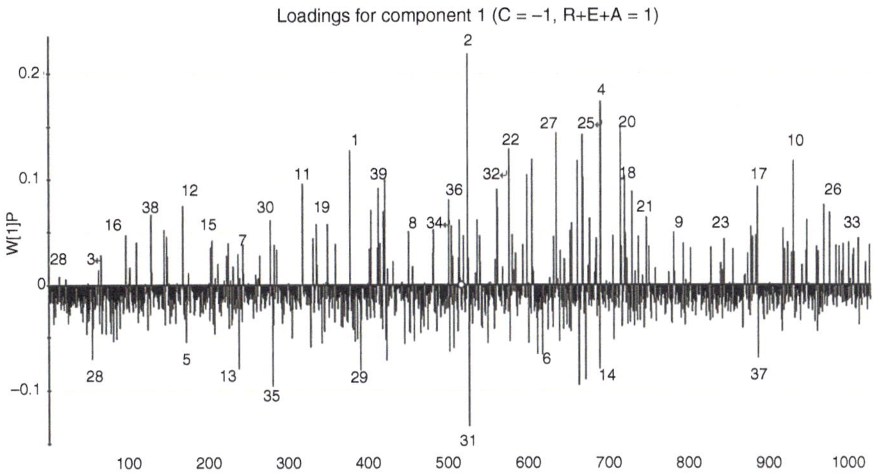

FIGURE 7.5 OPLS loading plot of urine metabolites of Yinhuang syndrome.

7.3.5 Effects of YCSNT on YIS Based on Metabolite Profile

Using the presented method, the urinary metabolic profiles of the YCSNT-treated group were obtained. From the score plot of the PCA, a clear separation among the control, YCSNT-treated (low, medium, and high dose) group was easily seen, which was represented by an ellipse in Fig. 7.6. The score plot of the PCA showed that the control, YIS group were separated clearly, and

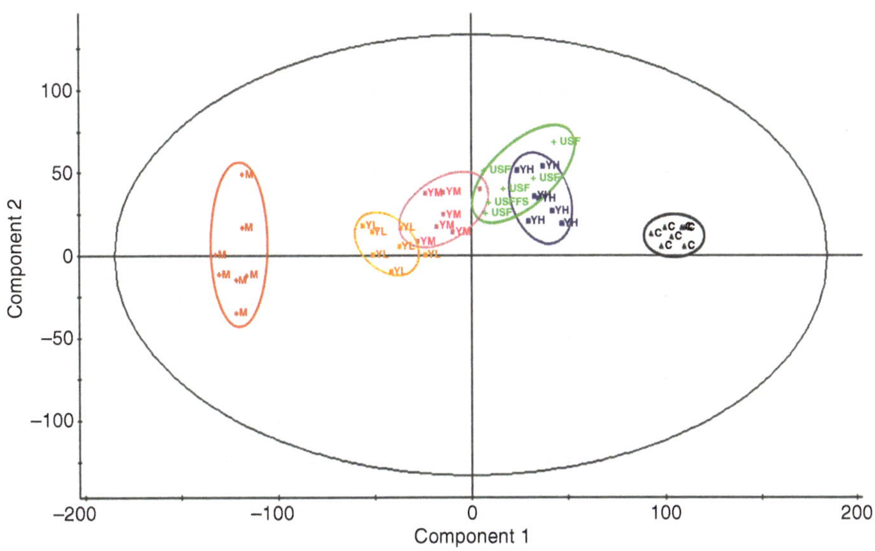

FIGURE 7.6 Effects of Yinchensini Tang against Yinhuang syndrome based on metabolite profile.

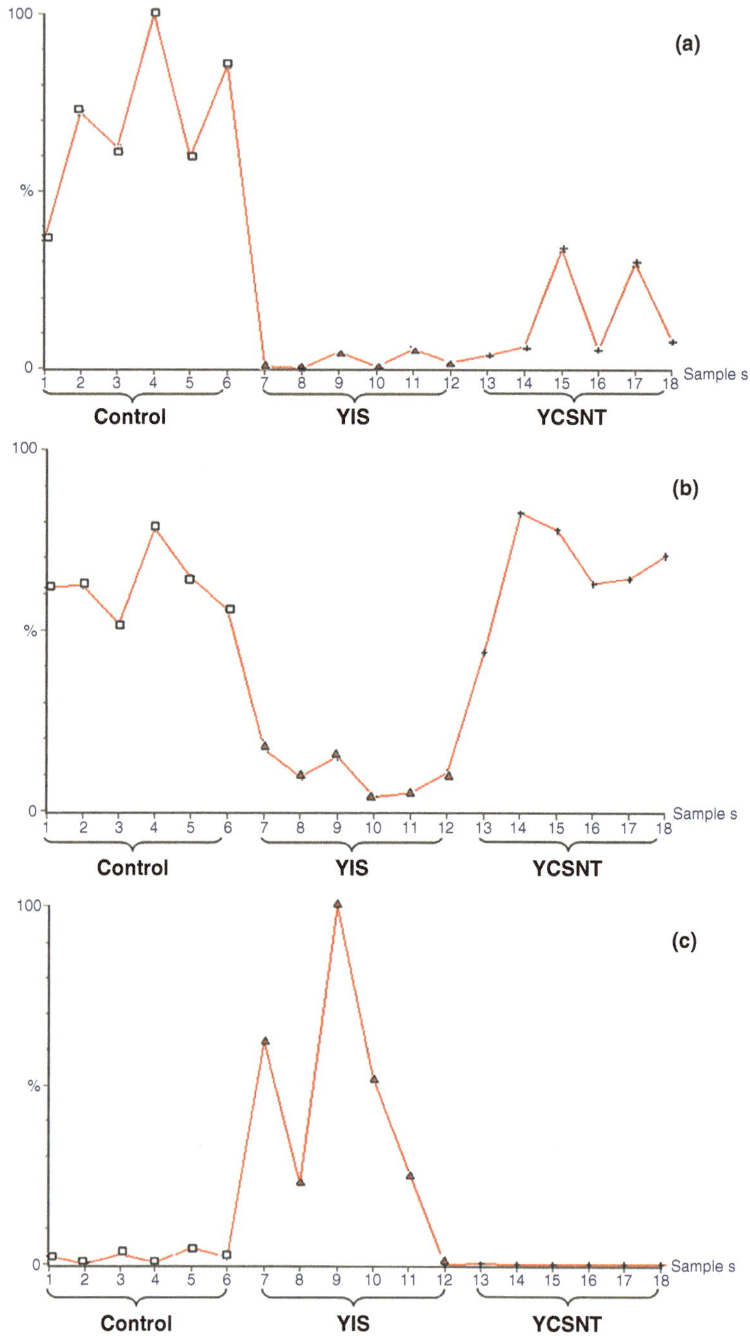

FIGURE 7.7 Change trends of the relative content of the key biomarkers.
(A) Glycocholic acid; (B) 3-methyldioxyindole; (C) 5-hydroxy-6-methoxyindole glucuroinde.

the YCSNT high dose group was closer to the control group, which suggested that YCSNT could reverse the pathological process of YIS. By comparing the identified biomarker level in the control, the YIS and the YCSNT high dose groups, most of them were reset to a normal state after YCSNT administration (Fig. 7.7), and the other metabolites were also reversed at different degrees (data not shown).

In this study, metabolomic technique was used for evaluating the YIS rats established by the triplet factors of R, E, and A, and nineteen biomarkers were identified with approval of their metabolizing pathways and pathogenetic mechanisms. Importantly, some indices that were capable of reflecting the integral condition of organism were selected depending on a holistic view, which might make up the deficits of indices commonly used at present, and provide a completely new path for the research on TCM syndrome model establishment.

References

Sun, K.W., Chen, B., Huang, Y.H., 2006. Clinical observation on chronic severe hepatitis B treated by principles of cooling-blood and detoxicating combined with clearing-heat and resolving-damp or combined with strengthening-Pi and warming-yang. Zhongguo Zhong Xi Yi Jie He Za Zhi 26 (11), 981–983.

Sun, H., Zhang, A., Yan, G., Han, Y., Sun, W., Ye, Y., Wang, X., 2013. Proteomics study on the hepatoprotective effects of traditional Chinese medicine formulae Yin-Chen-Hao-Tang by a combination of two-dimensional polyacrylamide gel electrophoresis and matrix-assisted laser desorption/ionization-time of flight mass spectrometry. J. Pharm. Biomed. Anal. 75, 173–179.

Tong, X., Sun, H., Yan, G.L., Dong, W., Sun, W.J., Yu, Y., Wang, P., Han, Y., Wang, X.J., 2011. Evaluation study on urine metabolomics in yinhuang rat model induced by triplet factors of rhubarb, ethanol, and α-nephthylisothiolyanate. Chin. J. Integr. Med. 17 (5), 369–375.

Wang, X., Zhang, A., Han, Y., Wang, P., Sun, H., Song, G., Dong, T., Yuan, Y., Yuan, X., Zhang, M., Xie, N., Zhang, H., Dong, H., Dong, W., 2012. Urine metabolomics analysis for biomarker discovery and detection of jaundice syndrome in patients with liver disease. Mol. Cell. Proteomics 11 (8), 370–380.

Wang, X., Zhang, A., Wang, P., Sun, H., Wu, G., Sun, W., Lv, H., Jiao, G., Xu, H., Yuan, Y., Liu, L., Zou, D., Wu, Z., Han, Y., Yan, G., Dong, W., Wu, F., Dong, T., Yu, Y., Zhang, S., Wu, X., Tong, X., Meng, X., 2013. Metabolomics coupled with proteomics advancing drug discovery toward more agile development of targeted combination therapies. Mol. Cell. Proteomics 12 (5), 1226–1238.

Yan, G., Sun, H., Sun, W., Zhao, L., Meng, X., Wang, X., 2010. Rapid and global detection and characterization of aconitum alkaloids in Yin Chen Si Ni Tang, a traditional Chinese medical formula, by ultra performance liquid chromatography-high resolution mass spectrometry and automated data analysis. J. Pharm. Biomed. Anal. 53 (3), 421–431.

Zhang, A., Sun, H., Wang, X., 2014. Potentiating therapeutic effects by enhancing synergism based on active constituents from traditional medicine. Phytother. Res. 28 (4), 526–533.

CHAPTER 8

Metabolite Profiling and Biomarkers Analysis of Jaundice Syndrome-Related Animal Models

Hui Sun, Wenjun Sun, Aihua Zhang, Xijun Wang

Subchapter 8.1

Jaundice Syndrome Rats Induced by D-Galactosamine

8.1.1 INTRODUCTION

Hepatitis, one of the manifestations of Jaundice syndrome (JS), is a disease with a high incidence worldwide, and is mainly induced by alcohol, lipid peroxidative products, and viruses (Wang et al., 2010). The chronic type will leads to liver fibrosis and end stage cirrhosis. It has been a key public health issue, derived from the health-threatening complications of portal hypertension, liver degeneration, and the increased incidence of hepatocellular carcinoma. D-Galactosamine (GalN) is one of the known hepatotoxic agents, and has been proven to induce liver injury that is comparatively close to human viral hepatitis in pathophysiological phenotypes (Li et al., 2013). The experimental acute hepatitis has been successfully induced by GalN administration in the rat (Sato et al., 2013). For the enhanced elucidating mechanisms of GalN inducing hepatitis, the purpose of this work is design and to investigate the use of UPLC/MS-based metabolomic method to characterize the metabolic profiling associated with JS rats, to validate the special biomarkers, and changes in the metabolic mechanisms of the hepatitis rat, and to provide a rational basis for the methodological development of noninvasive prognosis, and nonsurgical medical therapies for hepatitis (Wang et al., 2014). The overall experimental process is shown in the flowchart (see Fig. 8.1).

CONTENTS

(Continued)

Chinmedomics. http://dx.doi.org/10.1016/B978-0-12-803117-9.00008-1

CONTENTS

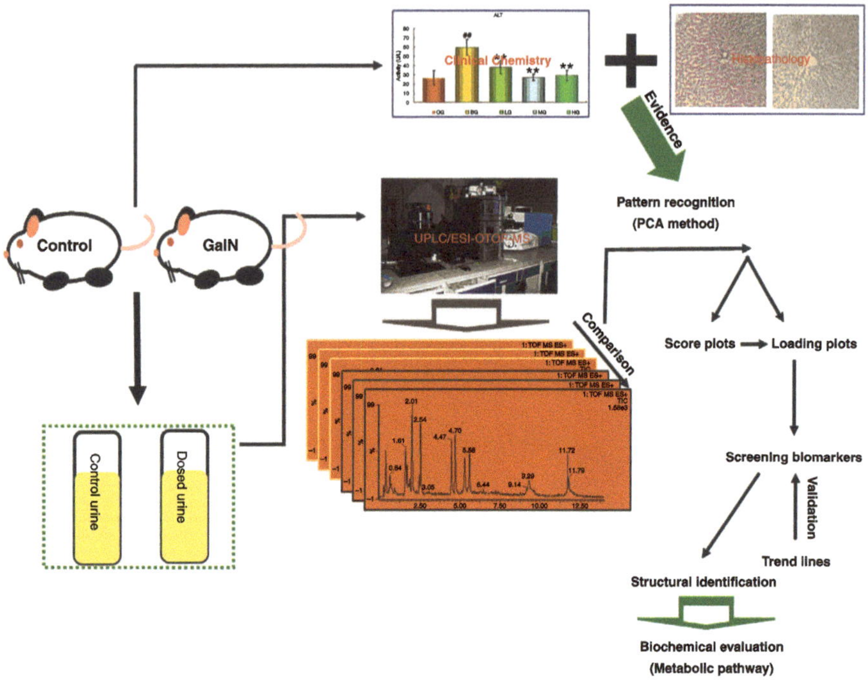

FIGURE 8.1 The overall flowchart of the application of metabolomic method in this study.

8.1.2 MATERIALS AND METHODS

8.1.2.1 Animals and Treatments

Twenty-four male Wistar rats (age: 6 weeks, body weight: 200 ± 20 g) were purchased from the Shanghai Slac Laboratory Animal Ltd Corp. (Shanghai, China). The animals were allowed to acclimatize for 7 days prior to dosing, and put in the metabolism cages during the urine collection periods specified below. The rats were randomly assigned to four dose groups: control-1, control-2, GalN dosed-1, and GalN dosed-2. Each rat in the GalN dosed groups was administered GalN in 5 mL/kg of body weight of saline (i.p., 400 mg/kg, $n = 6$) and in the control group the dose was saline (i.p., 0.9% w/v, $n = 6$). All rats in control-1 and GalN dosed-1 group were killed by exsanguinations under isoflurane anaesthesia 24 h postdosing, the collected blood samples, and liver tissues were prepared, and stored for determining clinical chemistry parameters and histopathological evaluation. In control-2 and the GalN dosed-2 group, urine samples were collected predose (−24 to 0 h), and 0–24, 24–48, 48–72, 72–96 h postdose. Urine samples were centrifuged at 13,000 rpm (5°C) for 10 min and frozen at −20°C prior to analysis.

8.1.2.2 Clinical Chemistry

The blood samples for the clinical chemistry study were taken with heparin containers from every rat of control-1 and GalN dosed-1 group, and centrifuged at 4000 rpm (5°C) for 10 min and separated plasma. The tissues for biochemical studies were washed with cold physiological saline and the connective tissue was removed. Liver homogenate was prepared with physiological saline by the ratio of 1/9 (*w*/*v*). The plasma ALT, AST, ALP, and CHO activities, and the liver SOD and MPO activities, were determined by the detection kits according to the manufacturer's protocols.

8.1.2.3 Histopathological Examination

Liver tissue for histopathological analysis was fixed in 10% buffered formalin saline, subsequently dehydrated and embedded in wax. The tissue wax was cut into 5 mm sections. Fixed sections were then stained with both hematoxylin and eosin (HE). The histopathological characters were observed and recorded under HPLAS-1000 colorized pathology image analyzer.

8.1.2.4 Chromatography

Chromatographic analysis was performed using a Waters ACQUITY UPLC system using a 2.1 mm ACQUITYC18 column (Milford, MA, USA) at a flow rate of 0.4 mL/min (back pressure: 11,000 psi) and a 5 μL injection volume. The column temperature was maintained at 45°C. The column eluent was directed to the mass spectrometer without split. All samples were eluted using a bilinear gradient, 0–20% B over 10 min, 20–80% B from 10 min–15 min, and then 80–100% B from 15 min–16 min (A: 0.1% formic acid in water; B: 0.1% formic acid in ACN). After every 10 sample injections, a pooled sample followed by a blank was injected in order to ensure the stability and repeatability of the LC–MS systems.

8.1.2.5 Mass Spectrometry

MS analysis was performed on a Micromass QTOF (Waters MS Technologies, Manchester, UK) with an ESI source operating in positive and negative ionizations mode. The desolvation gas rate was set to 550 L/h at a temperature of 300°C, the cone gas rate was set at 60 L/h and the source temperature at 110°C. The capillary voltage and the cone voltage were set at 3200 V and 35 V in positive ionization mode, and 2800 V and 35 V in negative ionization mode, respectively. All analyses were acquired using the lockspray to ensure accuracy and reproducibility, the leucine–enkephalin was used as the lock-mass (*m*/*z* 556.2771 for positive ion mode; *m*/*z* 554.2771 for negative ion mode) at a concentration of 50 fmol/L and flow rate 30 L/min. Data were collected in centroid mode from *m*/*z* 50 to *m*/*z* 900 with a lockspray frequency of 0.48 s, and data averaging over 10 scans.

CONTENTS

8.1.2.6 Data Processing and Pattern Recognition

The raw data were analyzed by the Micromass MarkerLynx applications manager (Waters, UK); the retention time and m/z data pair for each peak was detected. The ion intensities for each peak detected were then normalized, within each sample, to the sum of the peak intensities in that sample. The resulting normalized peak intensities were then multiplied by 10,000. Then the data were exported and analyzed by PCA.

8.1.2.7 Statistical Analysis

Quantitative data were expressed as mean ± SD and all statistical comparisons were made by means of the one-way ANOVA test followed by the Tukey's test. p-Values less than 0.05 were considered statistically significant while p-values less than 0.01 were considered extremely significant.

8.1.3 RESULTS

8.1.3.1 Clinical Chemistry and Histopathological Examination

Almost all the rats highlighted retarded actions, slow reactions, and obviously tumescent livers, 24 h after being GalN dosed. The plasma levels of ALT, AST, ALP, and CHO, liver levels of SOD and MPO were within the normal ranges in the saline-treated control-1 group, all of which were significantly elevated respectively, in the GalN dosed-1 group (Fig. 8.2A–C). Compared with the normal liver histology of the control-1 rats (Fig. 8.2D), the liver inflammation, severe liver congestion, and hemorrhage were indicated in the GalN dosed-1 group rats (Fig. 8.2E).

8.1.3.2 Validation of UPLC Conditions

The flow rate is limited within 0.4 mL/min in order to obtain the ideal ionization efficiency and avoid ion suppression that will seriously influence the analytical sensitivity of the mass spectrometer. Considering sensitivity and resolution meanwhile, the final flow rate was optimized at 0.4 mL/min in this study. Furthermore, the effects of an acid modifier (e.g., formic acid) with different concentrations (1, 0.1, and 0.01%) were characterized, the optimal peak resolution and shape was obtained at 0.1 % formic acid. The UPLC column temperature was set at 35°C, also this reduces the column pressure from a higher flow rate. The gradient elution was adopted to reach a more efficacious separation. Six batches of urinary samples from the same rat were injected into the UPLC column for assessing the reproducibility of the chromatography. The retention time of visible peaks varied with the RSD values within the range of 0.05–0.6% during the analysis, suggesting that the retention time reproducibility of the system was good for large-scale sample analysis. For testing the ability of the UPLC system to differentiate the control and GalN dosed rats, the urine

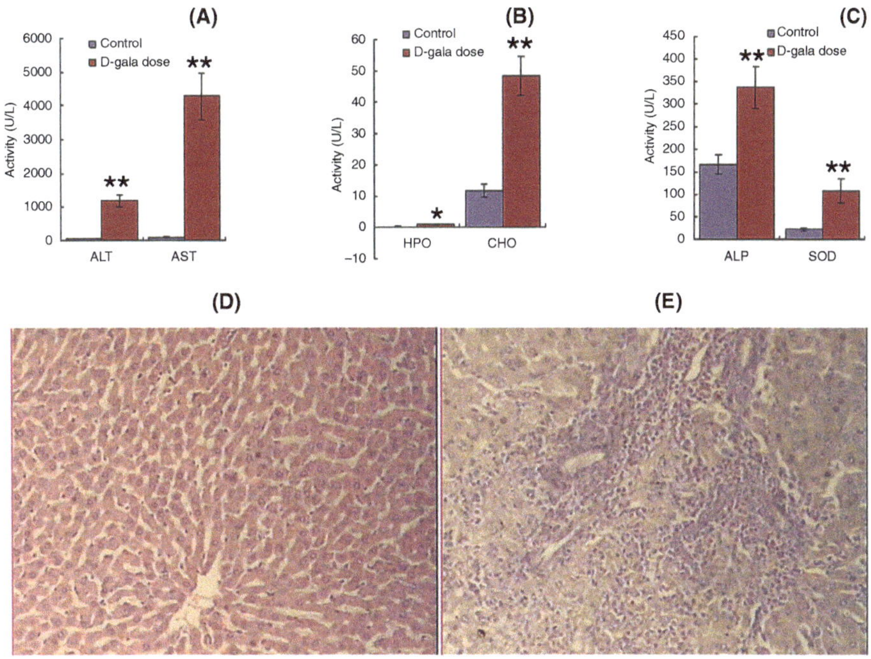

FIGURE 8.2 Clinical chemistry and histopathological comparison of the GalN dosed group and controls.

(A) Activities of ALT and AST in GalN dosed group and controls; (B) activities of MPO and CHO in GalN dosed group and controls; (C) activities of ALP and SOD in GalN dosed group and controls; (D) the normal liver morphology; (E) the liver morphology of GalN dosed. *$p < 0.05$ vs. control; **$p < 0.01$ vs. control.

taken from the different rats were analyzed, typical MS chromatograms of these samples were shown in Fig. 8.3. Visual examination of these chromatograms demonstrated distinctly among different samples.

8.1.3.3 Optimization of MS Conditions

The desolvation gas flow rate was set at 600 L/h, and the desolvation temperature at 300 °C, respectively, in order to remove redundant solvent resulting from a higher flow rate of 0.4 mL/min for the mass spectrometer. Cone voltage was optimized to the corresponding values ranging from 15–90 V owing to the wide range of molecular weight metabolites (50–900 Da). The results indicated that a cone voltage of 35 V was sufficient to produce product ions in the ES^+ or ES^- mode for metabolites. As LC peaks become narrower for the UPLC system, the mass spectrometer is required to acquire data faster. The faster duty cycle of the TOF instrumentation should allow rapid data collection and good quality MS/MS analysis. In this study, the LC eluent was analyzed in positive or negative ion electrospray without splitting eluent. The data were collected with alternating collision energy, at 5 eV for parent ion information and a range of 15–25 eV for

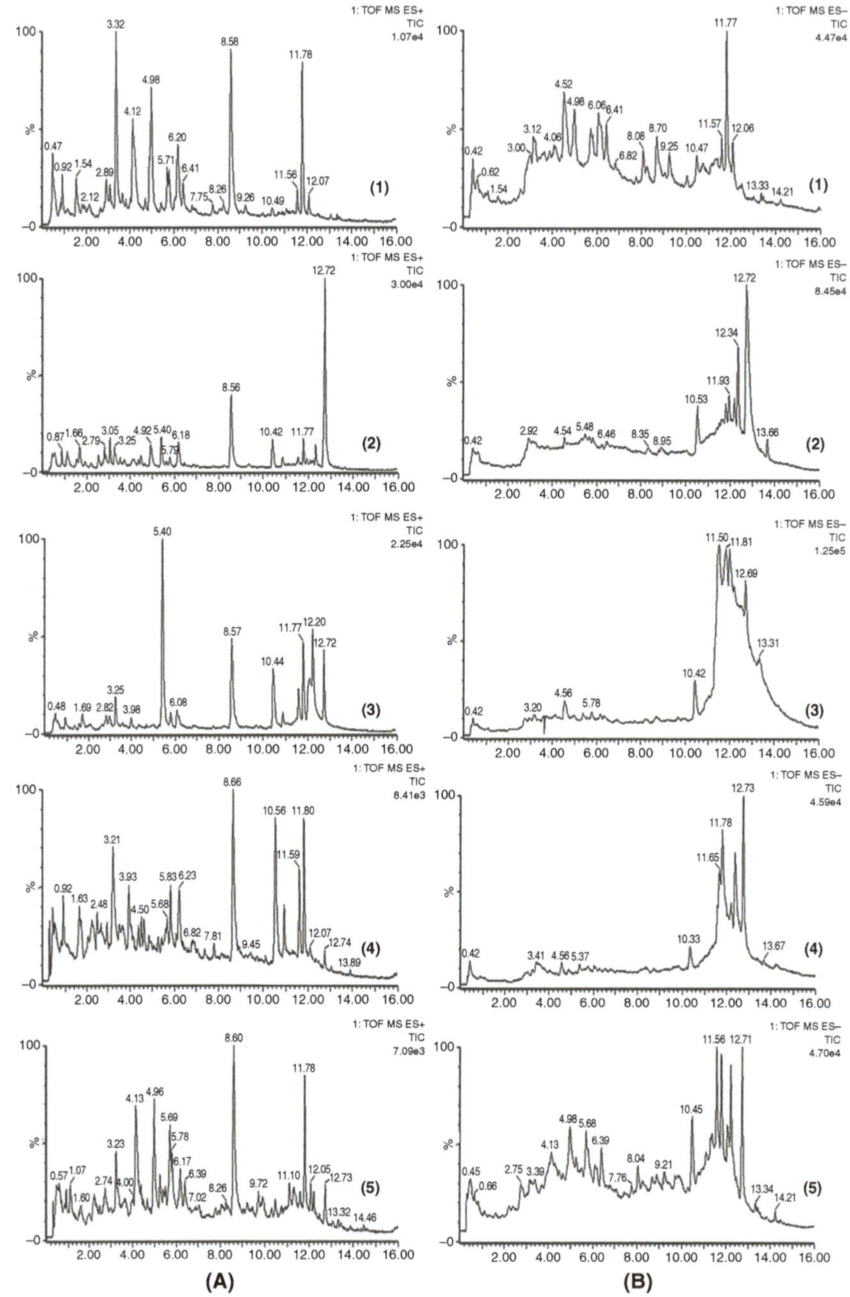

FIGURE 8.3 **Representative TIC chromatograms of positive (A) and negative (B) ions scanning of urine samples taken from control and GalN postdosed of different delaying time.**
A (1) and B (1): −24 to 0 h predosed GalN; A (2) and B (2): 0–24 h postdosed GalN; A (3) and B (3): 24–48 h postdosed GalN; A (4) and B (4): 48–72 h postdosed GalN; A (5) and B (5): 72–96 h postdosed GalN.

fragment ion information. The data acquisition rate of 100 ms per channel with a 20 ms interchannel delay was used. The resolution was greater than 5000 in both high and low collision energy modes. The Q-TOF instrument was switching with a frequency of 100 ms and we can see that the frequency was rapid enough to keep the chromatographic information while still obtaining good MS/MS data.

8.1.3.4 MS Analysis of Urine Samples

Both a positive and negative ion TIC chromatograms were shown in Fig. 8.3. According to peak heights data, about 6900 peaks of positive ions (ESI^+) and 4200 peaks of negative ions (ESI^-) were detected by Micromass MarkerLynx using the same acquiring method from 0.1 min to 16 min of retention time, and then the mean and normalized data were good for PCA analysis, respectively. The mass spectrometry of urine collected from the GalN dosed rats varied in metabolite profiling according to the severity of the hepatic lesions (illustrated in Fig. 8.3A and B), which revealed the fluctuated endogenous metabolite levels post-GalN-dosing.

8.1.3.5 PCA of Data from GalN Dosed Rats

In order to better visualize and describe the subtle similarities and differences amongst these complex data sets, PCA was applied to data analysis. The multivariate data analysis tool has many advantages including the ability to reduce large data sets to a graphical form where trends are easily demonstrated. By virtue of the mathematics of PCA, those samples that lie close to one another will have similar spectral profiles. The loadings plot indicates which variables (concatenated *m*/*z* values and retention times in our analysis) are responsible for the observed clustering in the scores plot. The differences between groups can be highlighted. Figure 8.4 showed the scores and loading plot of PCA, in positive and negative ions scan modes, the GalN dosed group and control group significantly can be separated, the separation between control and the GalN dosed, varied with postdosing time prolonging, the obvious returning was founded during postdose 72–96 h from the score plots of both ion scan modes. As can be seen from Table 8.1, the markers selected from the ESI^+ and ESI^- scans were very different, thus, using two different scan modes gives us fuller information about the urine metabolomes. By comparing the retention time of the markers, it is found that most of the markers from the ESI^+ were most of peaks appearing from 3 min–6 min which varied significantly with postdose delaying, compared with other time ranges, and most of the markers from the ESI^- were selected appearing from 11 min–13 min. Ten potential biomarkers were selected according to separation contribution from loading plots, similarly, the trends of each biomarkers during 0–24, 24–48, 48–72, and 72–96 h again demonstrated postdosing that the GalN induced significant separation of groups between the control and the GalN dosed which is attributed to the variation of urine metabolomes

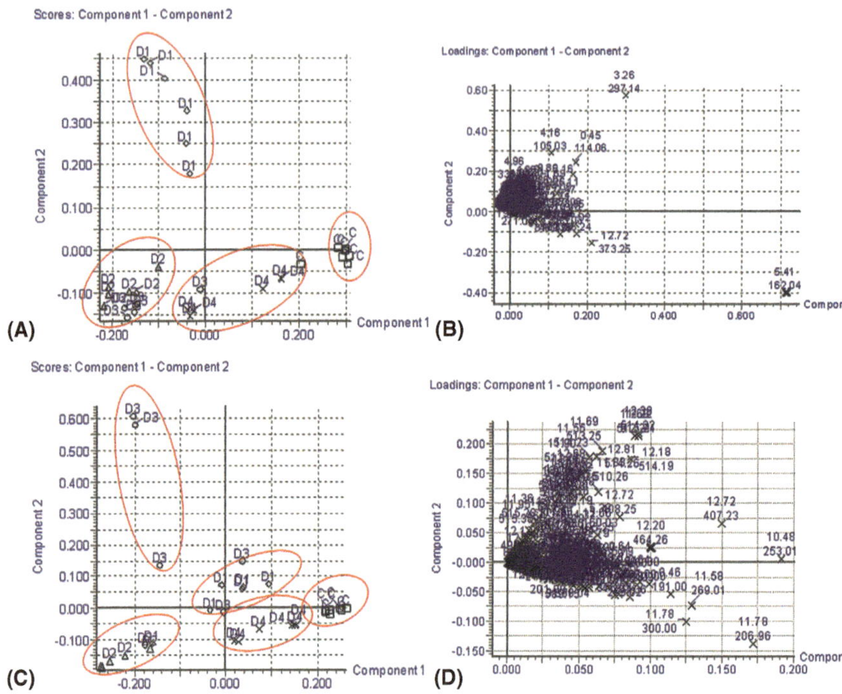

FIGURE 8.4 **PCA of data from GalN dosed rats.**
(A) Score plot of PCA, ESI^+; (B) loading plot of PCA, ESI^+; (C) score plot of PCA, ESI^-; and (D) loading plot of PCA, ESI^-. (□) control group (predose); (◇) 0-24 h postdosed GalN; (Δ) 24-48 h postdosed GalN; (○) 48-72 h postdosed GalN; (×) 72-96 h postdosed GalN.

resulting from the GalN inducing disorders of the normal metabolic pathway, and the levels of most of the biomarkers gradually recovered to normal in the control group during 72–96 h (Table 8.1) this supports the conclusion that the score plots demonstrate and validate a selection of potential biomarkers.

Here, the *m*/*z* 191 (ESI^-, Fig. 8.5) as an example to elucidate the biomarker identification process. First, the molecular weight (MW) and proposed formula of the potential biomarker was to be determined, the extract mass was acquired (Fig. 8.5A and B), ensuring the MW is 192, subsequently the accurate molecular weight was also measured through the Q-TOF micro MS system (191.23) and the proposed formula was determined by extract mass and element composition analyses ($C_6H_8O_7$). Second, further information about fragmentation patterns of the marker were searched for in the mass spectrum, and the MS/MS scan was also employed in this step (Fig. 8.5C). Finally, the online database was retrieved subjected to the information obtained from the above process. As a result, the biomarker of 191 was identified as citric acid, its level varied with postdose time (as shown in Fig. 8.5D). The standard sample of citric acid

Table 8.1 Markers Selected Indicated a Difference Between ESI^+ and ESI^- Scans When Global Urine Metabonome Data

No.	R_t_m/z	Formula	Identified Metabolites	0–24 h Postdose	24–48 h Postdose	48–72 h Postdose	72–96 h Postdose
Positive ion mode							
1	0.45_114.06	$C_4H_7N_3O$	Creatinine	↑	↑	↑	↑
2	3.26_297.14		Unknown	↓	↓	↓	↑
3	4.16_105.03	$C_4H_8O_3$	β-Hydroxybutanoic acid	↓	↓	↓	↑
4	5.41_174.18	$C_{10}H_9NO_2$	Indole-3-acetic acid	↑	↑	↑	↓
5	12.72_373.25	$C_{16}H_{28}N_4O_4S$	Unknown	↓	↓	↓	↑
Negative ion mode							
1	12.20_464.26	$C_{26}H_{43}NO_6$	Glycocholic acid	–	–	–	↑
2	11.78_300.00	$C_{18}H_{39}NO_2$	Sphinganine	↑	↑	↑	↑
3	12.72_407.23	$C_{24}H_{40}O_5$	Cholic acid	–	–	↑	–
4	11.78_206.96	$C_{11}H_{13}NO_3$	*n*-Acetyl-L-phenylalanine	↑	↑	↑	↓
5	0.46_191.23	$C_6H_8O_7$	Citric acid	↓	↓	↓	↑
6	11.58_269.01	$C_{11}H_{12}NOS_3$	Unknown	↓	↓	↓	↑

Notes: ↑ *represents that the level of compound in urine was increased in GalN postdose, compared with controls;* ↓ *represents that the level of compound in urine was increased in GalN postdose, compared with controls; – represents that the level of compound in urine was equal in GalN postdose with controls.*

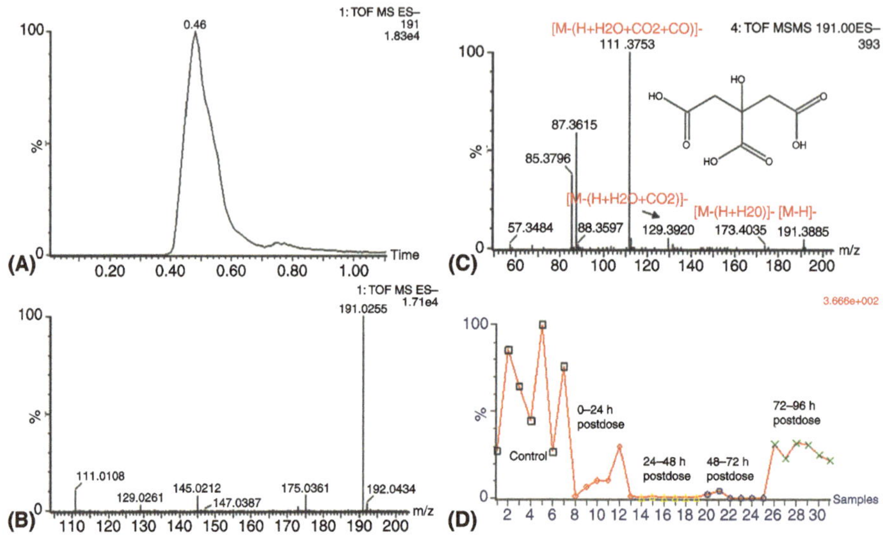

FIGURE 8.5 **Typical identification of potential biomarker.**
(A) Extract mass chromatogram of m/z 191 of ESI$^-$; (B) parent ion scanning chromatogram of m/z 191 in ESI$^-$; (C) product ions scanning chromatogram of m/z 191 of ESI$^-$; (D) the trends plot of m/z 191 pre- and postdosed GalN.

confirmed this result. According to the above process, 8 out of 11 potential biomarkers have been preliminarily identified.

8.1.4 DISCUSSION

GalN is known to cause reversible acute hepatitis, the mechanism of development is not now uncertain, the reported results, however, only partly address this topic. One of the major strengths of the use of high definition mass spectrometry of biofluids in toxicological studies lies in the conferred ability to evaluate simultaneously a whole series of low-molecular-weight metabolites, thus allowing characterization of a lesion in terms of a combination of bioactive markers. This study developed a validated and rapid UPLC/MS-based metobolomic method for characterizing metabolic profiling of GalN inducing hepatitis in rats, elucidating potential biomarkers of validating development, the returning of hepatitis, and basing information of method development on noninvasive prognosis, and nonsurgical clinical therapies for hepatitis.

8.1.4.1 Energy Metabolism

Citric acid (citrate) is a key metabolite of energy metabolism, which is formed in the TCA cycle, or may be introduced with diet. The concentration of citrate will rise to appreciable levels only under conditions of substantive metabolic imbalances. The levels of citrate in urine is increased in the GalN dosed rats, this suggest that GalN hepatitis has more energy consumption, compared with that of the control rats, resulting in the citrate biosynthesis being inhibited, which may indicate a more weak immunity system in the GalN dosed rats to xenophobic intervention. Simultaneously, the elevated levels of creatinine in urine involved the disturbance of energy metabolism and oxidative stress induced by GalN. The creatine–phosphocreatine system is vital for cellular energy transportation. It is well known that the larger part of creatine is phosphorylated by enzyme creatine kinase into phosphocreatine in the cell. This reaction is reversible and can act as an energy buffer to maintain ATP homeostasis. In addition, about 2% creatine and the produced phosphocreatine are nonenzymatically converted to creatinine (Wyss and Kaddurah-Daouk, 2000) increased excretion of it in urine has been correlated with oxidative stress (Wang et al., 2005). All creatine kinase isoenzymes are easily susceptible to oxidative damage because creatine kinase isoenzymes invariably have a highly reactive sulfhydryl group in the active site (Cys-283), oxidative modification of which could result in pronounced inhibition of the enzyme (Schlattner et al., 2006). Whereas, it is very likely that GalN dosed could inhibit the creatine kinase activity, and consequently creatine was diverted into creatinine production rather than phosphocreatine synthesis.

8.1.4.2 Lipid Metabolism

8.1.4.2.1 Bile Acid Metabolism

In this study, some perturbations were observed up to 72 or 96 h postdose, namely, the increasing excretion of bile acids. The accumulation of hydrophobic bile acids plays a role in the induction of the apoptosis and the necrosis of hepatocytes during cholestasis (Rost et al., 2003). Glycocholic acid is one of the bile acids involved in the emulsification of fats and cholestasis. At this point, the level of glycocholic acid in urine was increased. Cholic acid is also one of major primary bile acids produced in the liver and usually conjugated with glycine or taurine. It facilitates fat absorption and cholesterol excretion. The correlation between both cholic acid and cholic acid/chenodeoxycholic acid levels with histological liver injury, and fibrosis progression suggests a potential monitoring role for these bile acids in cystic fibrosis-associated liver disease, thus, the high level of cholic acid in urine existed during 48–72 h postdosed GalN which may indicate that the cystic fibrosis is a development stage of GalN hepatitis, it may be summarized that the bile acid metabolism was seriously effected by GalN resulting in the increase levels of bile acid, e.g., glycocholic acid and cholic acid, they finally induced cholestasis and cystic fibrosis of liver.

8.1.4.2.2 Ketone Metabolism

Beta-hydroxybutanoic acid is one of the ketone bodies produced in the liver and occurs at high levels in the blood and urine in ketosis. Some studies revealed that ketonic compounds showed protection from acetaminophen-induced liver necrosis and modified the chronic liver injury (cirrhosis) induced by haloalkanes (Price and Jollow, 1983). The data showed that the level of beta-hydroxybutanoic acid was significantly reduced during 0–72 h postdosed GalN, the result was demonstrated during 72–144 h, on the one hand, which suggests development of GalN hepatitis will exhaust β-hydroxybutanoic acid based on above data (Charbonneau et al., 1986), when the recovery of GalN hepatitis in rats during 72–144 h, the production of beta-hydroxybutanoic acid in the liver was acclimated to a normal level; on the other hand, GalN may also induce ketosis when the hepatitis occurred, resulting in high level of beta-hydroxybutanoic acid in urine that will further demonstrate modulation to hepatitis and the accelerated recovery of hepatitis. Thus, the level of fluctuation of urine beta-hydroxybutanoic acid observed post-GalN dosing cannot be attributed to an immediate effect of GalN, but is the result of a delayed change in intermediary metabolism characterized by disturbing hepatic ketone production from fatty acids.

8.1.4.2.3 Sphingolipid Metabolism

Sphinganine is metabolized from L-serine and palmitoyl-CoA by serine palmitoyltransferase, it a blocker postlysosomal cholesterol transport by inhibition of low-density lipoprotein-induced esterification of cholesterol, and causes unesterified cholesterol to accumulate in perinuclear vesicles, thus obviously increasing levels of sphinganine in the urine of GalN dosed rats may indicate that the abnormal accumulation of esterification of cholesterol is a property of GalN hepatitis caused by perturbing sphingolipid metabolism.

8.1.4.3 Amino Acid Metabolism

8.1.4.3.1 Phenylalanine Metabolism

There were increased levels of *n*-acetyl-L-phenylalanine in the urine of GalN dosed rats from 0–72 h postdose, and returning during 72–96 postdose, the *n*-acetyl-L-phenylalanine is a hazardous amphipathic metabolite of phenylalanine. It appears in large amounts in the urine of patients with phenylketonuria that is a human genetic disorder that is caused by a lack of phenylalanine hydroxylase, the enzyme necessary to metabolize phenylalanine to tyrosine. Comprehensive studies on the structure and the function of the liver were performed in children with phenylketonuria (PKU). The results illustrated that the liver is still involved in the pathological process of PKU, so that GalN inducing hepatitis may enhance phenylalanine metabolism to increase in the levels of urine *n*-acetyl-L-phenylalanine, and suggest that GalN may induce PKU in this study.

8.1.4.3.2 Tryptophan Metabolism

Indole-3-acetic acid is a normal urinary indolic tryptophan metabolite and has been found elevated in patients with liver diseases, the increase of urine level of indole-3-acetic acid was induced by GalN during 0–72 h postdose, and the recovery of production was revealed during 72–96 h postdose, it is likely that

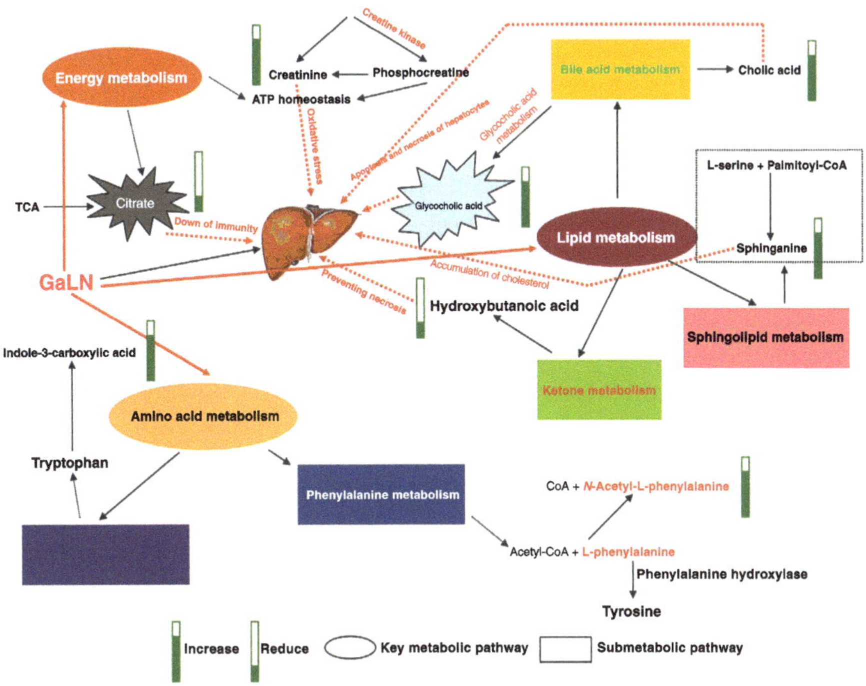

FIGURE 8.6 **Effects of GalN induced hepatitis on metabolic pathways.**
Putative metabolic pathways were inferred in the GalN dosed rats from changes in the urine levels of intermediates during substance metabolism.

GalN could activate tryptophan metabolism so as to enhance the production of indole-3-acetic acid.

Based on the above evaluation, the putative metabolic pathways were deduced in the GalN hepatitis (Fig. 8.6) based on changes in the levels of intermediates during substance metabolism, and the mechanisms underlying GalN damage to the liver were inferred. The pathways included energy metabolism, lipid metabolism (bile acid metabolism, ketone metabolism, and sphingolipid metabolism) and amino acid metabolism (phenylalanine metabolism and tryptophan metabolism). The production of ketone bodies (beta-hydroxybutanoic acid) and the TCA cycle (citric acid) were inhibited in the GalN dosed rats, whereas the amino acids metabolism (indole-3-acetic acid and *n*-acetyl-L-phenylalanine) and lipid metabolism (glycocholic acid, sphinganine, and cholic acid) were enhanced in these rats.

In this study, the identified biomarkers were attributed to the separation of groups between the control and the GalN dosed rats which will be a key complementation for NMR or GC–MS-based metabolomic studies of GalN dosing in rats and mice, especially the novel identified biomarkers in this work, and would better assist the clinic in providing a full understanding of metabolic profiling and the mechanism of actions of GalN inducing hepatitis, and setting a basic information for the establishment of prognosis and rational therapy. However, there are still many potential biomarkers that were not elucidated, and the internal mechanisms of the changes of the metabolites are still unclear. Further elucidation of the total physiological changes, and mechanism of GalN inducing hepatitis requires progress in methodology (holistic analysis of all molecules) and its combination with more clinical data of hepatitis.

8.1.5 CONCLUSIONS

A metabolomic method based on UPLC/MS and a multivariate statistical technique has been developed to elucidate the changes in metabolic intermediates and to investigate the metabolic profile of acute hepatitis induced by GalN, one of the manifestations of JS. The significant differences in the urine levels of many metabolites were highlighted when compared to the controls with GalN hepatitis rats. According to the developed method of markers selection and identification, eight potential biomarkers were identified that significantly contributed to the development and the recovery of JS. It demonstrates that this metabolomic approach is a powerful tool with which to characterize changes in the levels of metabolic intermediates, and to facilitate the search for metabolic markers under certain physiopathological conditions. Markers linked to metabolic pathways may also be used for the prognosis and the treatment of hepatitis in of clinical practice.

Subchapter 8.2

Jaundice Syndrome Rats Induced by Alcohol

8.2.1 INTRODUCTION

Excessive alcohol consumption is the third most common cause of lifestyle-associated mortality in the world, and more than half these deaths were attributed to alcohol-induced liver disease (ALD). An early and reliable tool to assess ALD risk would be helpful for treatment intervention. Only an estimated 13% of people with identified ALD have ever received specialty treatment due to the lack of effective medications that ameliorate withdrawal syndrome and cure alcohol dependence. Metabolomics has brought much excitement to the field of life sciences as a potential translational tool, and offers a global analysis of low-molecular-weight metabolites in biological samples, attempts to capture global changes, and the physiological status in biochemical networks and pathways in order to elucidate the sites of perturbations. Common tools used for metabonomic studies include: nuclear magnetic resonance (NMR) spectroscopy and gas or high-performance liquid chromatography coupled to mass spectrometry. Ultraperformance liquid chromatography coupled to mass spectrometry (UPLC/MS) is the ideal tool with the highest resolution and this sensitive technique is considered a powerful tool in metabolomics because of its ability to obtain multiparametric metabolite profiles from biofluids rapidly and effectively. Here, it describes the results of LC–MS-based metabolomic investigations on the liver metabolomes of rats for the endogenous metabolites using alcohol-fed models, with metabolite identification via high accuracy MS analysis. The power of mass spectrometry-based metabolomics to capture and elucidate metabolic changes during alcohol consumption has been demonstrated in this study.

8.2.2 METHODS

8.2.2.1 Animal Handling and Sample Preparation

Male Wistar rats (weighing 220–260 g) were supplied by the GLP Center of Heilongjiang University of Chinese Medicine (Harbin, China). After acclimatization, animals were randomly divided into two groups with 10 rats in each: the control and the model groups. Rats model were orally administrated with 50% alcohol (5 mL/kg body weight) olive oil solution at 3 day (6:00 p.m.) to induce liver injury model for 5 consecutive days, and until day 8. The rats in the control group were administrated with 0.9% saline in the whole procedure for 7 consecutive days. Urine was collected daily (at 6:00 a.m.) from metabolism cages at ambient temperature throughout the whole procedure and centrifuged at 13,000 rpm at 4°C for 5 min, and the supernatants were stored frozen at −20°C until metabolomic analysis. All the experimental procedures were conducted according to the principles expressed in the Declaration of Helsinki.

8.2.2.2 Metabolic Profiling Platform

8.2.2.2.1 Chromatography

Chromatographic analysis was performed in a Waters ACQUITY UHPLC system controlled with Masslynx (V4.1, Waters Corp., Milford, USA). An aliquot of 6 μL of sample solution was injected onto an ACQUITY UHPLC HSS T3 C_{18} column (100 × 2.1 mm, 1.7 μm, Waters Corp., Milford, USA) at 40°C and the flow rate was 0.4 mL/min. The optimal mobile phase consisted of a linear gradient system of (A) 0.1% formic acid in acetonitrile and (B) 0.1% formic acid in water, 0–2 min, 98% A; 2–3 min, 98–80% A; 3–7 min, 80–5% A; 7–9 min, 5% A; 9–11 min, 5–98% A; 11–15 min, 98% A. In addition, the QC sample was used to optimize the condition of UPLC-Q-TOF/MS, as it contained the most information of whole urine samples.

8.2.2.2.2 Mass Spectrometry

The mass spectrometry was operated by electrospray ionization in the positive ionization mode. The eluent was introduced into the high-definition mass spectrometer (Waters Corp., Milford, USA) analysis, and the optimal conditions of analysis were as follows: the source temperature was set at 110°C, desolvation gas temperature was 300°C, cone gas flow was 100 h, desolvation gas flow was 600 L/h; the capillary voltage was 3.2 kV, the sampling cone voltage was 35 V, microchannel plate voltage was 2300 V, and extraction cone voltage was 3.0 V. The data acquisition rate was set to 0.14 s/scan, with a 0.1 s interscan delay. Data were collected in centroid mode from 100–900 Da.

8.2.2.2.3 Multivariate Data Analysis and Data Processing

The UPLC-MS data were processed using the MarkerLynx Application Manager (Waters Corp.). After the UPLC/ESI-QTOF/MS measurement, the raw data were imported into the MassLynx software for peak detection and alignment. The intensity of each ion was normalized with respect to the total ion count to generate a data matrix that consisted of the retention time, *m*/*z* value, and the normalized peak area. The multivariate data matrix was analyzed by EZ-info software 2.0 (Waters Corp., Milford, USA). All the variables were mean-centered and pareto-scaled prior to PCA and PLS-DA. If a separation between the control and the treatment groups was observed in the PCA scores plot, PLS-DA was performed to highlight the differences between the groups.

8.2.2.2.4 Biomarkers Identification and Reconstruction of Metabolic Pathway

Potential markers of interest were extracted from loading plots constructed following analysis with PLS-DA, and markers were chosen based on their contribution to the variation and correlation within the data set. The identities of the specific metabolites were confirmed by comparison of their mass spectra and chromatographic retention times. With regard to the identification of biomarkers, the ion spectrum of potential biomarkers was matched with the structure message of metabolites acquired from the available biochemical databases. The pathway analysis of potential biomarkers was performed with the MetPA software online database source.

8.2.2.3 Statistical Analyses

All statistical analyses were performed using the Student's t-test. Differences with a *p*-value of 0.05 or less were considered significant. Assays were performed in triplicate, and the results are expressed as mean ± SD.

8.2.3 RESULTS

8.2.3.1 LC–MS Analysis for Metabolic Profiling

Using the optimal reversed-phase UPLC/MS conditions described earlier, the representative total ion current (TIC) chromatograms of the urine samples obtained from UPLC/ESI-Q-TOF/MS analysis for a continuous 8 days are shown in Fig. 8.7. All the data containing the retention time, peak intensity, and exact mass were imported in the Masslynx™ software for multiple statistical analyses. Both PCA and PLS-DA can be often taken because of their ability to cope with highly multivariate, noisy, collinear, and possibly incomplete data. PCA score plots for the alcohol treatment in positive mode can reflect

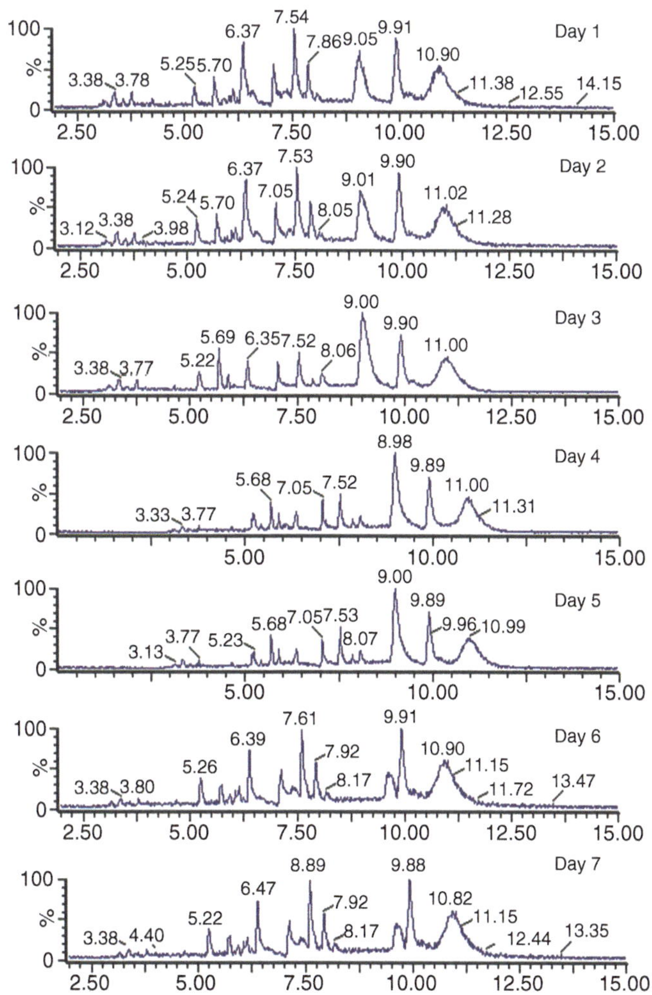

FIGURE 8.7 Typical total ion chromatograms of urine samples by UPLC/ESI-Q-TOF/MS analysis.

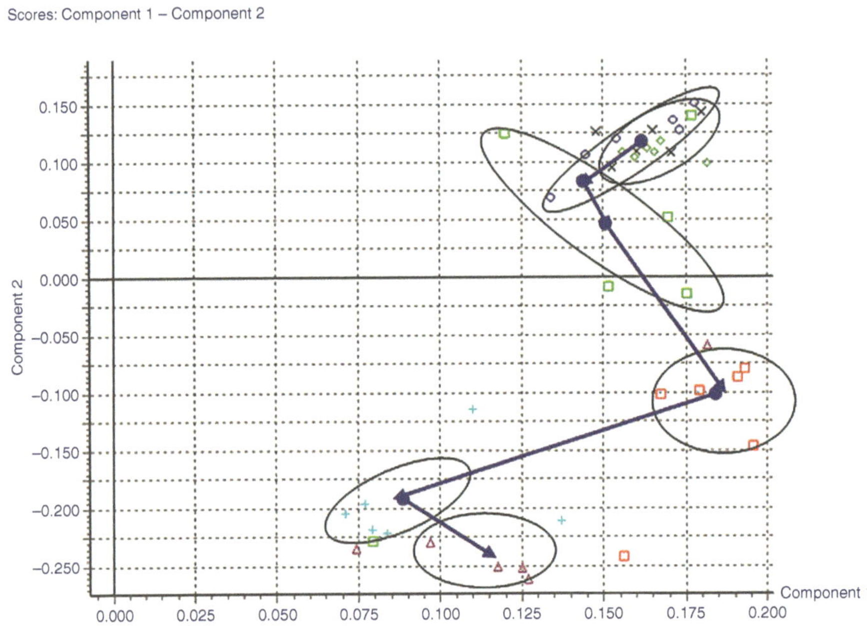

FIGURE 8.8 **Score plots of trajectory analysis for the alcohol treatment in positive mode.** (×: the 1st day; ○: the 2nd day; ◇: the 3rd day; □: the 4th day; □: the 5th day; +: the 6th day; Δ: the 7th day)

the differences between the model and the control groups, and the illustrated metabolic profiles in the control and the model groups were separated clearly (Fig. 8.8). PCA can effectively demonstrate the differences between the treated and the control groups. The PCA plot of the model group show similar behavior during the early stage of the experiment, but then gradually deviate from one another – on day 7 reaching the maximum trend. The tracks of the metabolic profiles at different time points also clearly demonstrate the time-dependent changes in the urine metabolites. The corresponding PLS-DA loadings plot indicated that differentiating metabolites were attributable to the clustering observed in the scores plot. The farther away from the origin, the higher the VIP value of the ions was. Four ions showed significant difference in abundance between the control and treated animals and contributed to the observed separation; these were selected from the loading plot of PLS-DA (Fig. 8.9).

8.2.3.2 Identification of Candidate Metabolites

The precise molecular mass was determined within a reasonable degree of measurement error using Q-TOF, and the potential element composition, degree

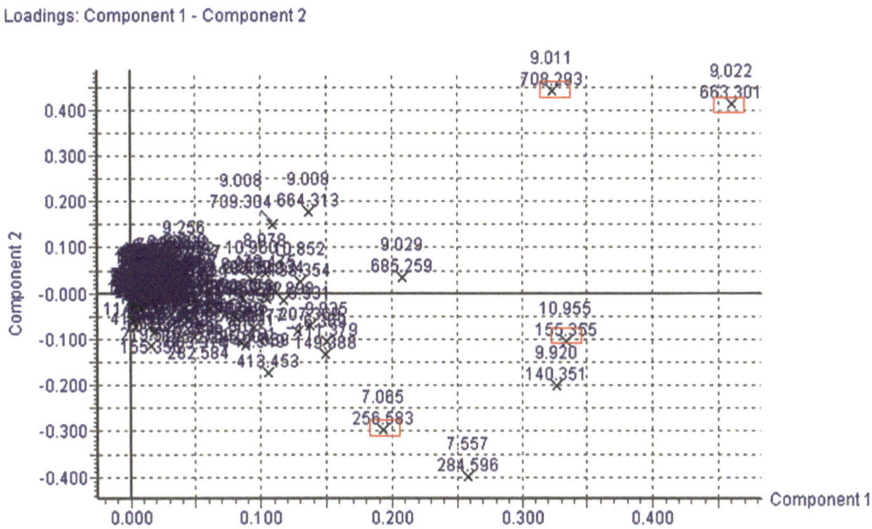

FIGURE 8.9 Loading plot of urine metabolome from model group.
Red data points indicate that ions most responsible for the variance in the score plot.

of unsaturation and fractional isotope abundance of the compounds were also obtained. The loading plot from the PCA based on UPLC/ESI-Q-TOF/MS data (10,297 variables) is shown in Fig. 8.9. We searched for the presumed molecular formula in the database to confirm possible chemical compositions, and the MS/MS data helped to identify the potential biomarkers. The collision-induced dissociation experiment was implemented to get fragmentation patterns of these potential biomarkers. Furthermore, metabolite identification was conducted with high-resolution MS and MS/MS fragments, as well as database analyses. For example, the postulated elemental compositions for the ions of 1,1′-(1,8-naphthylene)bis(1H-1,2,3-triazole-4,5-dicarboxylic acid) tetra-tert-butyl ester are given in Fig. 8.10. The metabolite, which gave an elemental composition of $C_{34}H_{42}N_6O_8$, and a molecular mass of $[M + H]^+$ 663.3013 were identified. The $[M + H]^+$ 663.3013 demonstrated even number nitrogen, the product spectra such as *m/z* 607, 551, 495, and 439 contributed to $[M + H]^+$ peak continuous loss $-C_4H_8$ when subjecting to MS/MS analysis, and retrieving corresponding literatures. Finally, it was speculated as 1,1′-(1,8-naphthylene)bis(1H-1,2,3-triazole-4,5-dicarboxylic acid) tetra-tert-butyl ester after searching in the database. Accordingly, the protocol described previously, 2-pyrocatechuic acid, 3-methoxy-4-hydroxyphenylglycol sulfate, glucosylceramide (d18:1/18:0) were identified and summarized, see Table 8.2.

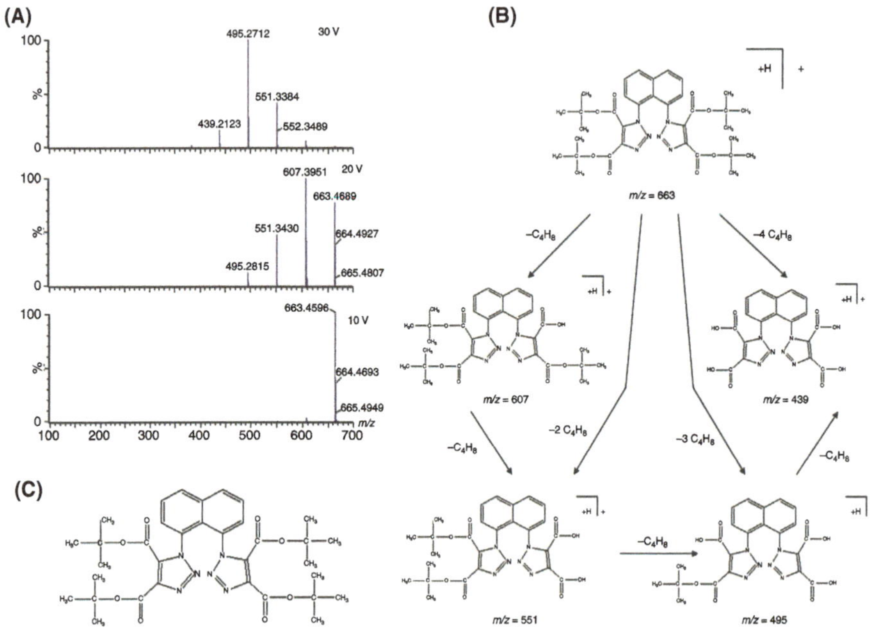

FIGURE 8.10 **Typical identification of potential biomarker 1,1′-(1,8-naphthylene)bis(1H-1,2,3-triazole-4,5-dicarboxylic acid) tetra-tert-butyl ester in urine using UPLC-ESI-QTOFMS-based metabolomics.**

(A) Mass spectrum of full scan and product ion scan of determined biomarkers; (B) possible fragmentation pathway; (C) the chemical structure.

Table 8.2 A List of Potential Urinary Biomarkers of Urine Samples from Rats Treated With Alcohol in Positive Ion Mode

No.	Retention Time	Calculated [M + H]$^+$	Proposed Molecular Formula	Identified Metabolites
1	9.02	663.3013	$C_{34}H_{42}N_6O_8$	1,1′-(1,8-Naphthylene) bis(1H-1,2,3-triazole-4,5-dicarboxylic acid) tetra-tert-butyl ester
2	7.06	256.5827	$C_{12}H_{18}NOS_2$	3-Methoxy-4-Hydroxyphenyl-glycol sulfate
3	10.95	155.3517	$C_7H_6O_4$	2-Pyrocatechuic acid
4	9.01	708.2913	$C_{46}H_{60}O_6$	Glucosylceramide (d18:1/18:0)

8.2.3.3 Metabolic Pathway Analysis

Metabolite profiling focuses on the analysis of a group of metabolites related to a specific metabolic pathway in biological states. To determine whether our observations of changes in the metabolites in the setting of liver injury in fact reflected coordinate changes in defined metabolic pathways, we used MetPA software to identify a network pathway. It revealed that potential biomarkers were identified from the citrate cycle, sphingolipid metabolism, taurine and hypotaurine metabolism that changed specifically in the setting of ALD. Of two distinct metabolites identified from these pathways, many were in various progress stages of liver injury. The detailed construction of the metabolism pathways with a higher score was shown in Fig. 8.11. Results suggested that these target pathways showed the marked perturbations over the time-course of liver injury and could contribute to development of liver injury.

8.2.4 DISCUSSION

Alcohol abuse is one of the main causes of liver disease worldwide and has become a social problem. Due to the increased frequency of drinking, the incidence of alcoholic liver disease has increased in the world, becoming another important risk factor for morbility and mortality in addition to ALD. Metabolomics is a rapidly evolving field that aims to identify and quantify the concentration changes of all the metabolites due to endogenous or exogenous perturbations. PCA revealed robust differences between profiles from the control and the alcohol-treated animals. The major metabolites seen to differ between the control and the alcohol-treated animals were identified using high accuracy MS^n data and verified using external search engines. The main metabolite classes to show major changes in the alcoholic liver-derived samples were 1,1′-(1,8-naphthylene)bis(1H-1,2,3-triazole-4,5-dicarboxylic acid)tetra-tert-butyl ester, 2-pyrocatechuic acid, 3-methoxy-4-hydroxyphenylglycol sulfate, glucosylceramide (d18:1/18:0). In order to more clearly capture and elucidate metabolic changes, network reconstruction has led to the integration of metabolites associated with the caused perturbation pathways including citrate cycle, sphingolipid metabolism, taurine, and hypotaurine metabolism. These metabolites demonstrated that abnormal metabolism occurred in the model animals and metabolic analysis of liver injury was inferred from changes in the intermediates during substance metabolism. Urinary excretion of these metabolites was also shown to have high specificity and sensitivity as markers. Glucosylceramide (d18:1/18:0) is a component of the cell plasma membrane that modulates cell signal transduction events.

A combination of UPLC/MS and chemometrics were used to identify urinary biomarkers associated with ALD. It will increase our understanding of the pathophysiological processes involved and this should help us to identify potential

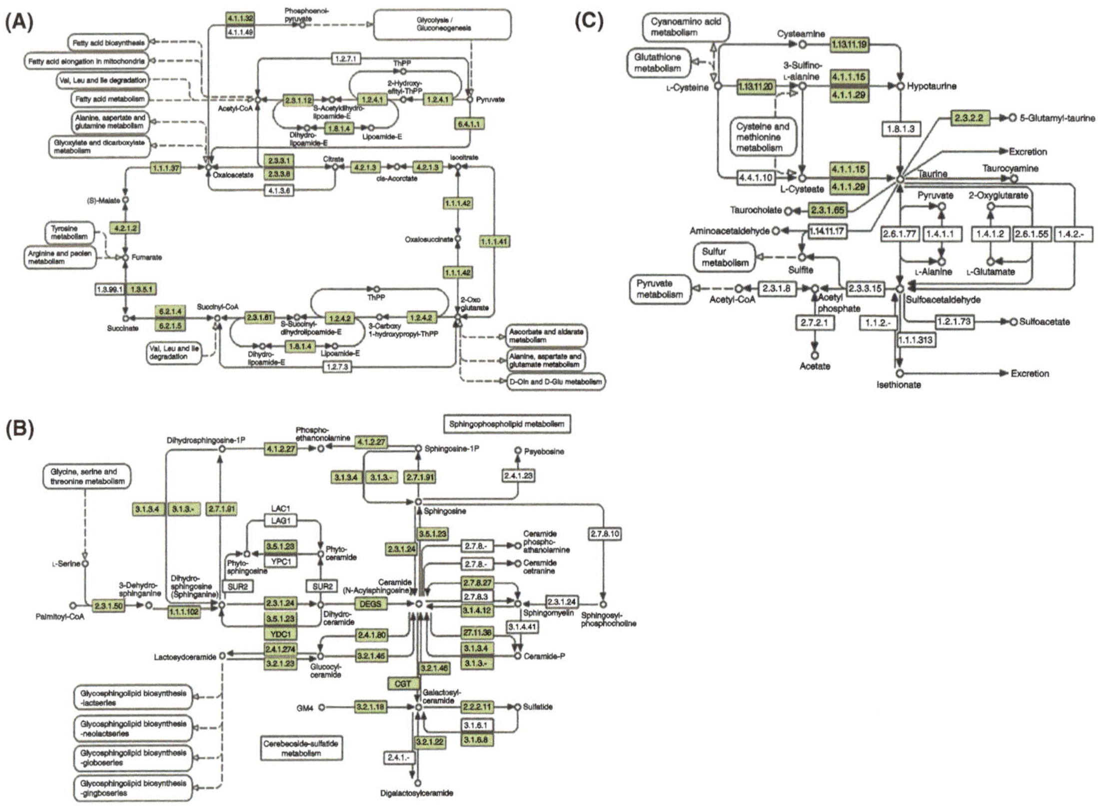

FIGURE 8.11

Related metabolic pathways of citrate cycle: (A) sphingolipid metabolism; (B) taurine and hypotaurine metabolism; (C) in jaundice syndrome rats induced by alcohol. The map was generated using the reference map by KEGG.

biomarkers to develop new therapeutic strategies. Novel metabolites and the metabolite-associated systems provide new insights into the mechanisms underlying pathogenesis. The findings demonstrate that the network-based methods are of importance for elucidating the interrelationships between complex diseases. Generalization of the proposed method for identifying markers will be the focus of future work. System analysis of metabolic networks will help us in generating more indepth understanding of the mechanism of ALD diseases, and thus provide better guidance for drug discovery. A discussion of the newly identified biomarkers, and implicated biochemical pathways is also presented. Future metabolomic studies in human populations with ALD will be needed to validate the biomarkers found in the study.

This study also demonstrated the ability of the metabolomics approach to identify early, noninvasive biomarkers of ALD pathogenesis in the rat model.

Endogenous metabolites were measured and identified using a combination of high accuracy switching MS/MS data, acquired on a UPLC/MS system combined with multivariate statistical analysis, and verified to internal and external databases. The results indicate four ions in the positive mode were characterized as potentially differentiating metabolites. The identified metabolites are mechanistically related to the molecular events associated with development of ALD in the alcohol-treated rats. They were found to encompass a variety of pathways related to the citrate cycle, sphingolipid metabolism, taurine and hypotaurine metabolism. It provides the first metabolite network maps and may offer deeper insights into the potential pathways of ALD. Taken together, LC–MS can clearly enhance the interpretation and enrich the biological discoveries of urinary metabolome data of the molecular mechanisms of disease.

Subchapter 8.3

Jaundice Syndrome Rats Induced by Alpha-Naphthylisothiocyanate

8.3.1 INTRODUCTION

Metabolomics represents an emerging and powerful discipline concerned with the comprehensive analysis of small molecules (<1 kDa) and provides a powerful approach to discovering biomarkers in biological systems. It has been shown to have enormous potential when applied to subjects as diverse as toxicological mechanisms, disease processes, and drug discovery, depending on a series analysis of different sample spectra and combination with chemical pattern recognition methods. Various analytical techniques, such as ultraperformance liquid chromatography/mass spectrometry (UPLC/MS), gas chromatography mass spectrometry (GC-MS) and nuclear magnetic resonance (NMR), and multivariate data analysis, such as independent component analysis (ICA), and partial least squares-discriminant analysis (PLS-DA) have been applied in metabolomic-based drug metabolism studies. Today, UPLC coupled with MS has proven to be an effective combination for metabolites identifications and quantifications due to its excellent resolution and sensitivity. The aim of the current study was to obtain a systematic view of the dissection of the mechanisms of liver injury induced by α-naphthylisothiocyanate (ANIT) that produce acute intrahepatic cholestasis in rodents. The robust UPLC/ESI-Q-TOF/MS system and BEH C_{18} column with a 1.7 μm particle size was utilized and designed to elucidate the disease pathogenesis.

8.3.2 MATERIALS AND METHODS

8.3.2.1 Animal Handling and Sample Preparation

After acclimatization, male Wistar rats (weighing 220 ± 5 g) were randomly divided into two groups with 10 rats in each: the control and the model. The rats in the control group were administrated with physiological saline (0.9%) in the whole procedure for 6 consecutive days and anesthetized by an intraperitoneal injection of 1% pentobarbital sodium (0.15 mL/100 g body weight) on day 6. Rats were orally administrated with 20% ANIT (1 mL/kg body weight) olive oil solution at day 4 (6:00 p.m.) to induce the cholestasis model, and on day 6 the rats were sacrificed, as that in the control group. Urine was collected daily (at 6:00 a.m.) from metabolism cages at ambient temperature throughout the whole procedure and centrifuged at 13,000 rpm at 4°C for 5 min, and the supernatants were stored frozen at −20°C until metabolomic analysis. The serum was used for biochemical indicators detection – according to the instructions of the commercial kits utilized.

8.3.2.2 Biochemical Analysis

Blood samples were collected from the hepatic portal vein, allowed clotting time, and then serum was obtained by centrifugation at 3000 rpm for 15 min at 4°C, flash frozen in liquid nitrogen and stored at −80°C until the liver function tests. The serum ALT, ALT, ALP, γ-GT activity, and T-BIL, D-BIL content was measured according to the instructions of the assay kits. Postblood collection, postcaval vein of liver was ligated using an artery clamp; the hepatic portal vein was needled by perfusing physiological saline (0.9%) for discharging the blood of the liver; simultaneously, the precaval vein of the liver was cut. Liver was picked and obtained to reach an exsanguine condition and rinsed using ice physiological saline. Subsequently, 4.5 mL of physiological saline was added to the second hepatic lobes (0.5 g) and hepatic tissue was misced as well as centrifuged at 3500 rpm for 15 min at 4°C. Finally, 10% of hepatic tissue homogenate was obtained. The MDA and β-G were calculated according to the instructions of the assay kits, respectively.

8.3.2.3 Metabolic Profiling Platform

8.3.2.3.1 Chromatography

Chromatographic analysis was performed in a Waters ACQUITY UHPLC system controlled with Masslynx (V4.1, Waters Corp., Milford, USA). An aliquot of 6 μL of sample solution was injected onto a BEH C_{18} column (50 × 2.1 mm, 1.7 μm, Milford, USA) at 40°C and the flow rate was 0.5 mL/min. The optimal mobile phase consisted of a linear gradient system of (A) 0.1% formic acid in acetonitrile and (B) 0.1% formic acid in water, 0–2 min, 98% A; 2–3 min, 98–80% A; 3–7 min, 80–5% A; 7–9 min, 5% A; 9–11 min, 5–98% A; 11–13 min, 98% A. QC sample was used to optimize the condition of the UPLC-Q-TOF/MS, as it contained the majority of the information from the whole urine samples. In addition, the eluent was transferred to the mass spectrometer directly, that is, without a split.

8.3.2.3.2 Mass Spectrometry

The mass spectrometry was operated by electrospray ionization in the positive ionization mode. The eluent was introduced into the high-definition mass spectrometer (Waters Corp., Milford, USA) analysis, and the optimal conditions of analysis were as follows: the source temperature was set at 110°C, desolvation gas temperature was 300°C, cone gas flow was 100 h, desolvation gas flow was 600 L/h. In positive ion mode, the capillary voltage was 3.2 kV, the sampling cone voltage was 35 V, and extraction cone voltage was 3.0 V. The data acquisition rate was set to 0.14 s/scan, with a 0.1 s interscan delay. Data were collected in centroid mode from 100 Da–1000 Da. For accurate mass acquisition, a lock-mass of leucine enkephalin at a concentration of 0.2 ng/mL was used via a lock spray interface at a flow rate of 100 μL/min monitoring for positive ion mode ($[M + H]^+ = 556.2771$) to ensure accuracy during the MS analysis.

8.3.2.3.3 Multivariate Data Analysis and Data Processing

The MS data were processed using the MarkerLynx Application Manager (Waters Corp.). Raw data were imported into the MassLynx software for peak detection and alignment. The intensity of each ion was normalized with respect to the total ion count to generate a data matrix that consisted of the retention time, *m/z* value, and the normalized peak area. The multivariate data matrix was analyzed by Metalab software. All the variables were mean-centered and paretoscaled prior to ICA and PLS-DA. If a separation between the control and treatment groups was observed in the ICA scores plot, PLS-DA was performed to highlight the differences between the groups. After the analysis of all samples was finished, low-molecular weight metabolites were presented as chromatographic peaks in the base peak intensity (BPI) chromatograms.

8.3.2.3.4 Biomarkers and Metabolic Pathway Identification

MassFragment™ application manager (Waters Corp., Milford, USA) was used to facilitate the MS/MS fragment ion analysis process by way of chemically intelligent peak-matching algorithms. With regard to the identification of biomarkers, the ion spectrum was matched with the structure message of metabolites acquired from available biochemical databases. The MetPA software was based above the database source to identify the metabolic pathways. All statistical analyses were performed using the Student's t-test. Differences with a *p*-value of 0.05 or less were considered significant.

8.3.3 RESULTS

8.3.3.1 Biochemical Analysis

The biochemistry parameters of the control and model groups are presented in the Table 8.3. Some of the clinical parameters in the ANIT treated groups were significantly changed compared with the control group. ANITinduced increases in hepatic ALT, ALT, ALP, and γ-GT activities, T-BIL and D-BIL contents were significant ($p < 0.01$), whilst MDA content in the liver tissue was reduced by ANIT. The model group had higher ALT, ALT, ALP, γ-GT, T-BIL, and D-BIL, β-G values but had lower levels of MDA than the control animals. The biochemical observations were supplemented by a metabolomic examination.

8.3.3.2 LC–MS Analysis of Metabolic Profiling

All the data containing the retention time, peak intensity and exact mass were imported in the Masslynx™ software for multiple statistical analyses. Typically, the trajectory analysis of ICA score plots for the ANIT treatment in positive mode can really reflect the differences between the model and the control groups, and showed metabolic profiles in the control and the model group were separated clearly. Five ions showed significant difference in abundance between the control and treated animals, and contributed to the observed separation, these were selected from the loading plot of PLS-DA (Fig. 8.12).

8.3.3.3 Identification of Metabolite Candidates

We searched for the presumed molecular formula in the database to confirm possible chemical compositions, and the MS/MS data were helped to identify

Table 8.3 Alterations in the Clinical Chemistry Parameters in Model Rats

	ALP (U/L)	ALT (U/L)	AST (U/L)	TBIL (μmol/L)	DBIL (μmol/L)	r-GT (U/L)	TBIL (μmol/L)	β-G (nmol/mL)	MDA (nmol/mL)
Control	324.25 ± 52.63	27.63 ± 4	97.38 ± 17.77	2.18 ± 0.95	1.39 ± 0.5	7.13 ± 2.36	2.18 ± 0.95	117.04 ± 23.48	1.37 ± 0.10
Model	702.13 ± 92.39**	523.5 ± 235.51**	1439 ± 658.77**	26.63 ± 7.28**	22.07 ± 5.13**	45 ± 23.04**	26.63 ± 7.28**	164.27 ± 31.74**	1.20 ± 0.41

*p < 0.05; **p < 0.01 vs. control group; #p < 0.05; ##p < 0.01 vs. model group.

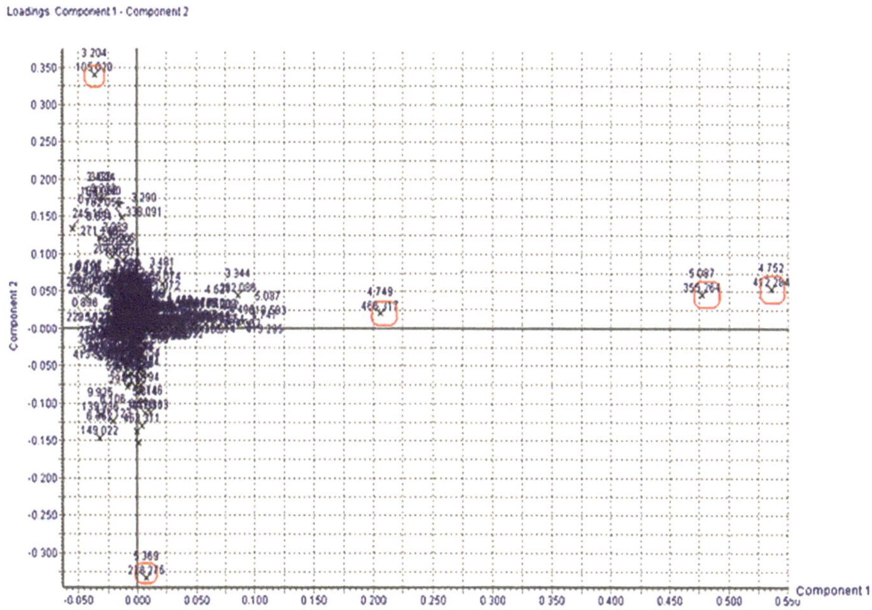

FIGURE 8.12 Loading plot of PLS-DA for the control and model groups.

the potential biomarkers. The collision induced dissociation experiment was implemented to obtain fragmentation patterns of these potential biomarkers. For example, the postulated elemental compositions for the ions of glycocholate are given in Fig. 8.13. The metabolite that gave an elemental composition of $C_{26}H_{43}NO_6$, and molecular mass of $[M + H]^+$ 466.5638 were identified. The $[M + H]^+$ 466.5638 demonstrated odd number nitrogen, the product spectra such as *m/z* 448, 430, 412, and 338 contributed to $[M\text{-}H_2O + H]^+$, $[M\text{-}2H_2O + H]^+$, $[M\text{-}3H_2O + H]^+$, and $[M\text{-}3H_2O\text{-}C_2H_4NO_2 + H]^+$ when subjected to a MS/MS analysis, and retrieved corresponding literatures. Finally after searching in the database, it was speculated as glycocholate. According the protocol described above, glycocholate, 2-hydoxybutanoic acid, deoxycytidie were identified and summarized in Table 8.4.

8.3.3.4 Metabolic Pathway Analysis

Metabolite profiling focuses on the analysis of a group of metabolites related to a specific metabolic pathway in biological states. In this instance it revealed that potential biomarkers were identified from pyrimidine metabolism, primary bile acid biosynthesis, and propanoate metabolism that changed specifically in the setting of liver cholestasis. Of the three distinct metabolites identified from these pathways, many were in various progress stages of liver injury.

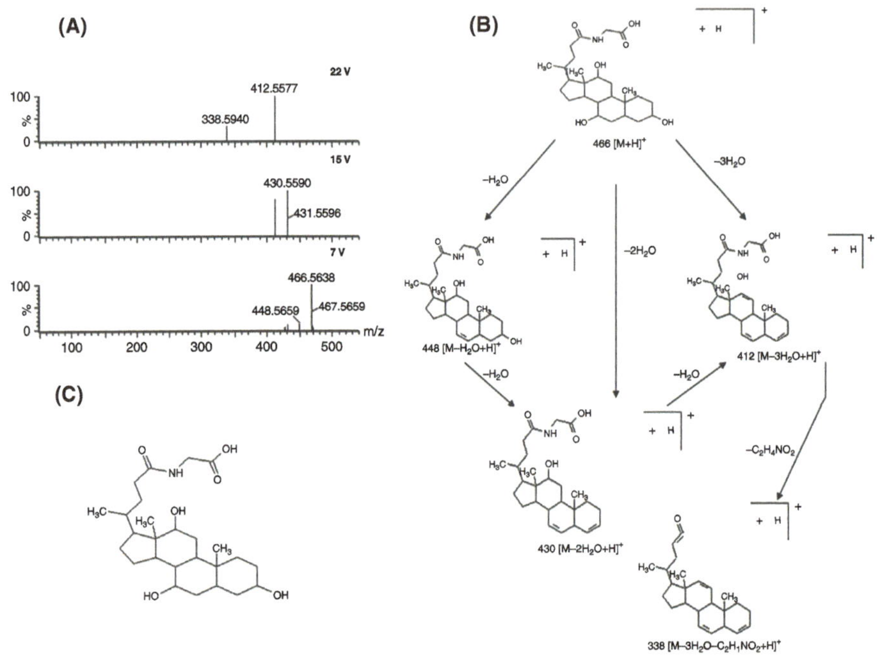

FIGURE 8.13 Typical identification of glycocholate using UPLC/MS.

(A) Mass spectrum of full scan and product ion scan; (B) possible fragmentation pathway; (C) chemical structure of glycocholate.

Table 8.4 A List of Potential Urinary Biomarkers of Urine Samples From Rats Treated With ANIT in Positive Ion Mode

No.	Elemental Composition	T_R (min)	Measured Mass	Calculated $[M + H]^+$/ $[M + Na]^+$	Identified Metabolites	Trend	Related Pathway
1	$C_4H_8O_3$	3.20	105.0203	105.0266	2-Hydoxybutanoic acid	↑	Pyrimidine metabolism
2	$C_{26}H_{43}NO_6$	4.72	466.3173	466.3169	Glycocholate	↑	Primary bile acid biosynthesis
3	$C_{25}H_{37}N_3O_2$	4.75	412.2837	412.2856	$C_{25}H_{37}N_3O_2$	↑	–
4	$C_{18}H_{19}O_6Na$	5.08	355.2644	355.2712	$C1_8H_{19}O_6Na$	↓	–
5	$C_9H_{13}N_3O_4$	5.36	228.2746	228.2982	Deoxycytidie	↑	Propanoate metabolism

Notes: '↑' and '↓' represent compounds that were up- and downregulated in the model group, compared with the control group.

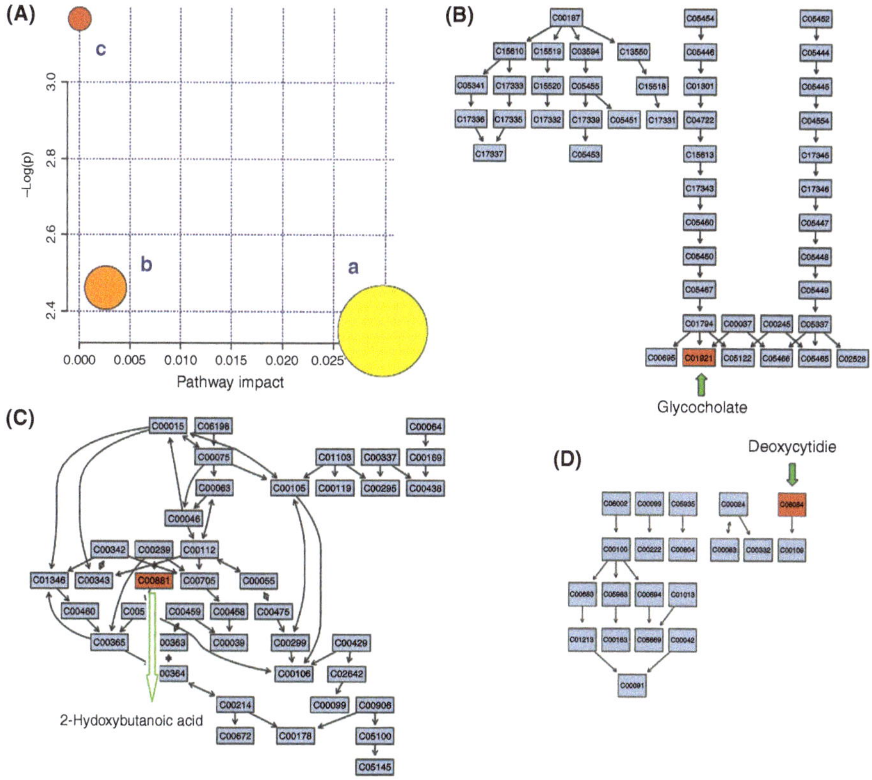

FIGURE 8.14 Biomarker network reconstruction disturbed in response to the ANIT treatment. Identifying network pathway by MetPA software. (A) primary bile acid biosynthesis: (a), pyrimidine metabolism (b) propanoate (c). The map was generated using the reference map by KEGG. Putative metabolic pathways of primary bile acid biosynthesis; (B) pyrimidine metabolism; (C) propanoate; (D) were inferred in the ANIT-dosed rats from changes in the urine levels of intermediates during substance metabolism. Red denotes the affected metabolites related to the pathway.

The detailed construction of metabolism pathways (with higher scores) are shown in Fig. 8.14. The results suggested that these target pathways showed the marked perturbations over the time-course of liver injury and could contribute to the development of liver injury.

8.3.4 DISCUSSION

Metabolomics adopts a "top-down" strategy to reflect the terminal symptoms of the metabolic network of biological systems in a holistic context. Here, we applied the metabolomics approach based on the UPLC/Q-TOF/MS system to study the mechanism of liver injury in rats induced by ANIT. Meanwhile, the combination of UPLC/Q-TOF/MS and multivariate data analysis, such as ICA and PLS-DA, was also predicted via ingenuity pathway methods, and was utilized to identify related potential biomarkers in the presence of urine samples from rats treated with ANIT. The biochemical observations were supplemented by metabolomic examination. We utilized a metabolomic approach further to delineate the metabolic changes of acute liver injury. Network reconstruction has led to the integration of metabolites associated with the causes of perturbation of multiple pathways including pyrimidine metabolism, primary bile acid biosynthesis, and propanoate metabolic pathways. These metabolites demonstrated that abnormal metabolism occurred in the model animals and the metabolic analysis of liver injury was inferred from changes in the intermediates during substance metabolism. The levels of the five metabolites exhibited the marked changes compared with that of the control rats. Urinary excretion of these metabolites was also shown to have a high specificity and sensitivity as markers. These markers and pathway components may prove useful as a non-invasive means of diagnosing liver diseases and to suggest novel therapeutic approaches.

The change in the metabolic profiles of the current study is consistent with the changes in the biochemical parameters. However, the changes in the metabolic profiles occurred earlier than those of the biochemical parameters. Compared with conventional clinical biochemical tests, metabolomics has unique superiority. In our study, the continuous, dynamic, and noninvasive detection of metabolites in the urine of rats treated with ANIT was successfully demonstrated in real time (Sun et al., 2014). Application of metabolomic technologies for the study of JS induced by ANIT will increase our understanding of the pathophysiological processes. System analysis of metabolic networks that are a central paradigm in biology will help us in identifying new biomarkers which will generate a more indepth understanding of the liver injury mechanism and thus provide better guidance for drug efficacy. Additionally, we suggest that the application of metabolite profiling tools can provide an efficient means for early diagnosis of liver injury and provide a practical method for monitoring therapeutic interventions.

This work was designed to study the metabonomic characters of the liver injury induced by ANIT, a classic jaundice and cholestasis. In this study, a model of liver injury rat was constructed, and dynamic metabolic profiles were investigated with UPLC-MS combined with multivariate statistical analysis, five potential biomarkers of significant contribution were characterizated in positive mode. By using a metabolomic approach, this study also exemplifies that metabolomics could provide a very promising way to elucidate the therapeutic mechanisms of TCM.

Subchapter 8.4

Jaundice Syndrome in Rats Induced by Carbon Tetrachloride

8.4.1 INTRODUCTION

Metabolomics is an emerging and powerful discipline concerned with the comprehensive analysis of low-molecular-weight metabolites and provides a powerful approach to discover biomarkers in biological systems, and attempts to elucidate sites of perturbations. Various analytical techniques, with multivariate data analysis have been applied in metabolomic-based disease metabolism studies. Now, UPLC/ESI-Q-TOF/MS has become one of the widely applied techniques in metabolomics studies owing to its high sensitivity and reproducibility (Wang et al., 2012). Carbon tetrachloride (CCl_4) is a well-known compound for producing chemical hepatic injury (Zhang et al., 2013a,b). To investigate the effect of the mechanism of acute liver injury on rats induced by CCl_4, the aim of the current study was to obtain a systematic view of Jaundice syndrome (JS) rats induced by CCl_4. Thus, a robust Waters UPLC/ESI-Q-TOF/MS system and Waters UPLC BEH C_{18} column with a 1.7 μm particle size was utilized and designed to elucidate CCl_4 induced liver injury in rats.

8.4.2 MATERIALS AND METHODS

8.4.2.1 Animal Handling and Sample Preparation

Male Wistar rats (weighing 220–260 g) were allowed to acclimatize for 7 days prior to dosing and put in the metabolism cages during the urine collection periods. After acclimatization, animals were randomly divided into 2 groups with 10 rats in each: the control and model groups. The rats in the control group were administrated with 0.9% saline in the whole procedure for 5 consecutive days. Rats were orally administrated with 20% CCl_4 (1 mL/kg body weight) olive oil solution at day 4 to induce the liver injury model for 5 consecutive days, and until day 8. The rats were administrated an oral accurate volume of 1 mL/100 g of each animal. Urine was collected daily from metabolism cages at ambient temperature throughout the whole procedure and centrifuged at 13,000 rpm at 4°C for 5 min, and the supernatants were stored frozen at −20°C until metabolomic analysis.

8.4.2.2 Metabolic Profiling Platform

8.4.2.2.1 Chromatography

UPLC/ESI-Q-TOF/MS was used for the global analysis of urine samples. Chromatographic analysis was performed in a Waters ACQUITY UHPLC system controlled with Masslynx (V4.1, Waters Corp., Milford, USA). An aliquot of 6 μL of sample solution was injected onto an ACQUITY UPLC C_{18} column (100 × 2.1 mm, 1.7 μm, Waters Corp., Milford, USA) at 40°C and the flow rate was 0.4 mL/min. The optimal mobile phase consisted of a linear gradient system of: (A) 0.1% formic acid in acetonitrile; and (B) 0.1% formic acid in water, 0–2 min, 98% A; 2–3 min, 98–80% A; 3–7 min, 80–5% A; 7–9 min, 5% A; 9–11 min, 5–98% A; 11–13 min, 98% A.

8.4.2.2.2 Mass Spectrometry

The optimal analysis conditions of mass spectrometry were as follows: the source temperature was set at 110°C, desolvation gas temperature was 300°C, cone gas flow was 100 h, desolvation gas flow was 600 L/h; the capillary voltage was 3.2 kV, the sampling cone voltage was 35 V, the microchannel plate voltage was 2300 V and extraction cone voltage was 3.0 V. The data acquisition rate was set to 0.14 s/scan, with a 0.1 s interscan delay. Data were collected in centroid mode from 100 Da–900 Da. For accurate mass acquisition, a lock-mass of leucine enkephalin at a concentration of 0.2 ng/mL was used via a lock spray interface at a flowrate of 100 μL/min monitoring for positive ion mode ($[M + H]^+ = 556.2771$) to ensure accuracy during the MS analysis.

8.4.2.2.3 Multivariate Data Analysis and Data Processing

The UPLC-MS data were processed using the MarkerLynx Application Manager (Waters Corp.). The intensity of each ion was normalized with respect to the total ion count to generate a data matrix that consisted of the retention time, the *m/z* value, and the normalized peak area. The multivariate data matrix was analyzed by EZinfo software 2.0 (Waters Corp., Milford, USA). Here, "Unsupervised" PCA and "Supervised" PLS-DA were used to process the acquired UPLC-MS data. In the PCA scores, each point represents an individual sample. The PCA results are displayed as score plots indicating the scatter of the samples that indicate similar metabolomic compositions when clustered together, and compositionally different metabolomes when dispersed. Further metabolite differences are indicated by the PCA loading plots. Each score plot has a loading plot associated with it that makes it possible to identify the spectral regions (metabolites) that are responsible for the observed sample clustering.

8.4.2.2.4 Biomarkers Identification and the Reconstruction of the Metabolic Pathway

Potential markers of interest were extracted from loading plots constructed following analysis with PLS-DA, and markers were chosen based on their contribution to the variation and correlation within the data set. A full spectral library, containing MS/MS data obtained in the positive ion mode, for all metabolites reported in this work are available on request from the authors. With regard to the identification of biomarkers, the ion spectrum of potential biomarkers was matched with the structure message of metabolites acquired from available biochemical databases. All statistical analyses were performed using the Student's t-test. Differences with a p-value of 0.05 or less were considered significant.

8.4.3 RESULTS AND DISCUSSION

8.4.3.1 Analysis of Metabolic Profiling

Using the optimal reversed-phase UPLC-MS conditions, as described previously, the representative BPI chromatograms of urine obtained from the UPLC/ESI-Q-TOF/MS analysis for a continuous 8 days are shown in Fig. 8.15. All the data containing the retention time, peak intensity and exact mass were imported in the Masslynx™ software for multiple statistical analyses. Both PCA and PLS-DA were taken because of their ability to cope with highly multivariate, noisy, collinear and possibly incomplete data. Typically, the trajectory analysis of the PCA score plots for the CCl_4 treatment in positive mode can reflect the differences between the model and the control groups, and illustrated that the metabolic profiles in the control and the model groups were separated clearly (Fig. 8.16). Therefore, PCA can effectively demonstrate the differences between the treated and the control groups. The PCA plot of the model group showed similar behavior during the early stage of the experiment, but then gradually deviated from one another, on day 6 reaching the maximum trend, and on day 7, and day 8 gradually going back to the normal group, coinciding with the classic CCl_4 induced liver injury. The tracks of the metabolic profiles at different time points, also clearly demonstrates the time-dependent changes in the urine metabolites. The corresponding PLS-DA loadings plot indicated that the differentiating metabolites were attributable to the clustering observed in the scores plot. The farther away from the origin, the higher the VIP value of the ions was. Five ions showed significant difference in abundance between the control and treated animals, and contributed to the observed separation, these were selected from the loading plot of PLS-DA (Fig. 8.17).

8.4.3.2 Identification of Metabolite Candidates

The loading plot from the PCA based on the UPLC/MS data (11,846 variables) was shown in Fig. 8.17. The distance of an ion from the origin represents the contribution to the clustering of different groups on the PCA. The collision-induced dissociation experiment was implemented to obtain the fragmentation patterns of these potential biomarkers. Furthermore, metabolite identification

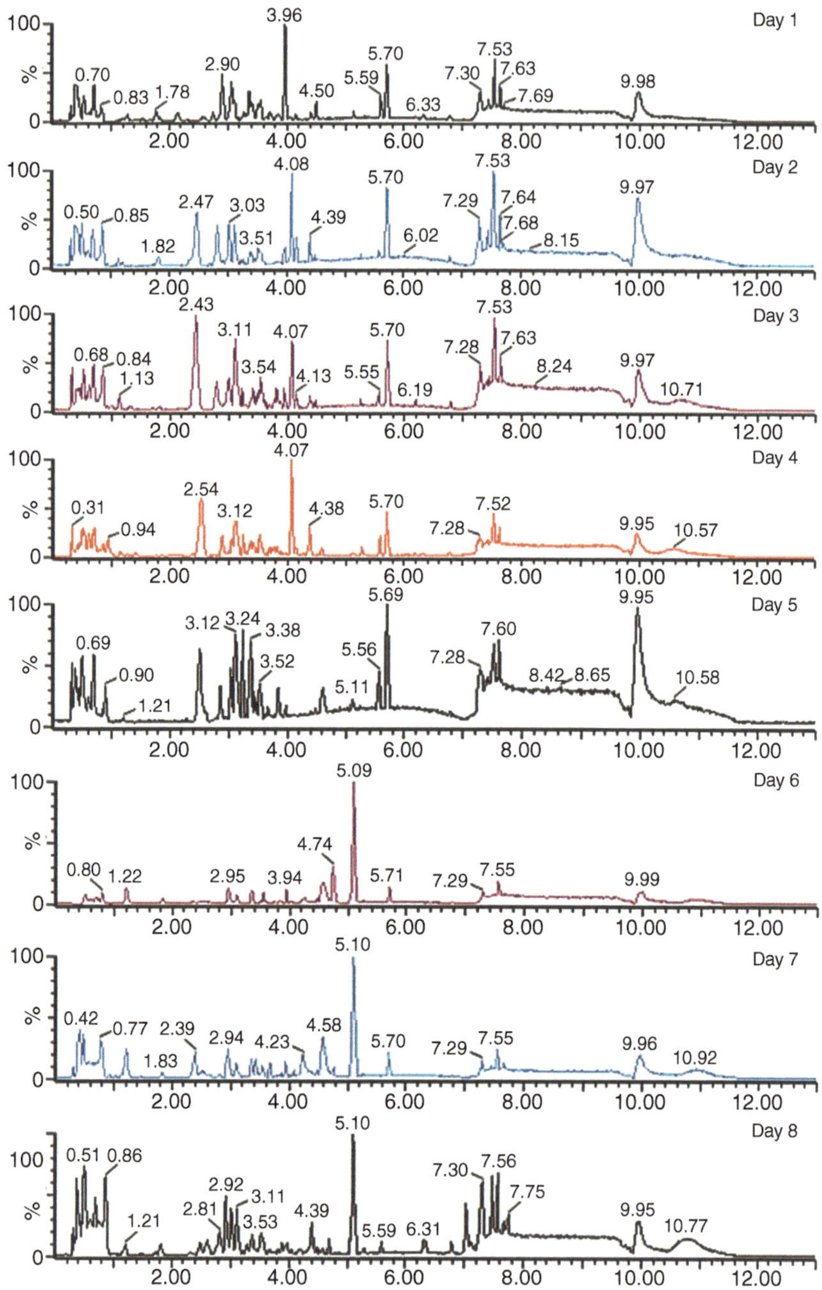

FIGURE 8.15 BPI chromatograms of urine obtained from UPLC/ESI-Q-TOF/MS analysis for a continuous 8 days.

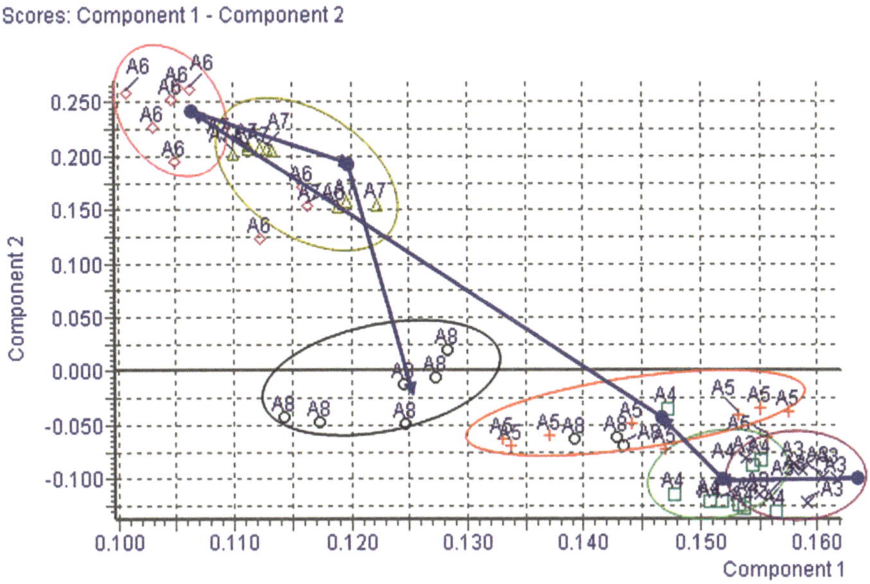

FIGURE 8.16 **Score plots of trajectory analysis for the CCl_4 treatment in positive mode.**
(×: the 1-3rd day; □: the 4th day; +: the 5th day; ◇: the 6th day; Δ: the 7th day; ○: the 8th day).

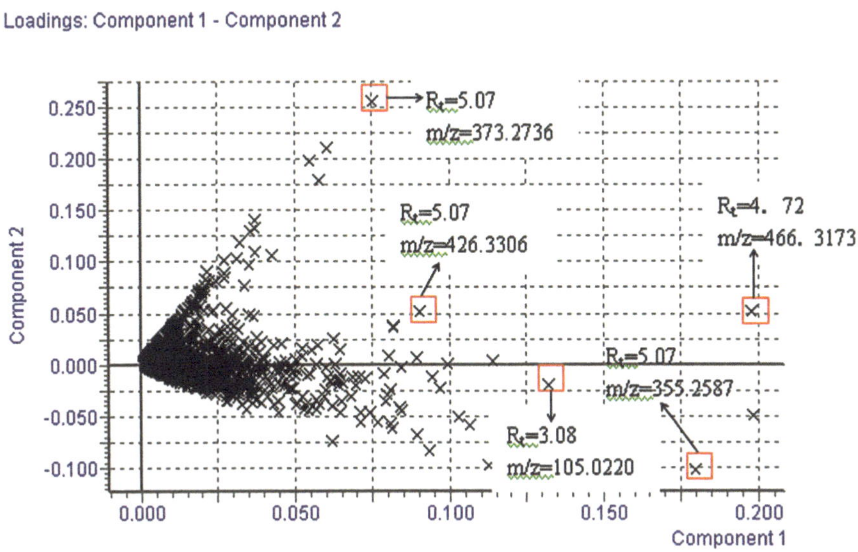

FIGURE 8.17 **Loading plot of urine metabolome from model group.**
Red data points indicate that ions most responsible for the variance in the score plot.

Table 8.5 A List of Potential Urinary Biomarkers of Urine Samples From Rats Treated With CCl_4 in Positive Ion Mode

No.	R_t	Measured Mass	Calculated $[M + H]^+$	Formula	Metabolites	Trend	Related Pathway
1	4.72	466.3173	466.3169	$C_{26}H_{43}NO_6$	Glycocholate	↑	Primary bile acid biosynthesis
2	5.07	354.240631	355.2597	$C_{19}H_{34}N_2O_4$	–	↑	–
3	3.08	105.0220	105.0188	$C_3H_4O_4$	2-Hydoxybutanoic acid	↓	Pyrimidine metabolism
4	5.07	425.3505	426.3372	$C_{28}H_{43}NO_2$	–	↑	–
5	5.07	372.150757	373.2743	$C_{24}H_{36}O_3$	–	↑	–

was conducted with high-resolution MS and MS/MS fragments, as well as database analyses. According the protocol described above, glycocholate and 2-hydoxybutanoic acid were identified and summarized in Table 8.5. Glycocholate is an acyl glycine and a bile acid–glycine conjugate. It is a secondary bile acid produced by the action of enzymes existing in the microbial flora of the colonic environment. In hepatocytes, both primary and secondary bile acids undergo amino acid conjugation at the C–24 carboxylic acid on the side chain, and almost all bile acids in the bile duct, therefore, exist in a glycine-conjugated form. More specifically, glycocholate or cholylglycine, is a crystalline bile acid involved in the emulsification of fats. It occurs as a sodium salt in the bile of mammals. As the glycine conjugate of cholic acid, this compound acts as a detergent to solubilize fats for absorption and is itself absorbed. Urinary excretion of these metabolites was also shown to have high specificity and sensitivity as markers.

8.4.3.3 Metabolic Pathway Reconstruction

Metabolic pathway analysis revealed that potential biomarkers were identified from glycolysis or gluconeogenesis, biosynthesis of unsaturated fatty acids, fatty acid biosynthesis and purine metabolism that changed specifically in the setting of liver injury. Of the two distinct metabolites identified from these pathways, many were in varied progress stages of liver injury. The detailed construction of the metabolism pathways with higher scores was not shown. Results suggested that these target pathways showed the marked perturbations over the time-course of liver injury and could contribute to the development of liver injury.

The liver has the important function of metabolizing and eliminating biological and xenobiotic compounds. Liver injury is considered to be a serious health problem leading to acute hepatitis, cholestasis, and cirrhosis. CCl_4 has been used extensively to result in liver injury in various animal models. In the present study, it was designed to study the metabolomic characters of JS induced by CCl_4. In this study, a model of liver injury rat was constructed, and the dynamic metabolic profiles were investigated with UPLC-MS combined with multivariate statistical analysis, five potential biomarkers of significant contribution was characterized in positive mode.

References

Charbonneau, M., Tuchweber, B., Plaa, G.L., 1986. Acetone potentiation of chronic liver injury induced by repetitive administration of carbon tetrachloride. Hepatology 6 (4), 694–700.

Li, Y., Cao, G., Zheng, X., Wang, J., Wei, H., Tian, Z., Sun, R., 2013. CRACC-CRACC interaction between Kupffer and NK cells contributes to poly I:C/D-GalN induced hepatitis. PLoS ONE 8 (9), e76681.

Price, V.F., Jollow, D.J., 1983. Mechanism of ketone-induced protection from acetaminophen hepatotoxicity in the rat. Drug Metab. Dispos. 11 (5), 451–457.

Rost, D., Herrmann, T., Sauer, P., Schmidts, H.L., Stieger, B., Meier, P.J., Stremmel, W., Stiehl, A., 2003. Regulation of rat organic anion transporters in bile salt-induced cholestatic hepatitis: effect of ursodeoxycholate. Hepatology 38 (1), 187–195.

Sato, K., Egashira, Y., Ono, S., Mochizuki, S., Shimmura, Y., Suzuki, Y., Nagata, M., Hashimoto, K., Kiyono, T., Park, E.Y., Nakamura, Y., Itabashi, M., Sakata, Y., Furuta, S., Sanada, H., 2013. Identification of a hepatoprotective peptide in wheat gluten hydrolysate against D-galactosamine-induced acute hepatitis in rats. J. Agric. Food Chem. 61 (26), 6304–6310.

Schlattner, U., Tokarska-Schlattner, M., Wallimann, T., 2006. Mitochondrial creatine kinase in human health and disease. Biochim. Biophys. Acta 1762 (2), 164–180.

Sun, H., Zhang, A.H., Zou, D.X., Sun, W.J., Wu, X.H., Wang, X.J., 2014. Metabolomics coupled with pattern recognition and pathway analysis on potential biomarkers in liver injury and hepatoprotective effects of yinchenhao. Appl. Biochem. Biotechnol. 173 (4), 857–869.

Wang, Y., Tang, H., Nicholson, J.K., Hylands, P.J., Sampson, J., Holmes, E., 2005. A metabonomic strategy for the detection of the metabolic effects of chamomile (Matricaria recutita L.) ingestion. J. Agric. Food Chem. 53 (2), 191–196.

Wang, S.B., Wang, J.H., Zheng, R.H., Zhong, B.H., Chen, M.H., 2010. Deep cholestatic jaundice as the predominant manifestation in autoimmune hepatitis. Hepatogastroenterology 57 (98), 326–329.

Wang, X., Zhang, A., Sun, H., Wu, G., Sun, W., Yan, G., 2012. Network generation enhances interpretation of proteomics data sets by a combination of two-dimensional polyacrylamide gel electrophoresis and matrix-assisted laser desorption/ionization-time of flight mass spectrometry. Analyst 137 (20), 4703–4711.

Wang, X., Lv, H., Zhang, A., Sun, W., Liu, L., Wang, P., Wu, Z., Zou, D., Sun, H., 2014. Metabolite profiling and pathway analysis of acute hepatitis rats by UPLC-ESI MS combined with pattern recognition methods. Liver Int. 34 (5), 759–770.

Wyss, M., Kaddurah-Daouk, R., 2000. Creatine and creatinine metabolism. Physiol. Rev. 80 (3), 1107–1213.

Zhang, A., Sun, H., Wu, G., Sun, W., Yuan, Y., Wang, X., 2013a. Proteomics analysis of hepatoprotective effects for scoparone using MALDI-TOF/TOF mass spectrometry with bioinformatics. OMICS 17 (4), 224–229.

Zhang, A., Sun, H., Sun, W., Ye, Y., Wang, X., 2013b. Proteomic identification network analysis of haptoglobin as a key regulator associated with liver fibrosis. Appl. Biochem. Biotechnol. 169 (3), 832–846.

CHAPTER 9

Metabolomic Evaluation of Hepatoprotective Effect of Yinchenhao Tang and its Major Bioactive Constituents

Xijun Wang, Aihua Zhang, Hui Sun

CONTENTS

9.1 INTRODUCTION

Yinchenhao Tang (YCHT), which was recorded in "Shanghanlun", and is one of the most famous Chinese herbal formulae. YCHT consists of *Artemisia annua* L., *Gardenia jasminoides* Ellis, and *Rheum Palmatum* L., and has been used for more than a thousand years to treat jaundice syndrome and liver disorders. Pharmacological studies and clinical practices have shown that it can be used to treat cholestasis, hepatitis C, primary biliary cirrhosis, liver fibrosis, and cholestatic liver diseases. Interestingly, 6,7-dimethylesculetin (6,7-DME), a major component of *Artemisia annua* L., has been shown to be effective in treating liver injury by exerting a hepatoprotective action, and contributes directly to the therapeutic effect of YCHT. The purpose of this present study is to develop a UPLC/MS-based method for investigating the urinary metabolome of liver injury rats induced with CCl_4. Furthermore, this study aims to confirm the biomarkers by demonstrating the effects of hepatoprotective drugs, and simultaneously to comprehensively understand the scientific connotation of the hepatoprotective effects of YCHT and 6,7-DME.

9.2 EXPERIMENTAL

9.2.1 Chemicals and Reagents

Acetonitrile, HPLC grade, was purchased from Sigma–Aldrich (MO, UK). Formic acid, analytical grade, was purchased from Beijng Reagent Company (Beijing, China). The deionized water, HPLC grade, was purified by Milli-Q system (Millipore, Bedford, MA, USA), Leucine enkephalin was purchased from Sigma–Aldrich (MO, UK). Carbon tetrachloride (CCl_4) was purchased from Sigma–Aldrich (MO, UK). The Assay Kits of plasma enzymes of aspartate aminotransferase (AST), alanine aminotransferase (ALT), alkaline phosphatase (ALP), and total bilirubin (T-BIL) were purchased from Zhongsheng BeiKong Biological Technology Limited Corp. (Beijing, China). The Assay Kits of malondialdehyde (MDA), superoxide dismutase (SOD), glutathione

Chinmedomics. http://dx.doi.org/10.1016/B978-0-12-803117-9.00009-3

peroxidase (GSH-PX) were purchased from Nanjing Jiancheng Bioengineering Institute (Nanjing, China).

The flower of *Artemisia capillaries* Thunb (Compositae) (YCH) was purchased from the China Branch Office of the Xinneitian Pharmaceutical Company of Japan (Tianjin, China); the fruit of *Gardenia jasminoides* Ellis (Rubiaceae)(ZZ) and the root of *Rheum officinale* Baill (Polygonaceae) (DH) were purchased from Harbin Tongrentang Drug Store (Harbin, China). All the crude drugs were authenticated by Prof. Xijun Wang of the Pharmacognosy Department at the Heilongjiang University of Chinese Medicine. Yinchenhao Tang (YCHT) was prepared according to the method recorded in "Shanghanlun": first, 18 g of YCH were boiled in 1400 mL of H_2O until the solution volume reduces to 700 mL, then adding 9.0 g of ZZ and 6.0 g of DH, and keep boiling for 10 min. The mixture was filtered through a five-layers bandage. Finally, the filtrate was prepared as freeze-drying powder. The dosing sample of Yin Chen Hao Tang was prepared with an accurate weight of 0.4 g of freeze-drying powder and was dissolved in 1 mL deionized water. The 6,7-dimethyesculetin (6,7-DME) was isolated from the flower of *Artemisia capillaries* Thunb (Compositae) in our laboratory, and the purity is > 99%. The dosing sample of 6,7-DME was prepared with an accurate weight of 1.0 mg of scoparone and was suspended and dissolved in 1 mL of 10% tweenum 80 deionized water.

9.2.2 Animals and Handling

All rats were randomly divided into four groups including a control group (orally administrating an equal volume of water to rats); a model group (orally administrating 20% CCl_4 in oil to rats at the dose of 1 mL/kg); a YCHT group (orally administrating freeze dried powders of YCHT to rats at the dose of 8 g/kg); and the 6,7-DME group (orally administrating a pure compound of 6,7-DME to rats at the dose of 10 mg/kg). The details are shown in Table 9.1. For each group, each male rat was housed in a metabolism cage and urine samples were collected from each animal over the course of 8 days. A standard diet was given to all animals and free access to water throughout the studies. Urine samples were collected continuously into tubes in ice every 24 h. The collected urine samples were centrifuged at 13,000 rpm for 5 min, to remove

Table 9.1 Treated Method of Different Rat Groups

Group	Treated Method							
	Day 1	Day 2	Day 3	Day 4	Day 5	Day 6	Day 7	Day 8
Control	H_2O	H_2O	H_2O	H_2O	H_2O	H_2O	H_2O	H_2O
Model	H_2O	H_2O	H_2O	H_2O	CCl_4	H_2O	H_2O	H_2O
YCHT	H_2O	YCHT	YCHT	YCHT	CCl_4 + YCHT	H_2O	H_2O	H_2O
6,7-DME	H_2O	6,7-DME	6,7-DME	6,7-DME	CCl_4 + 6,7−DME	H_2O	H_2O	H_2O

solid debris, and the supernatant was stored at −20°C, thawing and filtered through a 0.22 μm filter membrane before the UPLC analysis.

9.2.3 UPLC-MS Analysis of Urine Samples

Chromatography was performed on a Waters Acquity™ UPLC system (Waters Corp., Milford, USA) equipped with a column oven. The UPLC system was coupled with a Waters Micromass Q-TOF micro™ (Manchester, UK) equipped with an electrospray source operating in either positive ion or negative ion mode, and utilizing Makerlynx software. The source temperature was set at 110°C with a cone gas flow of 100 L/h, a desolvation gas temperature of 300°C and a desolvation gas flow of 600 L/h were employed. The capillary voltage was set at 3.2 kV in positive ion mode, and 2.6 kV in negative ion mode, and the cone voltage was set at 35 V. A scan time of 0.4 s with an interscan delay of 0.1 s was used throughout, with a collision energy of 4 eV. A lock-mass of leucine enkephalin at a concentration of 0.2 ng/μL, in acetonitrile (1% formic acid): water (0.1% formic acid) for positive ion mode ($[M + H]^+ = 556.2771$) and 1.0 ng/mL in acetonitrile (1% formic acid): water (0.1% formic acid) for positive ion mode ($[M + H]^+ = 554.2615$), was employed via a lock spray interface. An aliquot of urine (5 μL for negative or positive ion modes, respectively) was injected into an UPLC™ BEH C_{18} column (50 × 2.1 mm i.d., 1.7 μm, Waters Corp., Milford, USA) held at 45°C. The column was eluted with a linear gradient. A "purge–wash–purge" cycle was employed on the autosampler, with 90% aqueous methanol used for the wash solvent, and 0.1% aqueous formic acid used as the purge solvent, this ensured that the carry-over between injections was minimized. The mass spectrometric data was collected in full scan mode from m/z 100−900 from 0−10 min in positive ion mode with an accuracy of 5 ppm. The column elute was not split as a consequence approximately 400 μL/min were directed to the mass spectrometer.

9.2.4 Clinical Chemistry and Histopathology

Serum samples and liver tissue samples were characterized on an Automated Synchron LX20 Beckman–Coulter analyzer by following clinical chemical parameters: plasma enzymes of aspartate aminotransferase (AST), alanine aminotransferase (ALT), alkaline phosphatase (ALP), hepatocyte enzymes of superoxide dismutase (SOD) and glutathione peroxidase (GSH-PX), and content of malondialdehyde (MDA) in hepatocyte. Histological sections of control, CCl_4 treated, YCHT treated, and 6,7-DME treated rats were prepared and stained with hematoxylin and eosin (H and E) and examined by light microscopy.

9.2.5 Data Analysis for UPLC–MS

The UPLC/MS data were analyzed using the Micro-mass MarkerLynx Applications Manager (version 4.0). MarkerLynx incorporates a peak deconvolution package that allows detection and retention time alignment of the peaks eluting in each chromatogram. The data was combined into a single matrix by aligning peaks with the same mass/retention time pair together from each data file in the dataset, along with their associated normalized intensities.

9.3 RESULTS

9.3.1 Clinical Chemistry and Histopathology

The activity of the plasma enzymes: ALT, AST, and ALP all significantly increased after exposure to CCl_4 for 24 h compared to the control group, when YCHT or 6,7-DME treated, these increases were obviously elevated, compared to the CCl_4 treated group (Fig. 9.1A). The content of MDA significantly increased, the enzymes of SOD and GSH-PX in hepatocyte obviously reduced after exposure to CCl_4 for 24 h, compared to the control group. In the YCHT or 6,7-DME treated groups, those significant changes all were obviously recovered (Fig. 9.1B). Histopathological changes in the liver are shown in Fig. 9.1C. CCl_4 treated rats obviously showed feathery degeneration, periportal fibrosis, and vascular congestion; when compared with the control group rats. When YCHT or 6,7-DME treated, the pathological damages were significantly improved;

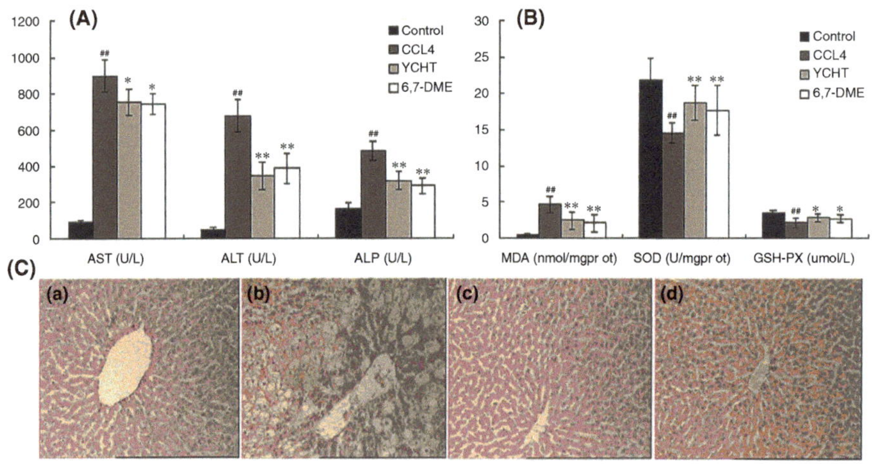

FIGURE 9.1 Effect of CCl_4 induced liver injury on the activities of AST, ALT, and ALP in rat serum. (A) The data were expressed as Means ± SD. # # Significant difference compared with control groups ($p < 0.01$); ** Significant difference compared with CCl_4 groups ($p < 0.01$). (B) Effect of CCl_4 induced liver injury on the content of MDA, the activities of SOD and GSH-PX in rat hepatocyte. The data were expressed as Means ± SD. Significant difference compared with control groups (# # $p < 0.01$; # $p < 0.05$); significant difference compared with CCl_4 groups (** $p < 0.01$; * $p < 0.05$). (C) Hepatohistological observation of different groups rat on day 8. (a) Control rat liver; (b) CCl_4 treated rat liver; (c) CCl_4 + YCHT treated rat liver; (d) CCl_4 + 6,7-DME treated rat liver.

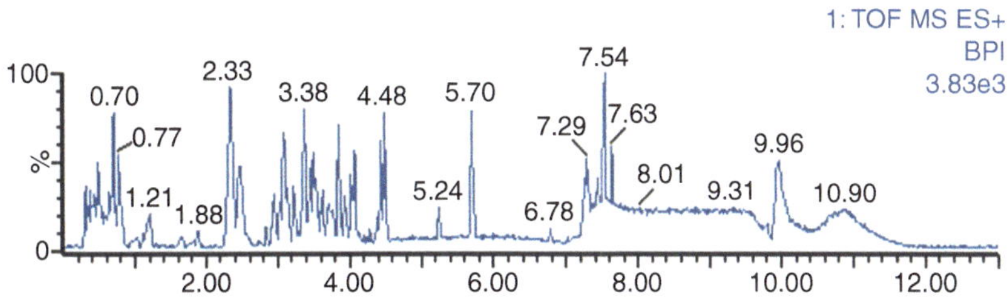

FIGURE 9.2 Typical BPI profile of metabolites in rat urine.

the liver demonstrated similar normal histology with mild congestion of the central vein. In conclusion, the above results revealed that the animal model of CCl_4 induced liver injury was successfully established and both YCHT and 6,7-DME have proven hepatoprotective effects.

9.3.2 Carbon Tetrachloride Treated Animals

The samples from both control and dosed animals were analyzed by reversed-phase gradient UPLC–MS in positive ion mode (as detailed in the experimental section). The typical MS profile of a urine sample was shown in Fig. 9.2. All of the control animals presented normal UPLC/MS spectra with only minor fluctuations in the concentrations of the endogenous metabolites over the time-course of the study. All of the CCl_4 treated animals also presented normal UPLC/MS spectra with only minor fluctuations in the concentrations of the endogenous metabolites in the initial 4 days before dosing. It was not until day 5 that the visual appearance of the urinary UPLC/MS profiles were seen to diverge from those of animals in the control animals group. The most marked changes were observed on day 6. From the day 7, the changes in the profiles decreased gradually, and the tendency of returning to those of animals in the initial 4 days was observed. Such tendency is fitted for the pathological process of CCl_4 inducing liver injury.

The score plot of PC1 vs. PC2 for the positive ion data after exclusion of the signals due to CCl_4 is shown in Fig. 9.3A. The most significant changes observed from day 6 or day 7 after dosing were a reduction in the intensity of the following ions in Fig. 9.3B of the loading plot. The principal ions responsible for the separation of the two groups in positive ion are *m/z* 466, 355, 180, 105, 426, 373, and 215 postdosing. The postulated elemental compositions for these ions are given in Table 9.2. Three of those metabolites, which gave an elemental composition of $C_{26}H_{43}NO_6$, $C_9H_9NO_3$, and $C_{10}H_{14}O_5$, and molecular mass of $[M + H]^+$ 466.3173, $[M + H]^+$ 180.0670, and $[M + H]^+$ 215.0929, were identified. The $[M + H]^+$ 466.3173 demonstrated odd number nitrogen, the product spectra such as *m/z* 448, 430, 412, and 338 contributed to $[M\text{-}H_2O + H]^+$,

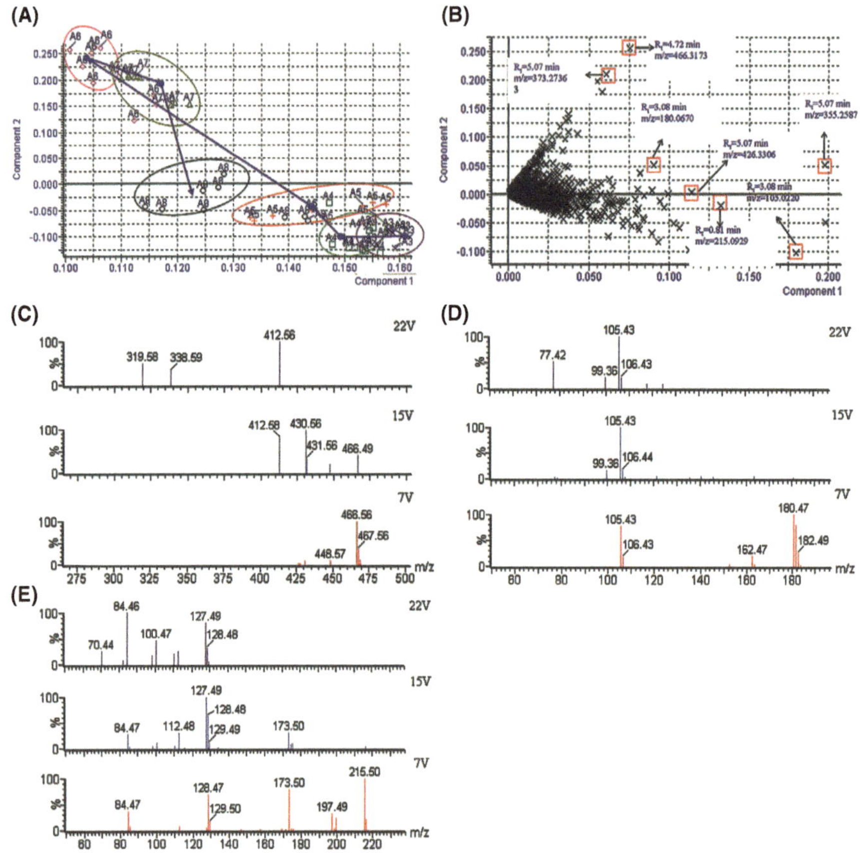

FIGURE 9.3
(A) The score plot for PCA of metabolites in model rats for the positive ion data obtained for the urine sample collection period for continuously 8 days; (B) distinct clustering is observed among different collection days. The loading plot for PCA of metabolites in model rats for the positive ion data obtained. (C) Seven metabolites were determined which markedly contributed to the separate tendency of score plot. Mass spectrum of full scan and product ion scan of determined biomarkers. (D) The basic information: R_t = 4.72 and $[M + H]^+$ = 466 mass spectrum of full scan and product ion scan of determined biomarkers. (E) The basic information: R_t = 3.08 and $[M + H]^+$ = 180 mass spectrum of full scan and product ion scan of determined biomarkers. The basic information: R_t = 0.81 and $[M + H]^+$ = 215.

$[M\text{-}2H_2O + H]^+$, $[M\text{-}3H_2O + H]^+$, and $[M\text{-}3H_2O\text{-}C_2H_4NO_2 + H]^+$ when subjecting to MS/MS analysis (Fig. 9.3C), and retrieving corresponding literatures, the metabolite was provisionally identified as glycocholate. The $[M + H]^+$180.0670 and $[M + H]^+$215.0929 were also preliminarily identified as methenamine hippurate and acetic acid (3-acetoxy-2-methyl-3,4-dihydro-2H-pyran-4-yl) ester, respectively (shown in Fig. 9.3D and E) when adopting the same identified mode. The identification of the metabolites of *m/z* 355, 105, 426, and 373 are in progress.

Table 9.2 The Basic Information of Determined Metabolites in Rat Urine

No.	Rt (min)	*m/z*	$[M + H]^+$ ion (accurate mass)	Metabolites
001	4.72	466.3173	$C_{26}H_{43}NO_6$ (466.3169)	Glycocholic acid
002	5.07	355.2587	$C_{19}H_{34}N_2O_4$ (355.2597)	–
003	3.08	180.0670	$C_9H_9NO_3$ (179.0582)	Methenamine hippurate
004	3.08	105.0220	$C_3H_4O_4$ (105.0188)	–
005	5.07	426.3306	$C_{28}H_{43}NO_2$ (426.3372)	–
006	5.07	373.2736	$C_{24}H_{36}O_3$ (373.2743)	–
007	0.81	215.0929	$C_{10}H_{14}O_5$ (214.0841)	Acetic acid (3-acetoxy-2-methyl -3, 4-dihydro-2H- pyran-4-yl) ester

9.3.3 YCHT and 6,7-DME Treated Animals

Examination of the trending plot of the above mentioned seven metabolites in different dosing animals revealed marked metabolic improvements in the urine of animals pretreated with YCHT and 6,7-DME, compared to model animals at all time points postdosing in positive ion mode, and the improvements reached approximately to the metabolomic profiles of nondosing animals (Fig. 9.4), which proved again the rationality of determining metabolites. Those endogenous metabolites could be used as special biomarkers for indicating occurrence, development, and recovery of CCl_4 inducing liver injury. Simultaneously, the score plots of YCHT and 6,7-DME interfering (Fig. 9.5A and B), indicates that the metabolic trajectories do not significantly change after dosing over the time-course of the study. The validation of groups pretreated with YCHT and 6,7-DME in interfering with CCl_4 inducing liver injury in rats was further elucidated. The results not only proved the validation of YCHT in preventing the liver injury induced by CCl_4, but also elucidated that 6,7-DME was a major hepatoprotective constituent of YCHT.

9.4 CONCLUSIONS

In the present work, similar changes in serum and liver enzymes, and the histopathological changes were observed in the experiment when rats were treated with CCl_4, and pretreated with YCHT and 6,7-DME. A metabolomic strategy based on a rapid UPLC coupled with a high resolution QTOF mass spectrometer (MS) method has been developed to analyze the metabolome of rats treated with CCl_4 and the intervention effects of YCHT and its major constituent 6,7-DME, which has demonstrated that the rats exposure to CCl_4 results in a marked separation in the urinary metabolic profile. Seven metabolites that mainly contributed to pathological disorder of liver in rats were found, they can be taken as special biomarkers to indicate the occurrence, development, and recovery of liver injury. Three metabolites concerning liver injury have

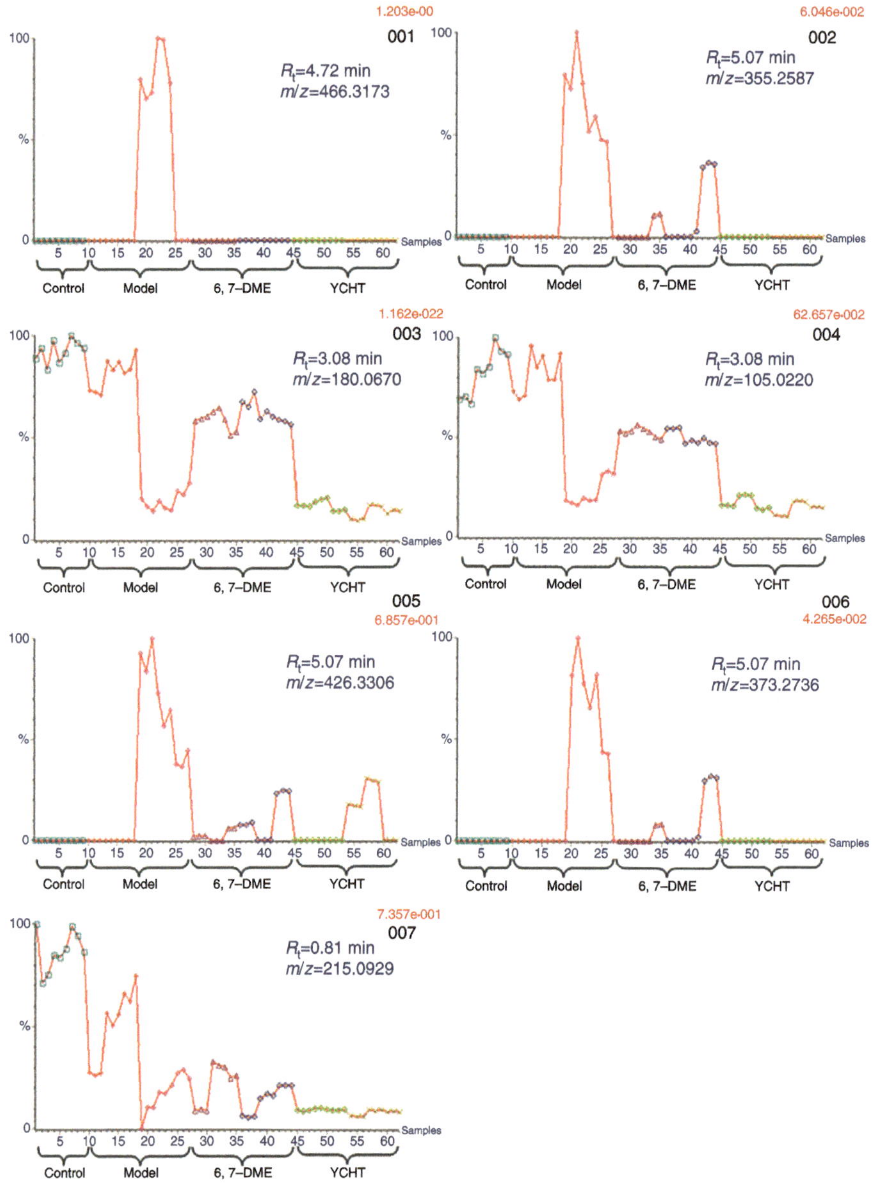

FIGURE 9.4 The trend plots of seven biomarkers contributing to the separation of metabomonic profiles in different dosing groups.

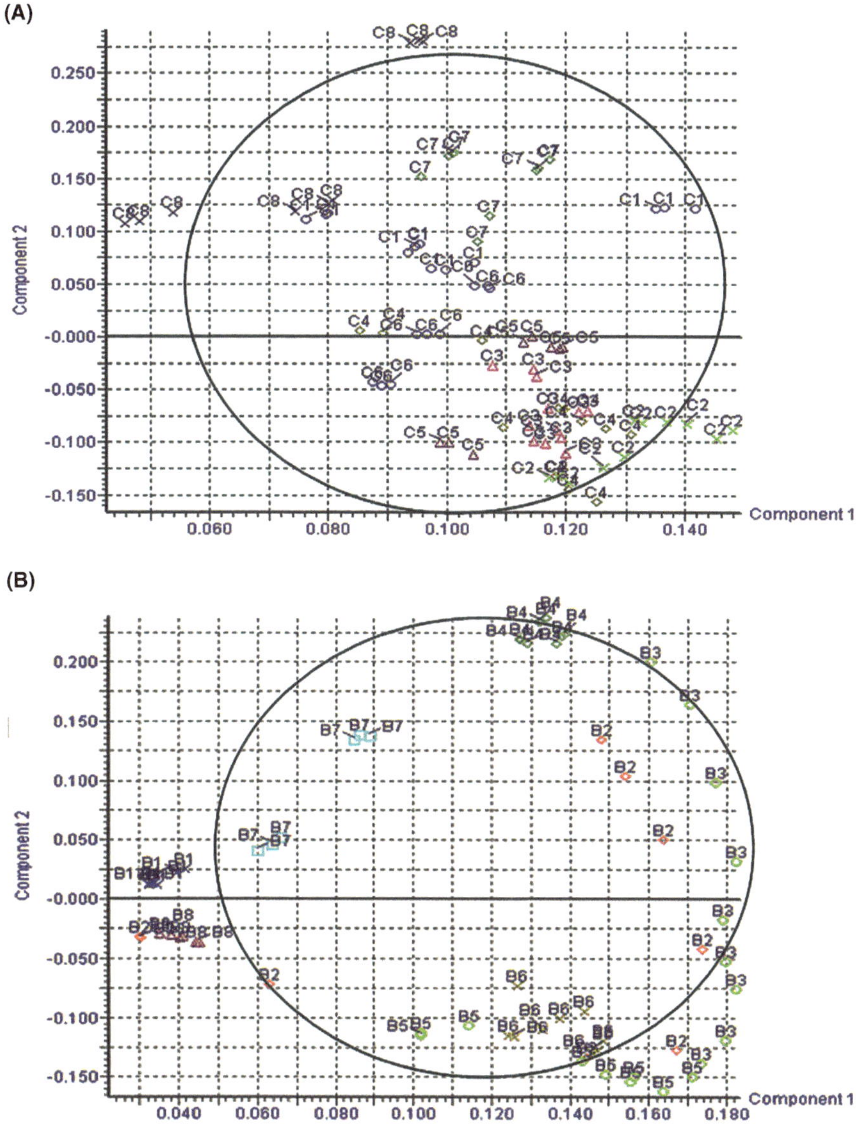

FIGURE 9.5 The score plot of metabolites in rat urine of pretreating with YCHT (A) and 6,7-DME (B).

been identified, the hepatoprotective effect of YCHT and 6,7-DME also can be obviously seen in the metabolic trajectory of seven biomarkers, these facts further authenticated the scientific rationality of the determined biomarkers related to liver injury. This study offered a rapid and efficient method for investigating and evaluating the hepatoprotective effect of TCM.

CHAPTER 10

Metabolomics and Proteomics Annotate Therapeutic Mechanisms of Geniposide

Aihua Zhang, Hui Sun, Wenjun Sun, Xijun Wang

CONTENTS

10.1 INTRODUCTION

"Omics" technologies are capable of measuring changes in large numbers of variables, often at a wide level, and building networks for analyzing drug action (Ioannidis and Khoury, 2011). Metabolomics combining with proteomics analyses will enable the development of predictive models of therapeutic efficacy, and for the identification of the target protein(s) for a compound with an unknown mechanism of action. Proteomic network analyses are emerging as a key tool in identifying potential drugs, and allow us to develop an initial understanding of the context within which molecular-level drug–target interactions occur (Price et al., 2012). As the newest of the "omics" sciences, metabolomics have been shown to have enormous potential when applied to subjects as diverse as drug mechanisms, disease processes, and drug discovery (Zhang et al., 2014).

Knowledge about the protein targets of natural products is critical for understanding drug action (Wang et al., 2008). To address drug effects in more physiological conditions, novel omics tools have recently been developed. Liver injury has become a severe health problem worldwide. Despite considerable and continuous efforts, effective treatment strategies against this disease resulting in fewer side effects are still lacking. Oriental herbal medicines, widely used for treatment of various diseases, have recently attracted the interest of the modern scientific community as alternative therapies (Zhang et al., 2013a). Zhizi (*Gardenia jasminoides* Ellis) is one important component of a famous Chinese medicinal Yinchenhao Tang that is one of the most popular traditional formulae for treatment of jaundice syndrome and liver disorders, and has been used more than a thousand years (Chinese Pharmacopoeia Commission 2010). Interestingly, geniposide (Fig. 10.1) is one of the biologically active ingredients of the gardenia fruit of Zhizi, and has been proven effective in treating liver diseases, and shows hepatoprotectivity and contributes directly to the therapeutic effect (Yang et al., 2011). While such a compound has distinct pharmacological effects; however, the precise mechanism of the action remains poorly understood.

Chinmedomics. http://dx.doi.org/10.1016/B978-0-12-803117-9.00010-X

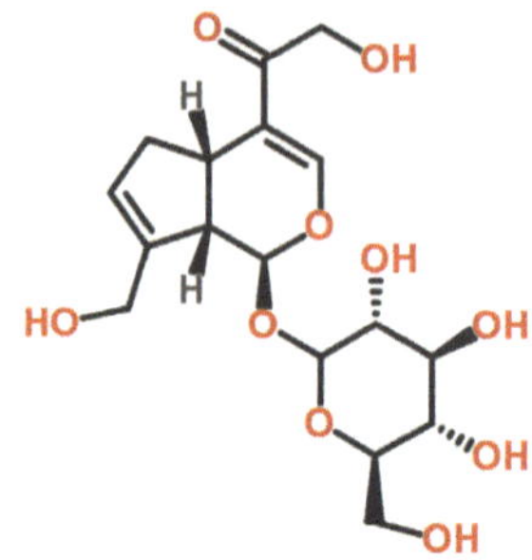

FIGURE 10.1 Chemical structures of geniposide.

Identification of the molecular targets of natural products is an important aspect of current drug discovery, as knowledge of the molecular targets will greatly aid drug development. The identification of their targets is required for a comprehensive understanding of their pharmacological role, and for unraveling their mechanism of action. Omics approaches combined with other biochemical methods can reconstruct regulatory networks, signaling cascades, and metabolic pathways upon drug treatment (Zhang et al., 2013b). Herein, we use the treatment of HI with geniposide as a working model from a systematic view of the dissection of mechanisms of geniposide as an effective treatment for liver injury, we conducted an investigation incorporating advanced metabolomic and proteomic technologies.

10.2 MATERIALS AND METHODS

10.2.1 Chemicals and Reagents

Geniposide (purity 99%) was purchased from the Sichuan Provincial Institute for Food and Drug Control (Sichuan, PRC), ultra-pure reagents for polyacrylamide gel preparation were obtained from Bio-Rad, 18 cm immobilized pH gradient (IPG) strips, and IPG buffer and dry strip cover fluid were purchased from Amersham Pharmacia Biotech (Uppsala, Sweden). Dithiothreitol (DTT), PMSF, iodoacetamide, urea, thiourea, α-cyano-4-hydroxycinnaic acid (CCA), and TPCK-Trypsin were purchased from Sigma Chemical (St. Louis, MO, USA).

10.2.2 Animal Handling and Sample Preparation

After acclimatization, male Wistar rats (weighing 240 ± 20 g) were randomly divided into three groups with eight rats in each: the control, HI model, and geniposide groups. The rats in the control group were administrated with 0.9% saline in the whole procedure for 5 consecutive days. Rats were orally administrated with 20% CCl_4 (1 mL/kg body weight) olive oil solution at day 4 (6:00 p.m.) to induce liver injury model for 5 consecutive days, and until day 8. Simultaneously, geniposide group was administrated with CCl_4 and 0.7 mg/m drug treatment for 5 consecutive days. The rats were administrated an oral accurate volume of 1 mL/100 g of each animal.

Urine was collected daily from metabolism cages at ambient temperature throughout the whole procedure and centrifuged at 13,000 rpm at 4°C for 5 min, and the supernatants were stored frozen at −20°C until metabolomic analysis. Blood was collected from the hepatic portal vein, kept on ice for 1 h and serum was separated via centrifugation at 4500 rpm for 5 min at 4°C, flash frozen in liquid nitrogen and stored at −80°C until use.

10.2.3 Biochemical Analysis and Histology Assay

We quantified the levels of serum ALT, AST, ALP, T-BIL, MDA, SOD, and GSH-PX content using assay kits according to the manufacturer's instructions. The rat livers were removed immediately after plasma collection and stored at -80°C until analysis. The livers were fixed in 4% neutral buffered formaldehyde at 4°C and embedded in paraffin. The liver tissue was stained with H&E for histopathological analysis.

10.2.4 Metabolic Profiling Platform

10.2.4.1 Chromatography

An aliquot of 6 μL of sample solution was injected onto an ACQUITY UPLC BEH C_{18} column (50 × 2.1 mm i.d., 1.7 μm, Waters Corp., Milford, MA, USA) at 35°C and the flow rate was 0.5 mL/min. The optimal mobile phase consisted of a linear gradient system of (A) 0.1% formic acid in acetonitrile; and (B) 0.1% formic acid in water, 0–8.2 min, 90–30% A; 8.2–8.5 min, 30–1% A; 8.5–10.0 min, 1% A; 10–10.5 min, 1–99% A; 10.5–12.5 min, 99% A. In addition, the QC sample was used to optimize the condition of UPLC-Q-TOF/MS, as it contained the most information of whole urine samples.

10.2.4.2 Mass Spectrometry

For the urine UPLC–HDMS (Waters Corp., Milford, USA) analysis, the optimal conditions for positive ion mode were as follows: capillary voltage of 2500 V, desolvation temperature of 350°C, sample cone voltage of 35 V, extraction cone voltage of 3 V, microchannel plate voltage of 1600 V, collision energy of 4 eV, source temperature of 110°C, cone gas flow of 50 L/h, and desolvation gas flow of 600 L/h. A lock-mass of leucine enkephalin at a concentration of 200 pg/mL in acetonitrile (0.1% formic acid): H_2O (0.1% formic acid) (50:50, v/v) for positive ion mode ($[M + H]^+ = 556.2771$) was employed via a lock-spray interface. The data were collected in centroid mode, the lock spray frequency was set at 5 s and the lock mass data were averaged over 10 scans. The mass spectrometry full-scan data were acquired in the positive ion mode from 100 Da–1000 Da with a 0.1-s scan time.

10.2.4.3 Multivariate Data Analysis

MS data were generated recorded using MassLynx V4.1 (Waters Corp., Milford, USA), MarkerLynx Application Manager (Waters Corp., Milford, USA) was used for peak detection, and EZinfo software (which is included in the MarkerLynx Application Manager and can be applied directly) was used for: the "unsupervised" and principal component analysis (PCA); "supervised" partial least squares-discriminant analysis (PLS–DA); and orthogonal projection to latent structures (OPLS) analysis.

10.2.4.4 Biomarkers Identification and Reconstruction of Metabolic Pathway

Potential markers of interest were extracted from S-plot (EZinfo software 2.0) constructed following analysis with PLS–DA, and markers were chosen based on their contribution to the variation and correlation within the data set. Statistical analysis was conducted via direct comparisons of biochemical parameters and metabolite concentrations. The MassFragment™ application manager (Waters Corp., Milford, USA) was used to facilitate the MS fragment ion analysis process by way of chemically intelligent peak-matching algorithms. With regard to the identification of biomarkers, the ion spectrum of potential biomarkers was matched with the structure message of metabolites acquired from available biochemical databases. The reconstruction, interaction, and pathway analysis of potential biomarkers was performed with the MetPA online database source to identify the metabolic pathways.

10.2.5 Proteomics Analysis

10.2.5.1 Two-Dimensional Gel Electrophoresis

The plasma obtained was disrupted in a homogenization buffer containing 30 mM Tris, 7 M urea, 2 M thiourea, 4% CHAPS, and 40 mM DTT, after which the lysates were centrifuged at 12,000 rpm for 60 min at 4 °C. The supernatant was precipitated using the TCA sediment method to recover the total proteins, and the interfering components were removed using the 2-D Clean-Up Kit (GE Healthcare, Piscataway, NJ). The supernatant was removed, divided into several aliquots and stored at −70 °C until use. The protein concentration of each sample was measured using the Bradford method. A total of 350 mg of protein was loaded onto an 18 cm linear IPG strip (pH 3–10, Amersham Biosciences, Piscataway, NJ) for first-dimension isoelectric focusing (IEF). The strips were placed into a Protean IEF cell (Bio-Rad) and rehydrated at 50 V for 12 h, after which the proteins were separated based on their pI according to the following protocol: 250 V with linear climb for 60 min, 1,000 V with rapid climb for 60 min, 10,000 V with linear climb for 5 h, and 10,000 V with rapid climb until a level of 60,000 V-h was reached. After IEF, the IPG strips were equilibrated for 15 min in a buffer containing 50 mM Tris-HCl, pH 8.8, 30% glycerol, 7 M urea, 2% SDS, and 1% DTT; they were then further treated in a similar buffer (containing 4% iodoacetamide) for 15 min and directly applied onto 12% homogeneous SDS-PAGE gels for 6 h electrophoresis using a Protean II xi cell system (Bio-Rad). Each experiment was performed in triplicate.

10.2.5.2 Image Analysis

The gels were then silver-stained using Bio-Rad Silver Stain Plus kit reagents (Bio-Rad) according to the manufacturer's instructions. The silver-stained gels were scanned using a GS-800 densitometer (Bio-Rad) and then analyzed using PDQuest software (Bio-Rad). Labscan Software (Applied Biosystems, Foster City, CA) was used for image analysis that included background abstraction, spot intensity calibration, spot detection, and matching. The intensity of each spot was quantified by calculating the spot volume after the gel image had been normalized. Each sample was processed in triplicate to ensure reproducibility, and the Student's t-test was used to evaluate the average change in protein abundance corresponding to each target spot across the gels. Protein spots with more than a threefold significant change in density ($p < 0.05$) were considered to be differentially expressed and were selected for further identification using MALDI-TOF MS/MS.

10.2.5.3 Protein Identification Using MALDI-TOF-MS and Database Searching

Protein spots of interest with differential expression levels were excised from the 2-D gels with an EXQuest spot cutter (Bio-Rad), washed three times in distilled water, destained with 15 mM potassium ferricyanide in 50 mM sodium thiosulfate for 2 min at room temperature, dehydrated in acetonitrile and finally dried in a centrifugal vaporizer. After a 30-min incubation, the gels were digested for more than 12 h at 37°C. The peptides were then extracted twice using 0.1% TFA (trifluoroacetic acid) in 50% CAN dried under the protection of N_2. For MALDI-TOF/TOF MS, the peptides were mixed with 0.5 μL MALDI matrix (5 mg/mL α-cyano-4-hydroxycinnamic acid diluted in 0.1% TFA and 50% ACN) and spotted onto the MALDI sample plates. The MS measurements were performed on a Proteomics Analyzer with delayed ion extraction (Applied Biosystems). The parameters of MALDI-TOF were set up as follows: accelerating voltage 20 kV; grid voltage 64.5%; delay 100 ns; the number of laser shots 50, and the acquisition mass range 900–3500 Da. MS/MS accuracy was calibrated against the MS/MS fragments of m/z, which is one of the peaks that is generated in the myoglobin peptide mass fingerprint (PMF). The parameters for peak matching were as follows: the minimum signal-to-noise ratio was 20, the mass tolerance was 0.2 Da, the minimum peak to match reference masses was 4, and the maximum outlier error was 50 ppm. Bioinformatics databases were searched for protein identification based on tryptic fragment sizes. PMF were searched in the SWISS-PROT database (www.expasy.org/) using Mascot software (www.matrixscience.com). The protein score is $-10*\text{Log}(P)$, where P is the probability that the observed match is a random event. The missed cleavage sites allowed were up to 1, and scores that were higher than 61 were considered significant.

10.2.5.4 *Bioinformatics Analysis of Proteomic Data*

Identified proteins were further analyzed using STRING, chosen as the source for protein–protein interactions, to statistically determine the functions and pathways most strongly associated with the protein list. Prior to upload and analysis using STRING, the mean ratio of each quantified protein in a group was calculated and the fold change between the groups calculated. The PANTHER classification system and KEGG (www.genome.jp/kegg/) pathway database were used to further examine the significantly expressed molecules.

10.2.6 Statistical Analyses

All statistical analyses were performed using the Student's t-test. Differences with a P-value of 0.05 or less were considered significant. Assays were performed in triplicate, and the results are expressed as mean ± SD.

10.3 RESULTS

10.3.1 Geniposide Reduces Histologic Changes

To investigate the protective effects of geniposide on liver tissue damage, histological analyses were performed in liver tissue that was obtained from HI rats and compared with tissue from control rats. Microscopic analyses of H&E-stained liver sections showed that geniposide significantly decreased hepatocyte necrosis, and fibrotic area, making them comparable to normal liver (Fig. 10.2A). The histopathological examination of liver sections that were stained with H&E revealed numerous apoptotic hepatocytes, and the accumulation of massive necrosis with intralobular hemorrhage in the livers of HI rats. Minimal hepatocellular necrosis and inflammatory cell infiltration and mild portal inflammation were observed in rats that treated with geniposide. The degree of necrosis was clearly lower in the CCl_4-treated rats that received geniposide, as compared with animals that were treated with the model. Furthermore, the H&E-stained sections indicate that hepatocyte size had changed in the model rats, and that geniposide restored normal cell size. These results suggest that the geniposide prevents the destruction of liver tissue and exerts therapeutic efficacy.

10.3.2 Therapeutic Efficacy of Geniposide as Indicated by Biochemical Indicators

There was significant variation of the biochemical indicators such as ALT, AST, ALP, MDA, SOD, and GSH-PX between the control and model groups

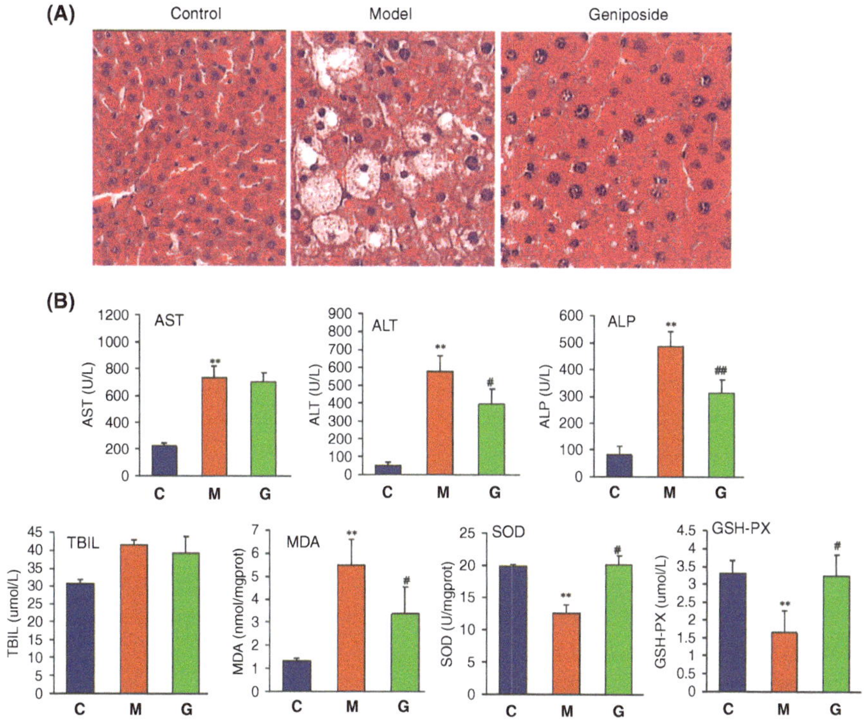

FIGURE 10.2

(A) Representative pictures of liver histopathology after CCl_4 with geniposide treatment in rats (original magnification: ×200); (B) therapeutic efficacy of geniposide as indicated by biochemical indicators. The values are expressed as mean ± SD and were compared by ANOVA. $^{*}p < 0.05$; $^{**}p < 0.01$; $^{\#}p < 0.05$, $^{\#\#}p < 0.01$.

(Fig. 10.2B). It indicated that the HI model was successfully replicated. The model group had higher ALT, AST, ALP, T-BIL, and MDA values but had lower levels of SOD and GSH-PX than the control animals. The indicators, SOD and GSH-PX results showed the recovery of the values by geniposide treatment, but values of AST and TBIL did not show any significant changes of values. Of note, the geniposide decreased the levels of ALT, AST, ALP, T-BIL, and MDA but increased the levels of SOD and GSH-PX. Geniposide group was treated back to baseline levels of the control group, which demonstrates that this drug had a therapeutic effect in the rat HI model.

10.3.3 Metabolomics Analysis

10.3.3.1 LC–MS Analysis of Metabolic Profiling

Using the optimal reversed-phase UPLC-MS conditions, all the data containing the retention time, peak intensity, and exact mass were imported in to the

Masslynx™ software for multiple statistical analyses. Both PCA and PLS-DA often can be taken because of their ability to cope with highly multivariate, noisy, collinear, and possibly incomplete data. Typically, the trajectory analysis of PCA in positive mode can really reflect the differences between the model and control groups, and showed metabolic profiles in the control and model group were separated clearly (Fig. 10.3A). 3D-PCA in Fig. 10.3B demonstrates a clear separation between the treated and control groups. Nine ions showed significant difference in abundance between the control and treated animals, and contributed to the observed separation, and were selected from the S-plot of PLS-DA (Fig. 10.3C). The farther away from the origin, the higher the VIP value of the ions was. Furthermore, PCA score plot for the liver injury rats after geniposide treatment in positive model showed significant differences between samples (Fig. 10.3D), gradually back to the normal group. Finally, according to the variable importance in the projection, the ions (listed in Table 10.1) were identified based on the metabolite identification strategy. The parallel analysis of the samples with the MetaboAnalyst allows for the ability to verify that ions, which are identified through both ways (i.e., Score plots, Heatmaps) are highly significant, as depicted through two completely different algorithms. Analysis of the control and geniposide groups utilizing the MetaboAnalyst's score plot revealed differences among groups (Fig. 10.3E). Using our metabolomics platform, interestingly, the parallels heatmap visualization (Fig. 10.3F) using Ward's method in computational systems analysis for the rats that treated with geniposide, and model groups could be achieved with distinct segregation.

10.3.3.2 Changes of Relative Intensity of Biomarkers

According to the protocol detailed above, nine endogenous metabolites (data not shown) contributing to the separation of the model group and control group were detected in the urine samples. Monitoring changes of these metabolites may predict the development of liver injury. Interestingly, the relative concentration of nine endogenous metabolites was significantly affected by geniposide treatment. Additionally, compared with the alterations of liver injury-related metabolites, most of them were reset to a normal state after geniposide administration. Among them, four metabolites in the control values, and two metabolites in the model values are zero (data not shown).

10.3.3.3 Biomarker Network and Metabolic Pathway Reconstruction

With pattern recognition analysis (PRA) of metabolites, a clear separation of the model and the control group was obtained, the geniposide group was located with the control group. To determine whether our observations of changes in the metabolites in the setting of liver injury in fact reflected coordinate changes in defined metabolic pathways, we used MetPA software to identify the network pathway (Fig. 10.4A). This software was based on the

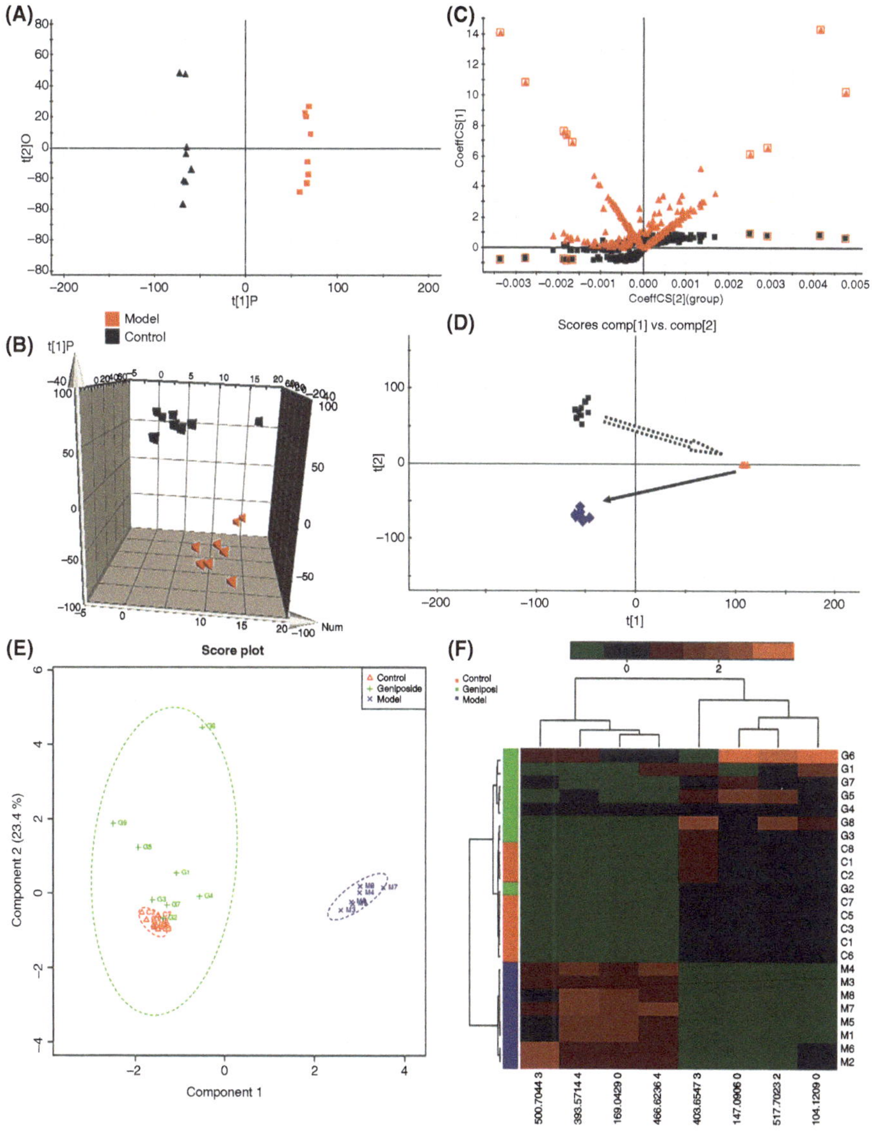

FIGURE 10.3 Metabolomic multivariate analysis.
(A) PCA score plot for the control and HI groups (■, control group; ■, model group); (B) 3D-PCA plot for the control and HI groups; (C) a combination plot of S-plot and VIP values (EZinfo software 2.0). The red point graph is the VIP value plot that represents the value of each metabolite. The farther from the origin, the higher the VIP value of the metabolites; (D) trajectory analysis of PCA for the CCl_4 treatment in positive mode; (E) score plots for acute livery injury after scoparone treatment in positive mode; (F) heat map visualization for the urine samples. The heatmap was constructed based on the potential candidates of importance that were extracted with PLS-DA analysis.

Table 10.1 Results of Identifications of Differentially Expressed Proteins Using MALDI-TOF/TOF MS

Database ID no.	Identified Protein Name	MASCOT Score	PI	MW
gi|125144	Ig kappa chain C	185	5.02	26051
gi|55628	Zinc finger protein	489	6.49	34886
gi|61556986	Serotransferrin	275	5.87	53532
gi|33086640	Haptoglobin	318	5.71	53590
gi|12084772	Macroglobulin	250	5.83	54878
gi|203063	α-1-antitrypsin	101	5.07	61662
gi|46237594	Complement factor	97	5.96	77040
gi|71824	Fibrinogen alpha chain precursor	131	6.43	38765
gi|510196	Glyceraldehyde-3-phosphate dehydrogenase	140	7.56	31613
gi|158138568	Albumin	208	6.75	34348
gi|17105350	2,4-Dienoyl-CoA reductase,	103	4.71	36531
gi|136467	Transthyretin	397	6.29	42190
gi|83816939	α-1-Inhibitor 3 precursor	303	5.84119	60934
gi|203941	Vitamin D-binding protein	126	5.44	62398
gi|135809	Prothrombin	156	5.26	91978

Notes: MW, relative molecular weight; PI, isoelectric point.

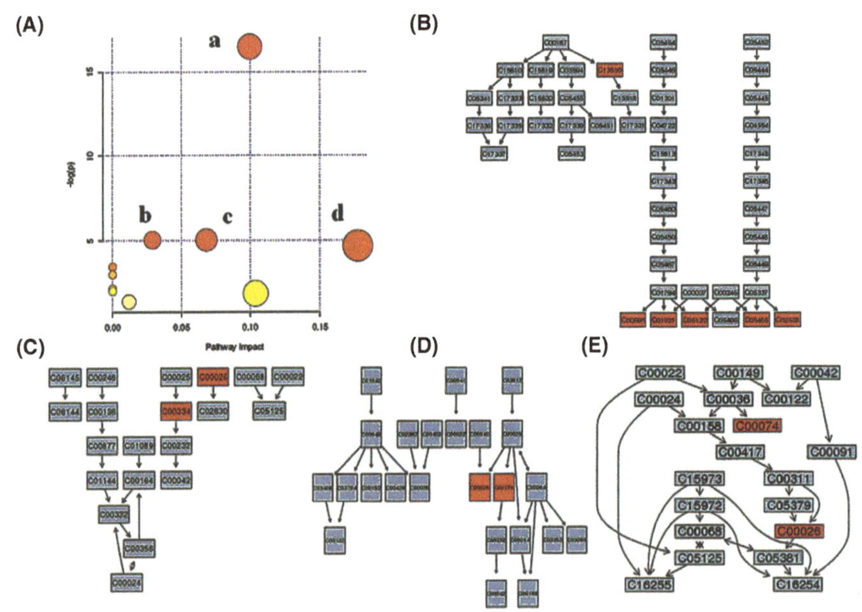

FIGURE 10.4 Summary of pathway analysis. Identifying network pathway by MetPA software. (A) (a) primary bile acid biosynthesis; (b) butanoate metabolism; (c) alanine, aspartate, and glutamate metabolism; (d) citrate cycle. (B) Putative metabolic pathways of primary bile acid biosynthesis; (C) butanoate metabolism; (D) alanine, aspartate, and glutamate metabolism; (E) citrate cycle metabolism were inferred in the CCl_4 dosed rats from changes in the urine levels of intermediates during substance metabolism. Red denotes affected metabolites related to the pathway. The map was generated using the reference map by KEGG (http://www.genome.jp/kegg/).

high-quality KEGG metabolic pathways as the backend knowledgebase to help researchers identify the most relevant pathways involved in the conditions under study. The detailed construction of the metabolism pathways with higher scores is shown in Fig. 10.4. MetPA revealed that the top four pathways including primary bile acid biosynthesis (Fig. 10.4B), butanoate metabolism (Fig. 10.4C), citrate cycle (TCA cycle, Fig. 10.4D), alanine, aspartate, and glutamate metabolism, (Fig. 10.4E) etc., that changed specifically in the setting of HI. Results suggested that these target pathways showed the marked perturbations over the time-course of liver injury and could contribute to the development of liver injury.

10.3.4 Proteomics Analysis

10.3.4.1 2-DE Pattern and Differential Analysis of Rat Serum

Proteins from the control, model and geniposide groups were respectively solubilized in the IEF-compatible lysis buffer, and separated on IPG strips and SDS-PAGE. Following staining with silver nitrate, the well-resolved and reproducible 2-D gel maps were obtained (data not shown). Using PDQuest 2-DE gels analysis software, approximately 1000 well-stained, clearly delineated protein spots were detected, and most spots were distributed in the range from pI 3−10. There were 15 spots that showed remarkable change (Student's t-test; $p < 0.05$) between the model and control groups observed.

10.3.4.2 Proteins Identified by MALDI-TOF-MS

Differentially expressed protein spots were excised from the 2-DE gel, and analyzed for protein identification by MALDI-TOF/TOF-MS. All PMFs were searched with Mascot software in the SWISS-PROT database to identify the protein spots. The result had high confidence if the protein was ranked as the best hit with a significant score and high sequence coverage. Finally, we identified 15 proteins in those spots, shown in Table 10.1. A representative PMF map and database query result of protein zinc finger protein 407, α-1-antitrypsin and transthyretin were shown in Fig. 10.5. The relative expression levels of the differentially expressed proteins are shown in Fig. 10.6. These proteins appear to be involved in antioxidation and signal transduction, energy production, immunity, metabolism, and chaperoning. The hepatoprotective effects of geniposide on acute hepatic injury Wistar rats were associated with the regulated expression of 6 proteins ($p < 0.05$) including zinc finger protein 407, haptoglobin, α-1-antitrypsin, glyceraldehyde-3-phosphate dehydrogenase, transthyretin, and prothrombin.

10.3.4.3 Visualization and the Analysis of Protein Interaction Networks

To facilitate access to the protein–protein interaction data, STRING integrates information about interactions from metabolic pathways, crystal

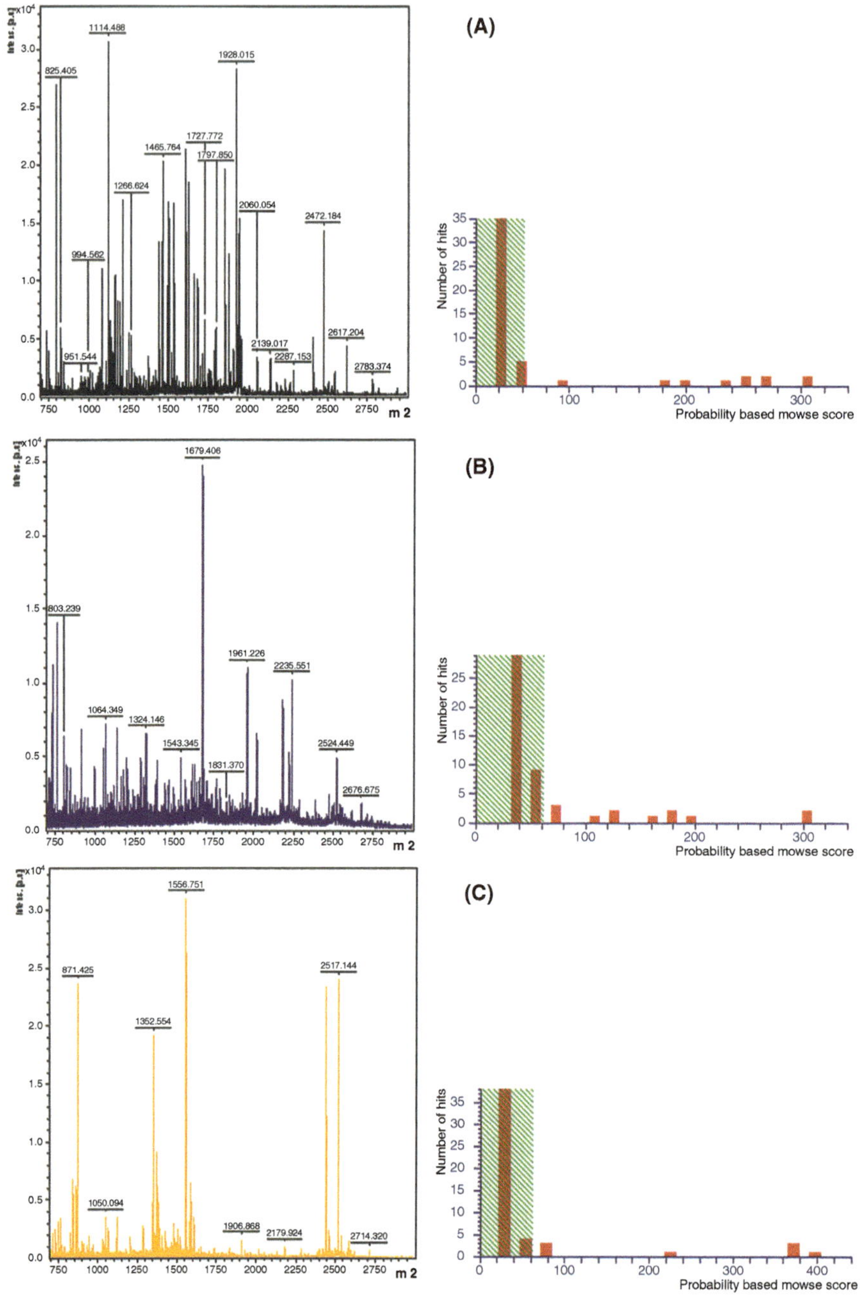

FIGURE 10.5 Peptide mass fingerprint spectrum of zinc finger protein 407, α-1-antitrypsin, transthyretin.

The spot was in-gel digested with trypsin. After being desalted, the peptide mixture was analyzed by MALDI-TOF-MS. The *x*-axis represents the mass-to-charge ratio (*m/z*), and the *y*-axis represents the relative abundance. All protein identifications are provided in supplementary Table 10.2. Down: the probability-based Mowse scores obtained using the Mascot search engine. Among predicted proteins with differential Mowse scores shown as multiple bars on the *x*-axis, only proteins with Mowse scores greater than 61 were considered significant, which were 305, 303, and 397 ($p < 0.05$) for zinc finger protein 407, α-1-antitrypsin, transthyretin, respectively.

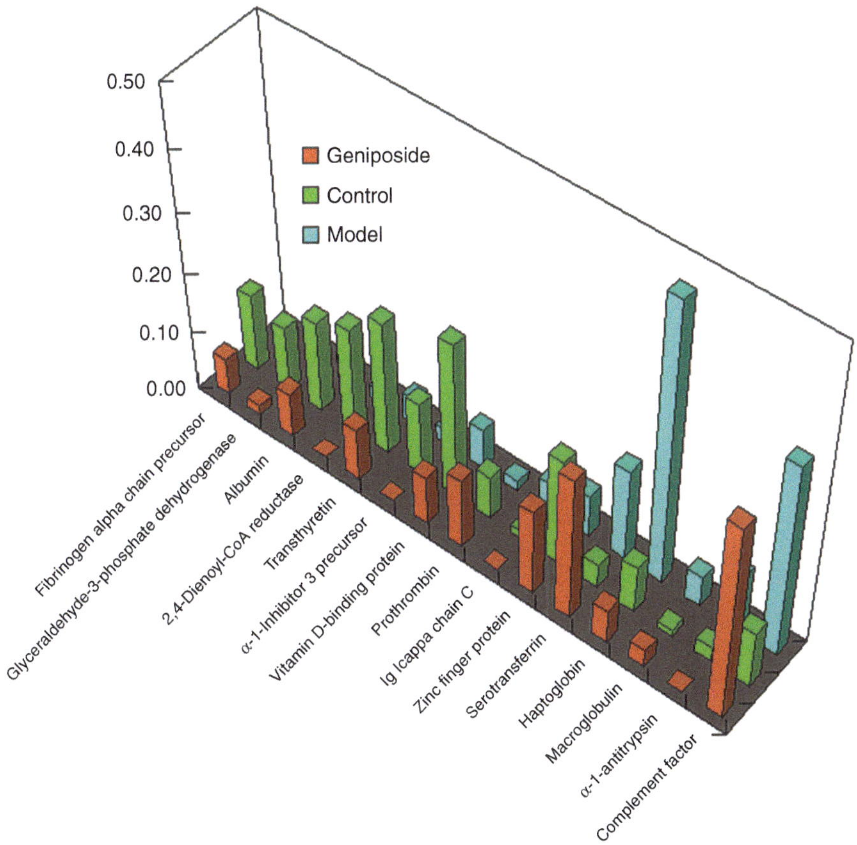

FIGURE 10.6 Relative level of the differentially expressed proteins obtained by MALDI-TOF/MS.

structures, binding experiments, and drug–target relationships. As seen in Fig. 10.7, association network of six differentially expressed proteins using STRING were constructed. To examine the protein functions and pathways of the differentially expressed proteins, the enriched GO terms were categorized for HI networks to identify the functional change. From all the GO categories covered by the differentially expressed proteins and their interactions, the identified proteins could be divided into several main groups according to the major biological process categories: proteins related to antioxidation and signal transduction, proteins involved in apoptosis, chaperone, energy generation, intermediary metabolism, regulation of cellular process, and molecular transport. These closely connected and coexpressed differentially expressed proteins in networks, are regarded as the signatures of the underlying targets.

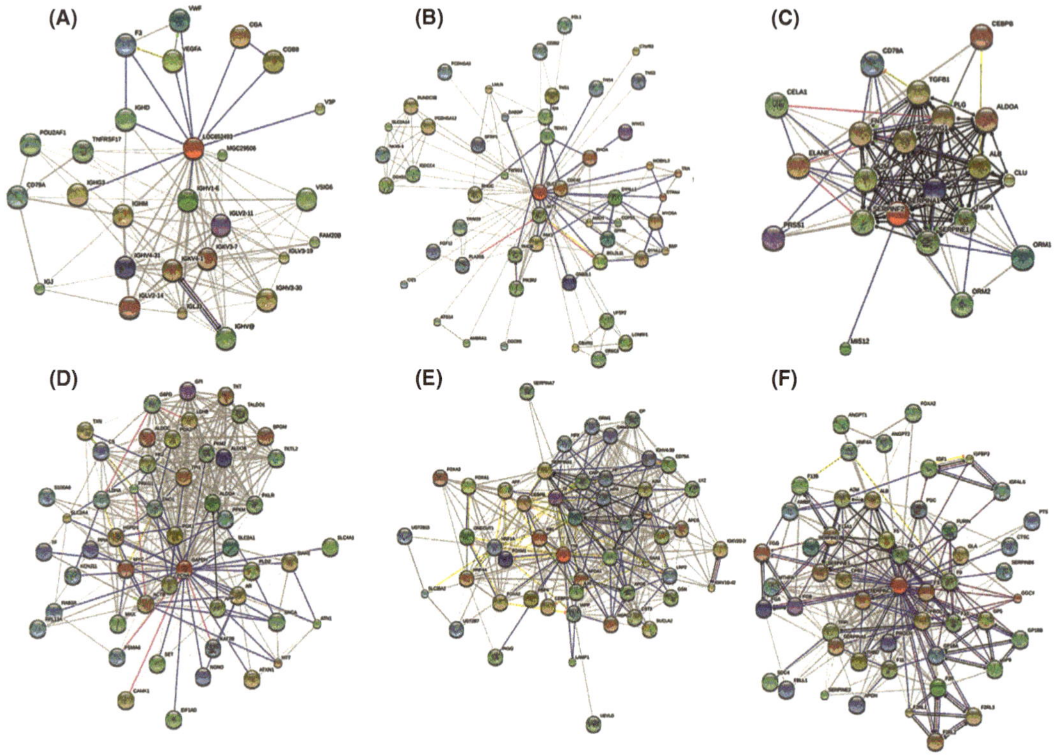

FIGURE 10.7 Interaction network generated for proteins identified.
Modes of action are shown in different colors. Differentially expressed proteins are represented as pill-shaped nodes, while other proteins are shown as spheres. Nodes that are associated to each other are linked by an edge. Thicker lines represent stronger associations. Lines and, for directed edges, arrows of different colors stand for different edge types in the actions view: binding (blue), activation (green), inhibition (red), catalysis (magenta), same activity (cyan) and reaction (black). (A) Zinc finger protein 407; (B) haptoglobin; (C) α-1-antitrypsin; (D) glyceraldehyde-3-phosphate dehydrogenase; (E) transthyretin; (F) prothrombin.

10.4 DISCUSSION

The tide of drug discovery has been turning back to Nature in the search for new drug candidates. Natural products have contributed to the development of important therapeutics to combat diseases over the past decades (Gersch et al., 2012). However, the targets and modes of action of these fascinating compounds are seldom known, hindering the drug development process. One example is geniposide, whose efficacy in treating HI has recently been well established (Ma et al., 2011). Since the means of action of geniposide on HI has not been sufficiently investigated, we used an "omics" platform and identified the key mechanisms that underlie the observed effects. Up to now, there have been no reports of the omics properties of geniposide in the relevant literature, although this information is important to our further exploration of geniposide as a drug candidate. To gain insight into the complex biochemical mechanisms of this effective HI therapy, we conducted an investigation that incorporated modern biochemical analyses, metabolomics and proteomics, to analyze the effects of the geniposide treatments on HI.

We employed UPLC-HDMS and MALDI-TOF MS/MS to systematically analyze the therapeutic properties of genipisode at the levels of the metabolome and proteome, thus exploring some of the key molecular mechanisms that underlie the effects. It was demonstrated that based on the UPLC-HDMS measurements of urine samples with statistically significant multivariate models, it could be constructed to distinguish between control and model subjects. The global metabolic profiling and subsequent multivariate analysis clearly distinguished HI rats from the matched controls. As shown in Fig. 10.2A and B, PCA revealed an evident separation between the model and control samples. By using our metabolomics platform, nine statistically important variables with VIP > 6 were defined. Interestingly, of the nine distinct metabolites identified from these pathways, many are in various stages of progress at the HI. Further study of these metabolites may facilitate the development of noninvasive biomarkers and more efficient therapeutic strategies for HI. Furthermore, primary bile acid biosynthesis, butanoate metabolism, citrate cycle, alanine, aspartate, and glutamate metabolism were the top functions listed by the Metaboanalyst tool. It is noteworthy that nine metabolites together are important for the host response to HI through target metabolism pathways. A significant separation using Heatmap analysis was obtained for the difference between HI vs. controls, as well as between genipisode vs. model. The HI rats were easily distinguished from control subjects by the powerful Heatmap approach when using autoscaling methods. Based on the KEGG, a detailed construction of the primary bile acid biosynthesis, butanoate metabolism, citrate cycle, alanine, aspartate, and glutamate metabolism pathways map with higher scores are shown in Fig. 10.4. Our results indicated that these target pathways show the marked perturbations over the entire time-course of HI and could contribute to the development of HI. We had also built metabolomic features that geniposide protects against liver injury. Interestingly, geniposide exhibited a hepatoprotective role in liver injury and kept animals in the normal situation because there were no distinct clustering differences between the control and the geniposide group. In order to more clearly characterize the treatment effects of geniposide, the intensity level of nine markers in the different groups were analyzed, and geniposide may regulate the metabolism of these markers efficiently for use in liver injury disorder. Urinary excretion of these metabolites was also shown to have high specificity and sensitivity as markers. It indicated that these metabolites may be the biomarkers which were related to the action mechanism of geniposide.

Development and use of protein regulatory networks to mechanistically interpret this data is an important development in molecular biology, usually captured under the banner of systems biology. Several networks have recently been constructed to help drug discovery (Wang et al., 2012; Rix and Superti-Furga, 2009). Additionally, predicting interactions between drugs and target proteins can help decipher the underlying biological mechanisms. Therefore, in this study, to identify the possible target proteins of geniposide, 2-DE-based comparative proteomics was performed and proteins altered in expressional level after geniposide treatment were identified by MALDI-TOF MS/MS. The hepatoprotective effects of geniposide on acute HI rats was associated with regulated expression of six proteins including zinc finger protein 407, haptoglobin, α-1-antitrypsin, glyceraldehyde-3-phosphate dehydrogenase, transthyretin, and prothrombin. The change in the levels of these proteins caused by geniposide treatment appear to be involved in antioxidation and signal transduction, energy production, immunity, metabolism, and chaperoning. As seen in Fig. 10.7, the association network of differentially expressed proteins in the geniposide-treated group was constructed using STRING. It is important to note that based on our results, geniposide targets not only immunity and metabolism, but also targets key regulatory pathways that are used in transport, signal transduction, cell growth and proliferation, thereby helping to restore the normal function of the liver.

Modern drug discovery is time-consuming and expensive, involving coordinated multidisciplinary research in multiple stages, each requiring intensive and specialized resources. The rapid advancement of "omics" approaches has provided a vast array of significant information in life science. Data relevant to drug discovery are not easily identified and recruited for application to pharmaceutical research. Understanding of the action mechanism of geniposide will be helpful to the study and the use of a likely promising, natural protector. Herein, we illustrate how omics can be utilized to explore the action mechanisms of geniposide that affect different "key pathway." It is the first demonstration of the omic approach to delineate metabolic changes in liver injury after dosing geniposide treatment. Geniposide has a hepatoprotective effect and keeps animals in the normal situation via multiple pathways including primary bile acid biosynthesis, and TCA cycle pathways, etc. The potential target identification problem can be solved effectively by our method, for extracting differentially regulated sub-networks from protein–protein interaction networks based on data from global quantification technologies, which may become an effective strategy for the underlying mechanism of disease. Six proteins whose expressions were significantly changed under geniposide treatment were identified and might be considered as the possible direct targets of geniposide. We conclude that omics is a powerful and versatile tool for both biomarker discovery, and for exploring the complex relationships between biological pathways and drug response, and is anticipated to make an important contribution for elucidating the underlying biological mechanism of actions at omics levels toward drug discovery research in this area.

References

Gersch, M., Kreuzer, J., Sieber, S.A., 2012. Electrophilic natural products and their biological targets. Nat. Prod. Rep. 29 (6), 659–682.

Ioannidis, J.P., Khoury, M.J., 2011. Improving validation practices in "omics" research. Science 334 (6060), 1230–1232.

Ma, T., Huang, C., Zong, G., Zha, D., Meng, X., Li, J., Tang, W., 2011. Hepatoprotective effects of geniposide in a rat model of nonalcoholic steatohepatitis. J. Pharm. Pharmacol. 63 (4), 587–593.

Price, J.V., Tangsombatvisit, S., Xu, G., Yu, J., Levy, D., Baechler, E.C., Gozani, O., Varma, M., Utz, P.J., Liu, C.L., 2012. On silico peptide microarrays for high-resolution mapping of antibody epitopes and diverse protein-protein interactions. Nat. Med. 18 (9), 1434–1440.

Rix, U., Superti-Furga, G., 2009. Target profiling of small molecules by chemical proteomics. Nat. Chem. Biol. 5 (9), 616–624.

Wang, L., Zhou, G.B., Liu, P., Song, J.H., Liang, Y., Yan, X.J., Xu, F., Wang, B.S., Mao, J.H., Shen, Z.X., Chen, S.J., Chen, Z., 2008. Dissection of mechanisms of Chinese medicinal formula Realgar-Indigo naturalis as an effective treatment for promyelocytic leukemia. Proc. Natl. Acad. Sci. USA 105 (12), 4826–4831.

Wang, X., Zhang, A., Sun, H., Wu, G., Sun, W., Yan, G., 2012. Network generation enhances interpretation of proteomics data sets by a combination of two-dimensional polyacrylamide gel electrophoresis and matrix-assisted laser desorption/ionization-time of flight mass spectrometry. Analyst 137 (20), 4703–4711.

Yang, D., Yang, J., Shi, D., Deng, R., Yan, B., 2011. Scoparone potentiates transactivation of the bile salt export pump gene and this effect is enhanced by cytochrome P450 metabolism but abolished by a PKC inhibitor. Br. J. Pharmacol. 164 (5), 1547–1557.

Zhang, A., Sun, H., Wang, X., 2013a. Recent advances in natural products from plants for treatment of liver diseases. Eur. J. Med. Chem. 63, 570–577.

Zhang, A., Sun, H., Wu, G., Sun, W., Yuan, Y., Wang, X., 2013b. Proteomics analysis of hepatoprotective effects for scoparone using MALDI-TOF/TOF mass spectrometry with bioinformatics. OMICS 17 (4), 224–229.

Zhang, A., Sun, H., Wang, X., 2014. Urinary metabolic profiling of rat models revealed protective function of scoparone against alcohol induced hepatotoxicity. Sci Rep 4, 6768.

CHAPTER 11

Metabolic Profiling and Potential Biomarkers of ShenYinXu Syndrome and the Therapeutic Effect of Liuweidihuang Wan

Xijun Wang, Ping Wang, Aihua Zhang, Hui Sun

11.1 INTRODUCTION

Liuweidihuang Wan (LW) a well-known pill for invigorating ShenYinXu (SYX, kidney-yin deficiency) Syndrome, was first recorded in "Xiao er Yao Zheng Zhi Jue," consisting of *Radix Rehmanniae Preparata, Fructus Macrocarpii, Rhizoma Dioscoreae Oppositae, Poria, Rhizoma Alismatis,* and *Cortex Moutan Radicis* with dose proportion of 8:4:4:3:3:3 (Yuhong et al., 2013). The first three crude drugs are known as Sanbu' (SB) that invigorate yin of the kidney, liver, and spleen, and the rest as Sanxie (SX), which attenuate the effects of invigoration. *R Rehmanniae Preparata*, as the principle drug in the famous formulae LW, greatly invigorates kidney yin. Both *F. Macrocarpii* and *R. Dioscoreae Oppositae* serve as the ministerial drugs. *F. Macrocarpii* invigorates yin of kidney and liver to assist *R. Rehmanniae Preparata* in invigorating kidney yin. *R. Dioscoreae Oppositae* invigorates kidney yin and spleen yin. Invigorating spleen yin could also assist *R. Rehmanniae Preparata* in invigorating kidney yin. The three drugs mentioned above could invigorate yin of three organs that are usually called Sanbu.

The kidney controls water metabolism and SYX leads to a deficiency-fire stirring up and malfunctioning in controlling water, so drugs with the effects of eliminating dampness and nourishing yin for lowering fire should be applied. *Poria, R. Alismatis,* and *C. Moutan Radicis* serve as the adjunctive drugs in the formulae. *Poria* could fortify the spleen and drain dampness, which could also assist *R. Dioscoreae Oppositae* in invigorating the spleen. *R. Alismatis* eliminates dampness and turbidity to attenuate the excessive invigoration effect of *R. Rehmanniae Preparata. C. Moutan Radicis* clears deficient heat to prevent ministerial fire from damaging yin. These three drugs could eliminate dampness and clear deficient heat and ministerial fire that are usually called SanXie. In a word, SanBu coordinates with SanXie to replenish kidney essence, and LW is applied to treat SYX in clinic. SYX is a common disease in China, especially for old people. However, with the increasing pace of life, many young people begin to suffer from this disease. Therefore, the present study was undertaken

CONTENTS

Chinmedomics. http://dx.doi.org/10.1016/B978-0-12-803117-9.00011-1

to investigate the metabolic profiling of rats with SYX induced by thyroxine and reserpine, as well as the therapeutic effects of LW, SB, and SX to mine the differentiating metabolites of this disease and to explore the therapeutic mechanism of LW on the metabolomic levels (Wang et al., 2010).

11.2 EXPERIMENTAL

11.2.1 Chemicals and Reagents

Six crude drugs: *R. Rehmanniae preparata*, *R. Dioscoreae Oppositae*, *F. Macrocarpii*, *Poria*, *R. Alismatis*, and *C. Moutan Radicis*, were purchased from the Harbin Tongrentang Drugstore, and were authenticated by Prof. Xijun Wang of the Department of Pharmacognosy, at the Heilongjiang University of Chinese Medicine. The crude drugs were made into honeyed pills (LW, SB, SX) according to the instructions recorded in the *Pharmacopoeia of the People's Republic of China* (2005 edn). The three types of pills were then cut into pieces, immersed 10 times in methanol for an hour each time, and ultrasonically extracted three times for 40 min each. Finally the filtrate was combined, and the solvents were recycled. The extracts of LW, SB, and SX pills were dissolved in distilled water with concentrations of crude drugs at 1.2 g/mL, 0.768 g/mL, and 0.432 g/mL respectively, and stored at 4°C for animal experimental usage. Comparison of fingerprint profiles of nine batches of LW samples by UPLC/ESI-HDMS is shown in Fig. 11.1.

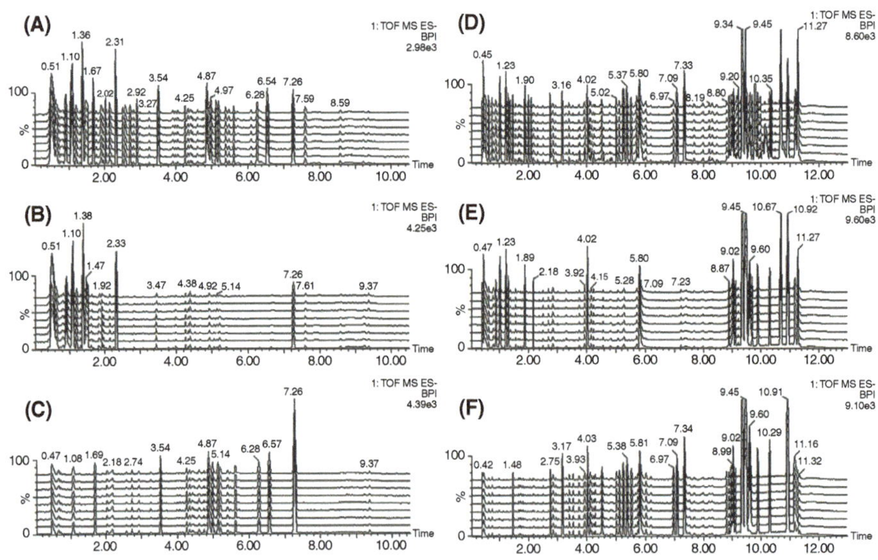

FIGURE 11.1 BPI chromatograms of LW, SB, and SX samples for reproducibility.
(A) QC-LW (ESI$^-$); (B) QC-SB (ESI$^-$); (C) QC-SX (ESI$^-$); (D) QC-LW (ESI$^+$); (E) QC-SB (ESI$^+$); (F) QC-SX (ESI$^+$).

Thyroxine tablets were purchased from Shanghai Industrial United Holdings Great Wall Pharmaceutical Co. Ltd (Shanghai, China); reserpine injections were purchased from Guangdong Bangmin Pharmaceutical Co. Ltd (Guangdong, China). The powder of the thyroxine tablets and the reserpine injection were dissolved in distilled water with concentrations at 16 mg/mL and 0.1 mg/mL, respectively. Pentobarbital sodium was purchased from the Shanghai Chemical Agent Company of China Medicine Clique (Shanghai, China); the assay kits for estradiol (E2), testosterone (T), thyroxine (T4), triiodothyronin (T3), and cortisol (Cort) were purchased from the Beijing North Biotechnology Institute (Beijing, China); the assay kits for cAMP and cGMP were purchased from the Shanghai University of Chinese Medicine (Shanghai, China); the assay kits for immunoglobulin M (IgM) and immunoglobulin G (IgG) were purchased from the Shanghai Sailor Biotechnology Ltd Co. (Shanghai, China).

11.2.2 Animal Handling and Sampling

Sixty male Wistar-derived rats (age, 6 weeks; bodyweight, 200 ± 20 g) provided by the GLP Centre of the Heilongjiang University of Chinese Medicine, were randomly divided into six groups with 10 rats in each: the control, model I, model II, LW, SB, and SX groups. All animals were kept in the standard environment in the GLP Centre and allowed to acclimatize in metabolism cages for 1 week prior to treatment. The rats in the control group were administrated with distilled water in the whole procedure for 20 consecutive days and anesthetized by intraperitoneal injection of 1% pentobarbital sodium (0.15 mL/100 g body weight) on day 21. Blood was collected from the hepatic portal vein and then was centrifuged at 3500 rpm for 15 min at 4°C. The supernatants were immediately frozen, stored at −20°C and thawed before analysis. The serum was used for biochemical indicator detection according to the instructions of commercial kits. The rats in the model I group were administrated with thyroxine and reserpine for 10 consecutive days, and on day 11 the rats were sacrificed, as in the control group. The rats in the model II group were administrated with thyroxine and reserpine for 10 consecutive days, and then administrated with distilled water for another 10 days, and on day 21 the rats were sacrificed, as in the control group. The rats in the LW, SB, and SX groups were administrated with thyroxine and reserpine for 10 consecutive days, then administrated with extracts of LW (12 g crude drug/kg body weight), SB (7.68 g crude drug/kg body weight), and SX (4.32 g crude drug/kg body weight) respectively for another 10 days, and on day 21 the rats were sacrificed, as in the control group.

Urine was collected daily (8.00 a.m.) from metabolism cages at ambient temperature throughout the whole procedure and centrifuged at 13,000 rpm at 5°C for 15 min, and the supernatants were stored frozen at −20°C until analysis. Although not all the urine would be analyzed using UPLC–HDMS, the urine would be collected every day in order to ensure that all the rats were in the same condition. The changes in environment will lead to changes in metabolic profiling. The "quality control" sample was used for the optimization of UPLC-MS conditions, as it contained most of the information of the urine samples. Every day, after the instrument was calibrated, the QC sample was first analyzed to test the stability of the instrument, making sure the instrument was in the same condition during the whole analytical procedure. The collected samples of rat serum of each group were characterized on a Beckman-Counter Access 2 Immune analyzer by following the clinical chemical parameters according to the corresponding method: T3, T4, Cort, T, E2, cAMP, and cGMP. IgM and IgG were analyzed on an Automated Synchron LX20 Beckman Coulter analyzer according to the instructions of the assay kits.

11.2.3 Metabolomic Analysis Techniques

Urine samples were centrifuged again at 13,000 rpm at 5 °C for 15 min, the supernatants were diluted 10 times with deionized water and 4 μL aliquots were transferred to autosampler vials for analysis. The autosampler was maintained at a temperature of 4 °C for the duration of the analysis. The UPLC-MS analysis was performed on a Waters ACQUITY UPLC system (Waters Corp., Milford, USA) coupled with a Waters Micromass Q-tof microTM Synapt High Definition Mass Spectrometer (Manchester, UK) equipped with electrospray ionization. For the reversed-phase UPLC analysis, the ACQUITY UPLC™ BEH C18 column (100× 2.1 mm i.d., 1.7 μm, Waters Corp., Milford, USA) was used. The column was eluted with a linear gradient of 1–20% B over 1.0–6.0 min, 20–35% B over 6.0–8.5 min, 35–43% B over 8.5–9.0 min, and 43–100% B over 9.0–9.5 min. The composition was held at 100% B for 1.0 min, returned to 1% B for 11.0 min, and then held for 2 min at an eluent flow rate of 500 μL/min (where mobile phase A consisted of 0.1% formic acid in deionized water, and mobile phase B consisted of 0.1% formic acid in acetonitrile).

For the UPLC-HDMS analysis, the optimal conditions were as follows: the source temperature was set at 110 °C with a cone gas flow of 50 μL/h, a desolvation gas temperature of 300 °C and a desolvation gas flow of 500 L/h. In negative ion mode, the capillary voltage was set at 2.4 V, the sample cone voltage was set 30 V and the extraction cone voltage was set 4.0 V. In positive ion mode, the capillary voltage was set at 2.6 V, the sample cone voltage was set 30 V and the extraction cone voltage was set 3.0 V. A scan time of 0.3 s with an inter-scan delay of 0.1 s was used through out, with Trap CE at 6.0 V and Transfer CE at 4.0 V. A lock-mass of leucine enkephalin at a concentration of 200 pg/mL in acetonitrile (0.1% formic acid): H_2O (0.1% formic acid) (50:50, v/v) for the positive ion mode ($[M + H]^+ = 556.2771$) and negative ion mode ($[M - H]^- = 554.2615$) were employed via a lockspray interface. Data were collected in centroid mode, the lockspray frequency was set at 5 s and the lock-mass data were averaged over 10 scans for correction. The mass spectrometric data were collected in full-scan auto mode from 0–13 min in both positive and negative ion mode.

11.2.4 Data Analysis

The data were recorded by MassLynx V4.1 (Waters Corp., Milford, USA), the MarkerLynx Application Manager (Waters Corp., Milford, USA) was used for the peak detection and the EZinfo software was used for the PCA, PLS-DA, and OPLS analyses. The EZinfo software was included in the MarkerLynx Application Manager, and can be applied directly. The biochemical data were analyzed using SPSS 16.0.

11.3 RESULTS

11.3.1 Biochemical Analysis

The sera marker (namely T4, Cort, and E2) of the model I group were increased compared with those of the control group, and levels of T and IgG were decreased significantly compared with those of the control group after thyroxine and reserpine treatment. These biochemical indicators of the model II group, namely T4 and E2, were increased significantly compared with those of the control group, and the levels of T and IgG were decreased significantly compared with those of the control group. These results indicated that the rodent model of SYX was successfully reproduced, and the rats had not recovered 10 days after administration of thyroxine and reserpine ceased, which is consistent with the clinical report that the potency of the two medicines lasted for several weeks after cessation. These biochemical indicators, namely T4, Cort, T, E2, and IgG were significantly regulated back toward their baseline levels as a result of the LW administration, demonstrating that LW might have some therapeutic effect on the rat model. However, SB and SX, components of LW, did not show significantly therapeutic effects in this rat model (only T and IgG were changed significantly in the SB group) indicating that the combination of the six drugs was reasonable (Table 11.1).

11.3.2 Metabolic Profiling Analysis

Using the optimal UPLC-HDMS conditions described in Section 11.2.3, the representative Based Peak Intensity (BPI) chromatograms for the reversed-phase UPLC-HDMS analysis of urine samples collected from representative rats for each of the different groups are presented in Figs 11.2 and 11.3. Urine samples on the last day from each group were used for UPLC-HDMS analysis, which was used for model I group, it was on day 11, for the others, on day 21. Some differences could be visually noted among these chromatograms. Low molecular mass metabolites could be separated well in the short time of 13 min due to the minor particles (sub-1.7_m) of UPLC.

Table 11.1 Blood Biochemical Examinations in Different Groups (mean + SD, *n* = 10)

	T3 (ng/mL)	T4 (ng/mL)	Cort (ng/mL)	T (ng/mL)	E2 (pg/mL)	cAMP (pM/mL)	cGMP (pM/mL)	IgM (g/L)	IgG (g/L)
Con.	0.87 ± 0.20	99.74 ± 16.93	303.15 ± 58.35	4.61 ± 2.06	64.84 ± 18.77	0.23 ± 0.03	0.032 ± 0.007	1.28 ± 0.63	8.47 ± 2.34
Mod I	0.90 ± 0.51	119.54 ± 27.20*	383.57 ± 48.03**	2.66 ± 0.93*	82.22 ± 20.16*	0.24 ± 0.02	0.034 ± 0.009	0.70 ± 0.22	4.78 ± 1.30**
Mod II	0.89 ± 0.40	117.75 ± 17.84*	323.96 ± 59.65	2.93 ± 1.61*	81.08 ± 14.96*	0.23 ± 0.03	0.034 ± 0.007	1.01 ± 0.71	5.19 ± 1.36**
LW	0.86 ± 0.16	99.97 ± 14.83#	306.17 ± 55.14#	5.02 ± 2.19#	65.43 ± 12.56#	0.23 ± 0.03	0.033 ± 0.007	1.41 ± 0.96	7.68 ± 1.18##
SB	0.81 ± 0.17	107.46 ± 21.84	345.87 ± 47.40	4.67 ± 1.80#	72.10 ± 14.08	0.22 ± 0.03	0.040 ± 0.009	1.12 ± 0.53	7.19 ± 1.64##
SX	0.92 ± 0.37	102.14 ± 23.47	316.27 ± 60.70	4.16 ± 1.37	77.06 ± 13.29	0.25 ± 0.03	0.040 ± 0.004	0.97 ± 0.34	5.59 ± 1.93

**Significant difference from control at* $p < 0.05$.
***Significant difference from control at* $p < 0.001$.
#Significant difference from model I at $p < 0.05$.
##Significant difference from model II at $p < 0.001$.

11.3.3 PCA Analysis

In order to clearly differentiate among groups, unsupervised pattern recognition PCA, a chemometric model that reduces a matrix of data to its lowest dimension of the most significant factors, was used for analyzing the chromatographic data. Before the urinary data was subjected to the PCA model using EZinfo software, the ions that were not contained in the urine sample of the control group were excluded. Therefore, the separations could not be derived by the metabolites of the three agents (LW, SB, SX). PCA score plots separated urine samples into different blocks, and samples subjected to the same treatment were located on the same trajectory, indicating that treatments have greatly disturbed the normal urine metabolic profiles of rats (Fig. 11.4). Urine from the model I group on day 11 was the furthest from the control group, suggesting the metabolic profiles have significantly changed as a result of thyroxine and reserpine administration. Rats in the model II group were administrated with thyroxine and reserpine for 10 consecutive days, then administrated with water for another 10 days and urine on day 21 was analyzed as the samples of model II group. The perturbation of metabolic profiles of the model II group shifted slightly toward baseline, however, it was still separated from the control group, demonstrating that the perturbation was not completely eliminated after stimulus from thyroxine and reserpine was removed, which was consistent with a clinical report that the potency of the two medicines lasts for several weeks after administration. Urine of the SX-treated group on day 21 was located near the

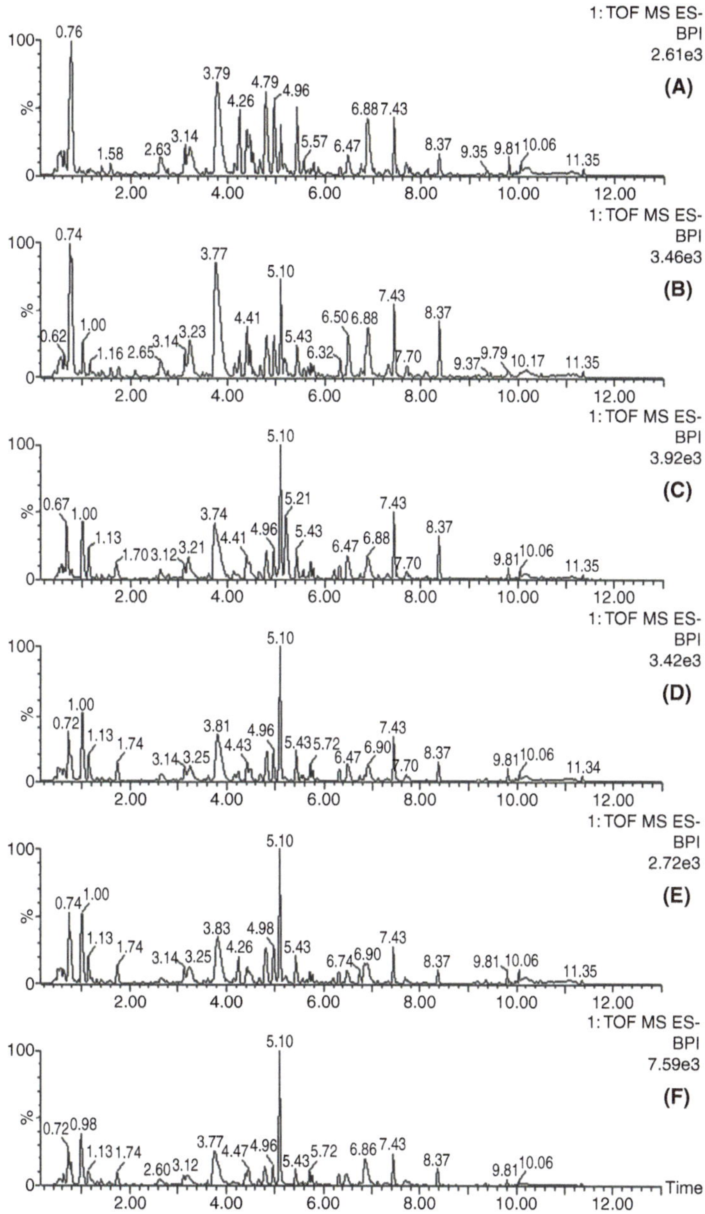

FIGURE 11.2 BPI chromatograms of: (A) control rats; (B) model rats; (C) LW-treated rats; (D) SB-treated rats; (E) SX-treated rats; (F) model II rats urine samples on the last day analyzed by LC-MS in ESI$^-$ mode.

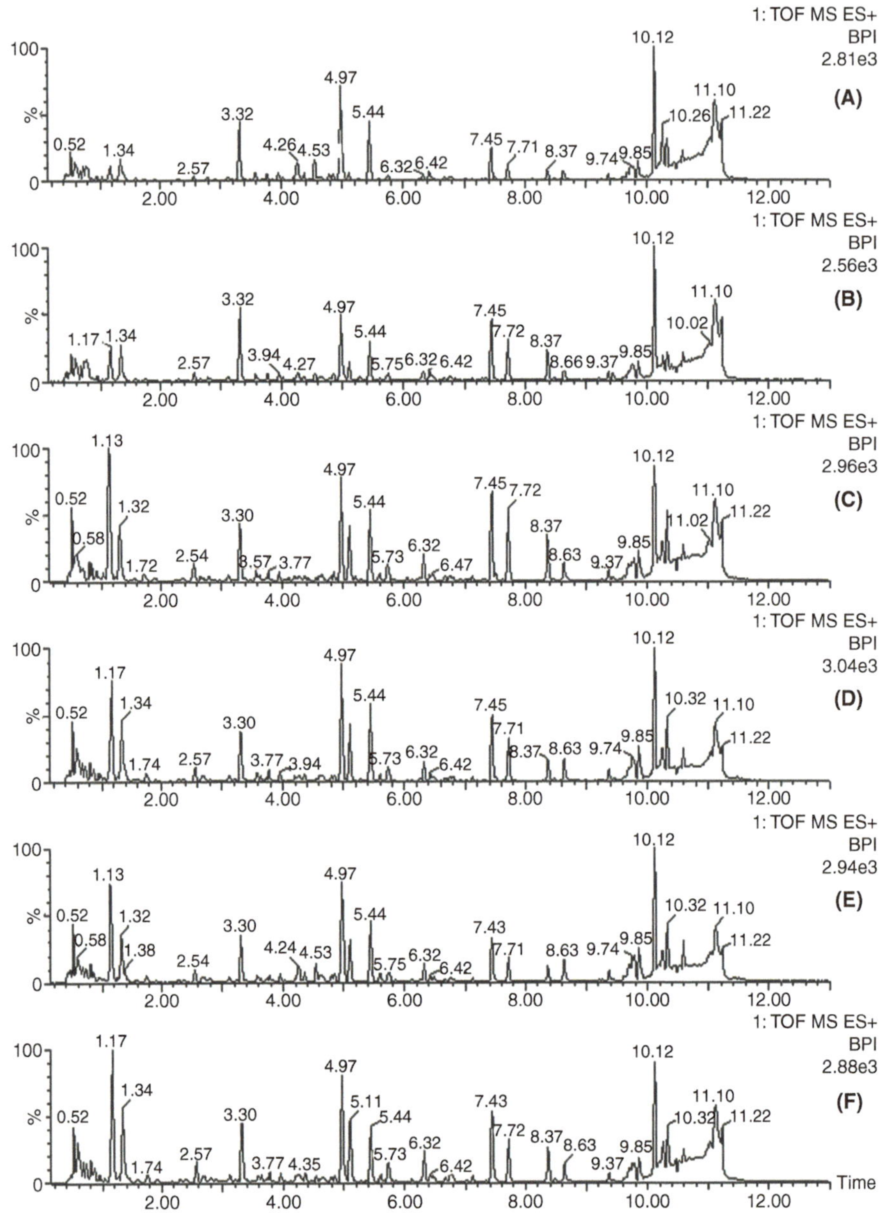

FIGURE 11.3 BPI chromatograms of urine samples of: (A) control rats, (B) model I rats, (C) LW-treated rats, (D) SB-treated rats, (E) SX-treated rats, and (F) model II rats on the last day analyzed by LC-MS in ESI$^+$ mode.

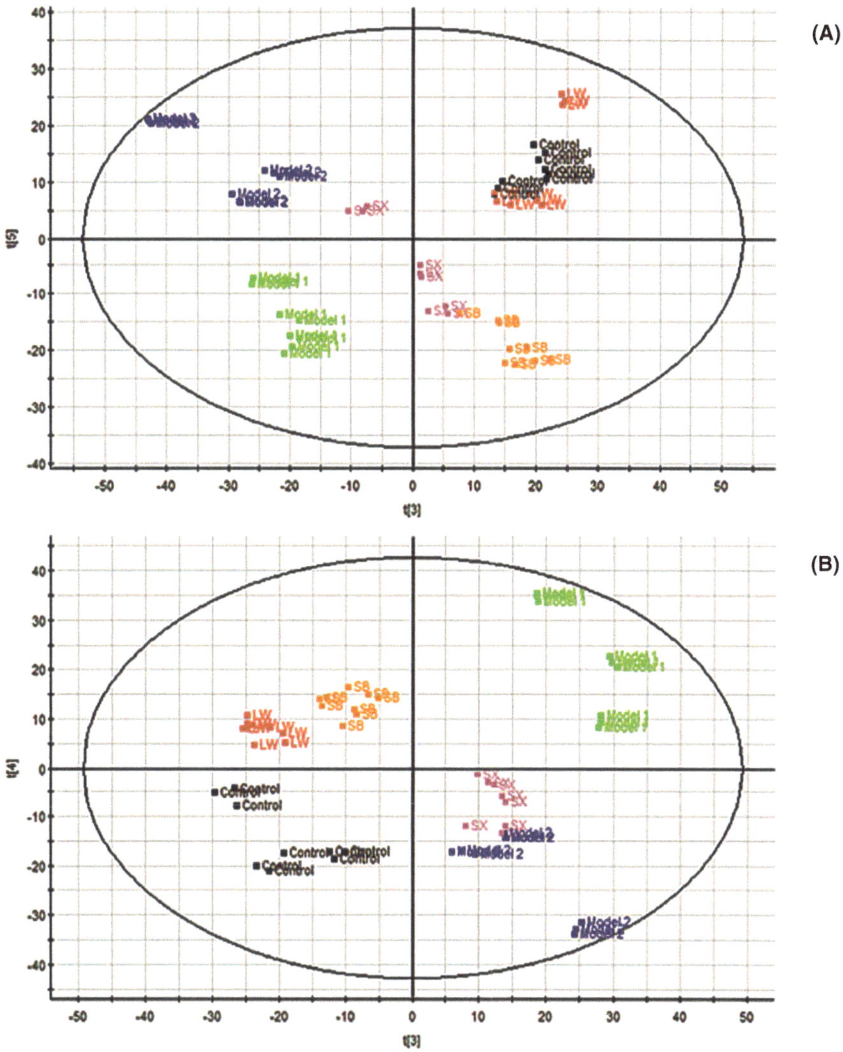

FIGURE 11.4 PCA score plots of urine samples collected from different treatment groups of rats: (A) in ESI⁻ mode; (B): in ESI⁺ mode. (■ control; ■ LW; ■ model 1; ■ model 2; ■ SB; ■ SX)

model I group, indicating they could have similar metabolic profiles. Urines of the LW and SB-treated groups on day 21 were located near the control group, especially the LW-treated group. Some of the samples in the control and LW groups clustered, suggesting that the LW, and control groups could have similar metabolic profiles, and that LW could assist recovery of the animals from earlier thyroxine and reserpine treatment. The PCA analysis of the chromatographic data identified the control, and the treated rats based on the differences in their metabolic profiles, demonstrating that a rodent model of SYX was successfully reproduced, and the classic formulae of LW showed significant therapeutic effects on the rodent model, which was consistent with blood biochemistry examinations. SB and SX as part of the classic formula did not show noticeable therapeutic effects on the rodent model, properly indicating that the combination of the six drugs was reasonable.

11.3.4 OPLS-DA Analysis

In order to obtain a better discrimination between the control and model I groups, the OPLS-DA approach was applied to the metabolic profiles. OPLS-DA score plots separated urine samples into two blocks, especially in the component P1 direction, and the component P2 properly explained individual variation in each group (Fig. 11.5). Loading plot of OPLS-DA indicated the relative content of each ion in the control and model I groups; the *Y*+ axis represented the model I group; the *Y*− axis represented the control group; the *X*-axis represented the number of detected ions (Fig. 11.6). To exhibit the responsibility of each ion for these variations more intuitively, S-plots and VIP-value plots were combined (Fig. 11.7). The black point graph is the S-plot, and most of the ions were clustered around the origin point; only a few of them scattered in the margin region, and just these few ions contributed to the clustering observed in the score plot and were also the differentiating metabolites. The red point graph is the VIP-value plot, which represents the value of each ion. The farther away from the origin, the higher the VIP value of the ions was. The black points and red points were in one-to-one correspondence in the combination.

11.3.5 Identification of Differentiating Metabolites

All the detected ions were arranged in descending order according to VIP values, and the highest VIP value was 7.84 in the negative mode. Combining the results of the OPLS analysis with the amount variation of ions in each group, 24 ions with VIP values exceeding 3 were selected in the negative mode. At the same time, the highest VIP value was 9.46 in the positive ion mode, and 19 ions with VIP values exceeding 3 were selected. The UPLC-HDMS segregation analysis platform provided the retention time, precise molecular mass, and MS/MS data for the structural identification of biomarkers. The precise molecular mass was determined within measurement errors (<5 ppm) by Q-TOF, and meanwhile, the potential elemental composition, degree of unsaturation, and fractional isotope abundance of compounds were obtained. The presumed molecular formula was searched in Chemspider, the Human Metabolome Database, and other databases to identify the possible chemical constitutions, and MS/MS data were screened to determine the potential structures of the ions.

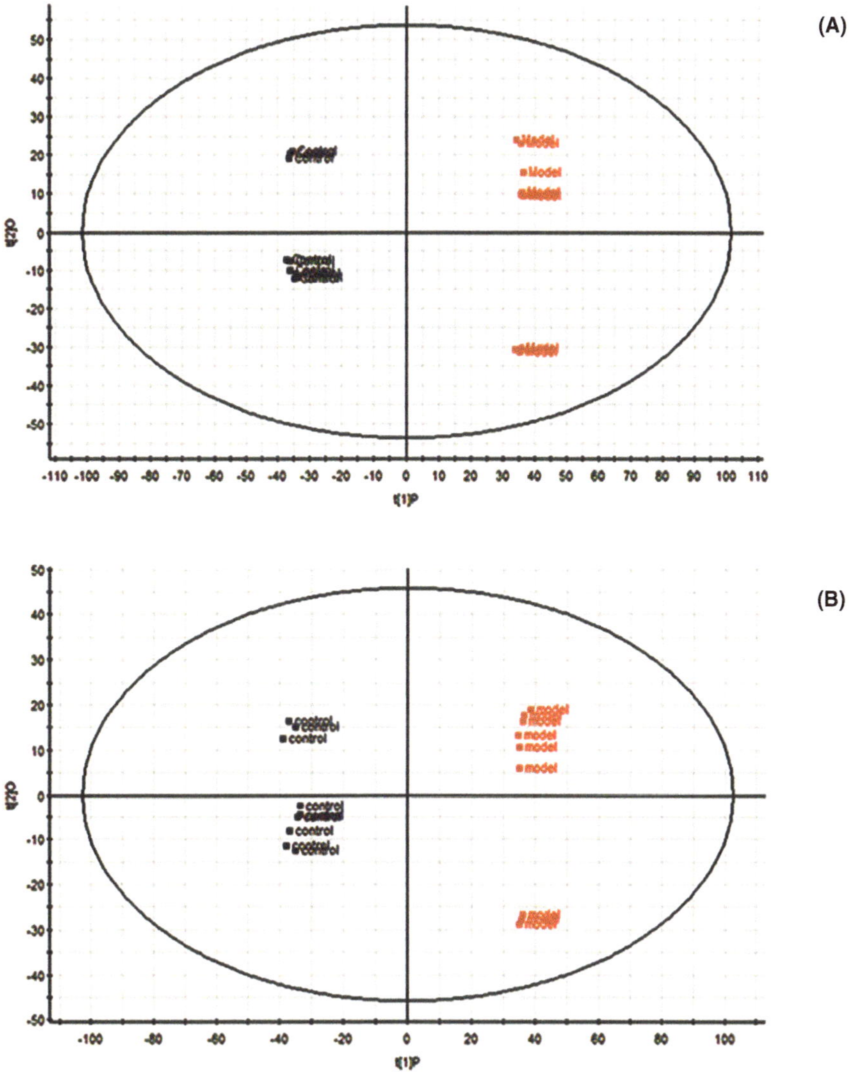

FIGURE 11.5 OPLS score plot of the control and model groups in: (A) ESI^- mode and (B) ESI^+ mode. (■ control; ■ model)

According to the protocol detailed above, the structures of 14 differentiating metabolites were preliminary identified in the negative mode and eight in positive mode (2 in both modes) (data not shown). Taking two ions as examples, the identification procedure was as follows. In the negative mode, the ion at $R_t = 5.46$ and $[M - H]^- = 162.0548$ has a high VIP value. This ion might contain an odd number of nitrogen atoms because its precise molecular weight

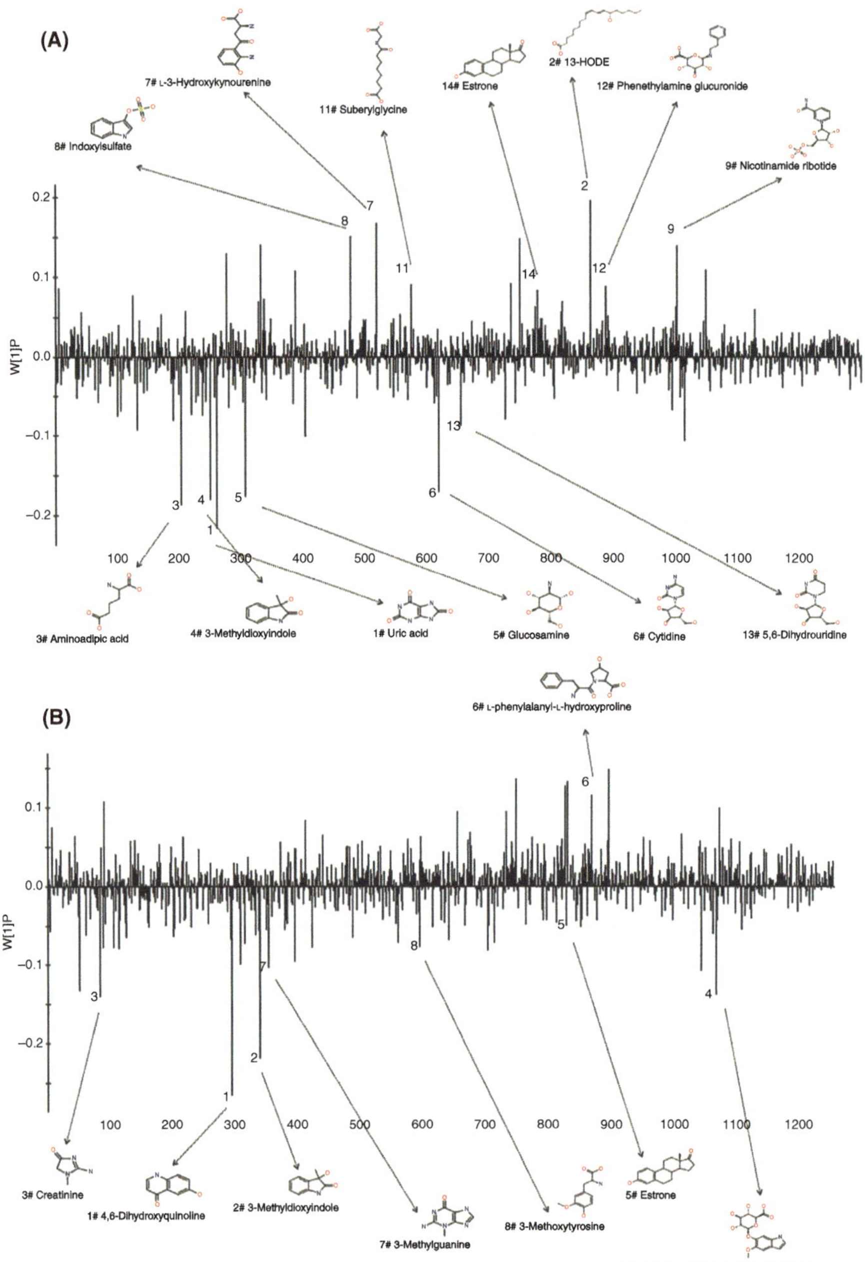

FIGURE 11.6 Loadings for component 1 in: (A) ESI^- mode and (B) ESI^- mode.

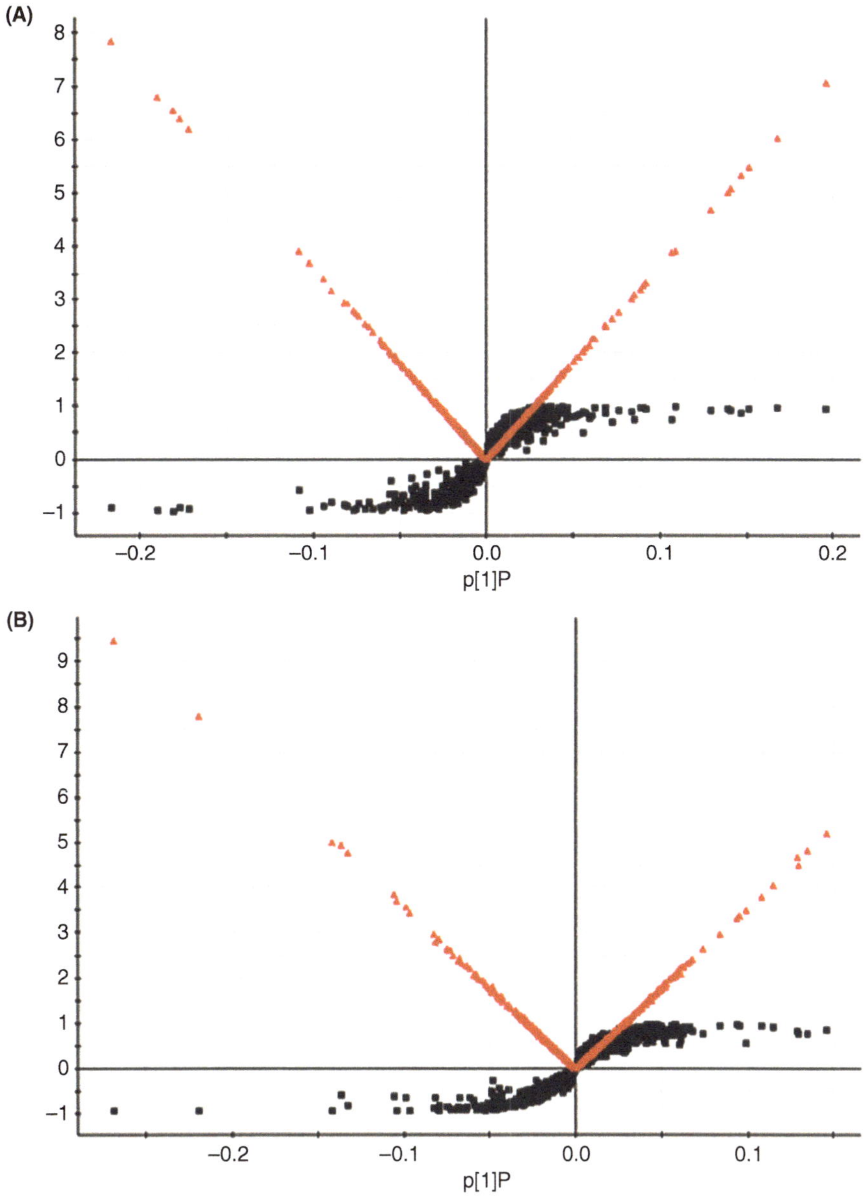

FIGURE 11.7 Combination of S-plot and VIP values in: (A) ESI$^-$ mode and (B) ESI$^+$ mode. (■ ions; ▲ VIP)

was 163.0633, and its molecular formula was speculated as $C_9H_9NO_2$ from the analysis of its elemental composition and fractional isotope abundance. The degree of unsaturation was calculated as 7, indicating that it was a ring compound. The main fragment ions analyzed by MS/MS screening were *m/z* 147, 144, 132, and 120, which could be the $[M - H]^-$ of lost $-CH_3$, $-H_2O$, $-CH_2O$, and –CNO, respectively. Finally, it was speculated as 3-methyldioxyindole when researched in databases, and its mass spectrum, and structure are displayed in Fig. 11.8. In the same way, the ion at $R_t = 4.96$ and $[M + H]+ = 162.0502$ has the highest VIP value, and it was primarily indentified as 4,6-dihydroxyquinoline

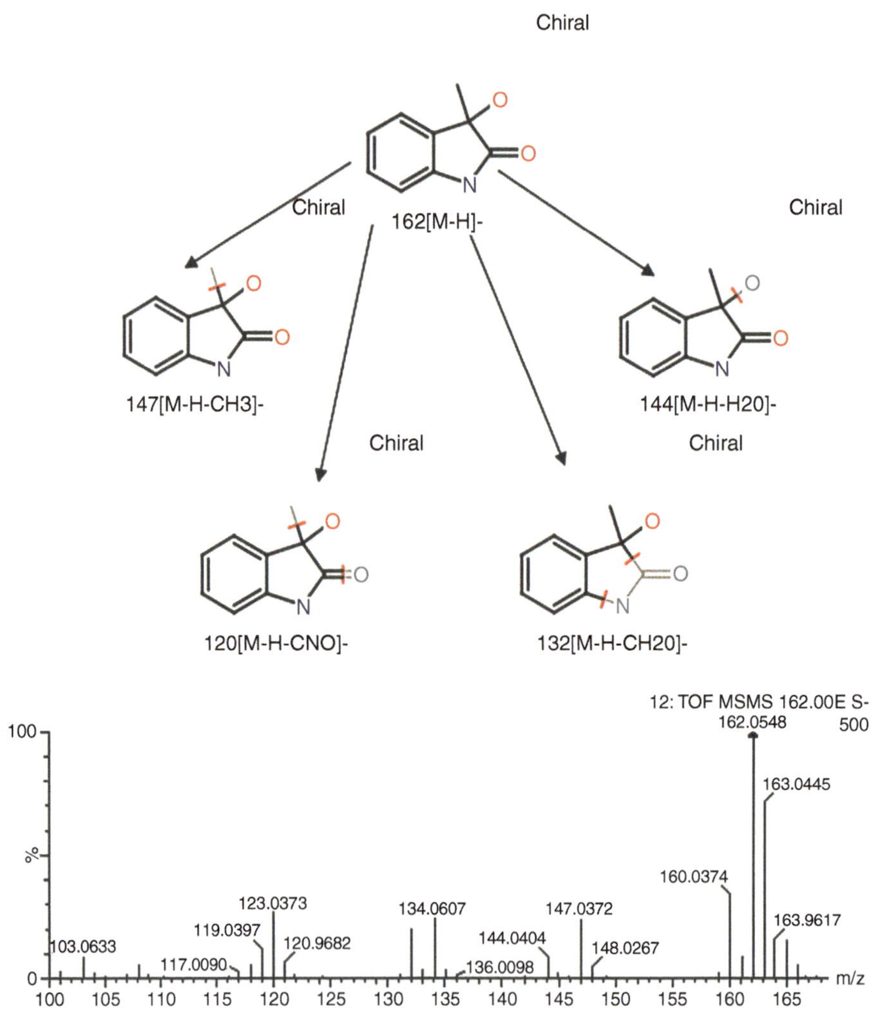

FIGURE 11.8 Chemical structure and mass fragment information of 3-methyldioxyindole in ESI^- mode.

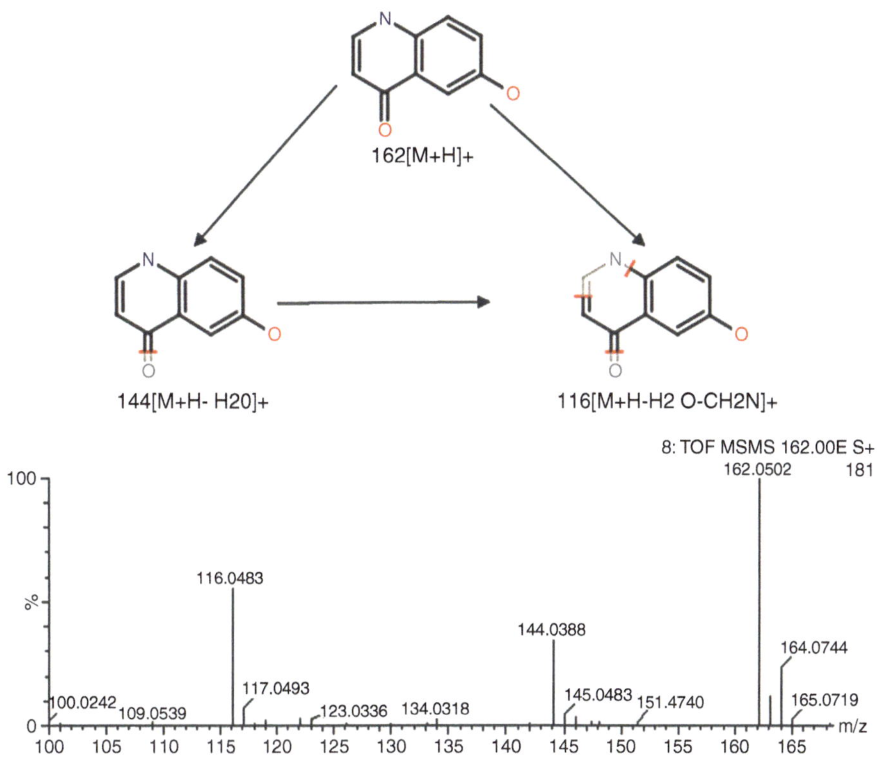

FIGURE 11.9 Chemical structure and mass fragment information of 4,6-dihydroxyquinoline in ESI$^+$ mode.

in the positive mode after searching in the database, and the main fragments were *m/z* 144 and 116, which could correspond to lost $-H_2O$ and $-CH_2N$, respectively. Its mass spectrum and structure are displayed in Fig. 11.9. The metabolic pathways of biomarkers related to SYX are explained in the discussion below.

11.4 DISCUSSION

The model of SYX is a classic symptom model in Chinese medical science, which was initially studied in the 1950s. To date there are eight methods to reproduce this animal model, such as thyroxine, thyroxine with reserpine, hypodesis of the renal artery, febricity herb medicine, and others (Wang, 2005) among which thyroxine with reserpine is the most common approach, and thus it was employed in the present study. The basic action of thyroxine is to induce synthesis of neonatal protein and regulate the metabolism of protein, fat, carbohydrates, water, salt, and vitamins. However, if thyroxine is administrated in higher doses, it can lead to dyssynchrony between energy metabolism and physical function, causing iatrogenic hyperthyroidism, whose clinical manifestation is impatience, trembling, hyperhidrosis, heat avoidance, dry mouth, polydipsia, polyphagia, magersucht, palpitations, and a powerful-sounding heart beat. Furthermore, it can also immunocompromise animals, causing pathological changes in the immune organs, hypothalamus-pituitary gland-adrenal glands axis, thyroid gland axis, and gonad axis functional disorder, and dysfunction in autonomic nerves. All of these symptoms are consistent with the clinical manifestation of SYX, and therefore, it was feasible to use thyroxine to replicate this disease model. As for reserpine, it could not influence the actions of thyroxine directly but causes hyperfunction of the parasympathetic nerve to assist in reproducing yin deficiency.

Blood biochemistry examinations revealed that sera markers (namely T4, Cort, E2, T, and IgG) of the model I group were significantly changed compared with those of the control group, and these biochemical indicators of the model II group were still different from those of the control group after thyroxine and reserpine treatment, indicating that the rodent model of SYX was successfully reproduced, and that the baseline was not restored 10 days after ending the administration of thyroxine and reserpine. These results are consistent with the clinical report that the potency of the two medicines lasted for several weeks after ending the administration. The marker sera (namely T4, Cort, E2, T, and IgG) returned to baseline when the model rats were subjected to LW; however, SB and SX as components of LW did not show significant therapeutic effects in this rat model, which properly demonstrated the efficiency of the LW formula.

In order to investigate the pathological essence of SYX induced by thyroxine and reserpine as well as the therapeutic effects of LW, SB, and SX, a metabolomic approach based on the UPLC–HDMS technique was employed to demonstrate the urinary metabolic characteristics. Hence, chemometrics such as PCA and OPLS were applied to metabolic data to mine the differentiating metabolites that contributed to the clustering observed in the score plots. Finally, 20 differentiating metabolites were preliminary identified in the negative and positive modes, which had relationships with the metabolic pathways of tryptophan, lysine, purine, pyrimidine, tricarboxylic acid cycle, and others. In the negative mode, uric acid was the most responsible ion for discrimination; as the terminal metabolite of purines, it is excreted by the kidney and the intestinal tract. The content of uric acid increased greatly in the rat model, suggesting that the kidney glomerulus suffered an inflammatory reaction (Kim et al., 2014) and SYX might have some relationship with the kidney organ. 13S-hydroxyoctadecadienoic acid (13-HODE) is synthesized in the body from linoleic acid, participates in cell multiplication in several systems, and has a close relationship with cell function (Spindler et al., 1997). The 13-HODE content decreased in the rat model, demonstrating that changes in cell function might affect the metabolism of 13-HODE directly. Lysine plays important roles in the absorption and utilization of protein, nutritional balance, improvement of nervous system function, enhancement of immune system function, prevention of osteoporosis, and other systems (Behr-Roussel et al., 2000). Normally, the metabolism of lysine is stable and cannot be affected by the dose. In the rat model of SYX, the content of aminoadipic acid, a metabolite of lysine, increased significantly, suggesting that the administration of thyroxine, and reserpine might have caused dysfunction that changed the metabolism of lysine. 3-Methyldioxyindole is the main metabolite of 3-methylindole and was increased in the rat model, but the relationship between this metabolite and SYX is still uncertain. Glucosamine, a metabolite of aminosaccharide, participates in energy metabolism. Its level greatly increased in the model group, indicating that the energy metabolism of the rat model was disordered, which is consistent with the clinical report of SYX. Cytidine is a nucleoside that is composed of the base cytosine linked to the 5-carbon sugar D-ribose. Cytidine is a pyrimidine that, besides being incorporated into nucleic acids, can serve as substrate for the salvage pathway of pyrimidine nucleotide synthesis. Metabolic disorders of pyrimidine suggest that congenital enzyme defects, acquired drugs or foods result in increases in the metabolites of pyrimidine in blood or urine. Cytidine as a metabolite of pyrimidine increased greatly in the urine of the model group, indicating a metabolic disorder of pyrimidine induced by thyroxine and reserpine. L-3-hydroxykynurenine (L-3-HK) is a metabolite in the kynurenine pathway, the major route of tryptophan degradation in mammals, and indoxyl sulphate is a metabolite in another tryptophan degradation pathway. Tryptophan is one of the eight essential amino acids in the body, playing a fundamental role in physiology and biochemistry, and can stabilize polyribosomes to regulate humoral immunity. The urine contents of L-3-HK and indoxyl sulphate decreased obviously, suggesting that the rats were immuno-compromised. Nicotinamide ribotide (NMN) is an important intermediate metabolite in the nicotinate and nicotinamide metabolic pathway. Mammals predominantly use nicotinamide rather than nicotinic acid as a precursor for NAD biosynthesis. Instead of the deamidation to nicotinic acid, nicotinamide is directly converted to NMN by nicotinamide phosphoribosyltransferase. NMN is an important metabolite for the maintenance of normal NAD biosynthesis, and circulating NMN levels may play an important role in regulating cell function in physiological and pathophysiological conditions. However, NMN decreased significantly in the urine of the rat model and resulted in changes to the normal cell function and substance metabolism. Aconitic acid is an intermediate metabolite in the tricarboxylic acid cycle. The decrease in its level in the model urine was a result of a functional disorder of mitochondria. Suberylglycine, an acyl glycine, is normally a minor metabolite of fatty acids. The decreased content in the model urine demonstrated that its metabolism was disordered, which was consistent with clinical reports of SYX. 5,6-Dihydrouridine is a pyrimidine that is the result of adding two hydrogen atoms to a uridine. Dihydrouridine is found only in tRNA molecules. The presence of 5,6-dihydrouridine in the urine suggests an increased degradation rate of tRNA (Topp et al., 1993). The main biological functions of tRNA are to transport the active amino acid and participate in the biosynthesis of proteins. The urine content of 5,6-dihydrouridine increased greatly in the model, indicating the degradation rat of tRNA was improved, which would affect the biosynthesis of proteins and the biological functions. Estrone is a major mammalian estrogen and can be detected in both blood and urine. The serum content of E2 increased in the SYX model, which was consistent with the elevated level of estrone in the urine.

In the positive mode, 4,6-dihydroxyquinoline was the most responsible ion for discrimination and is the product of the conversion of 5-hydroxykynurenamine by the enzyme monoamine oxidase, both metabolites from 5-hydroxytryptophan metabolism. 5-Hydroxytryptophan is a hormone secreted by the pars encephalica that can affect emotion, sleep, and the appetite of humans. 5-Hydroxytryptophan is also the precursor of 5-HT, an inhibitory substance of neural transmission. If the levels of 5-HT decrease, it can cause depression, anxiety, and insomnia. In the present study, the content of 4,6-dihydroxyquinoline increased in the urine of the rat model, so the metabolism of 5-HT might be enhanced, which would affect neural regulation to induce changes in emotion. Creatinine is a breakdown product of creatine phosphate in muscle. The loss of a water molecule from creatinine results in the formation of creatinine. Creatinine is transferred to the kidneys by blood plasma, whereupon it is eliminated from the body by glomerular filtration and partial tubular excretion. Creatinine is usually produced at a fairly constant rate by the body. Measuring serum creatinine is a simple test, and it is the most commonly used indicator of renal function. An increase in blood creatinine levels is observed only with marked damage to functioning nephrons (Kikuchi et al., 1981). It has been reported that thyroxine can affect kidney function, and on the other hand, kidney insufficiency can affect the action of thyroxine (Łebkowska et al., 2006). Accordingly after administration of thyroxine and reserpine, the urine content of creatinine increased greatly. However, to date there is no report about using urinary creatinine as an indicator to evaluate renal function. If urinary creatinine did have some relationship with renal function, it could alleviate patient distress to some extent. L-Phenylalanyl-L-hydroxyproline is a proteolytic breakdown product of collagen. Collagen, a structural protein, is an essential component of tissues, and accounts for 25–33% of total proteins and 6% of body weight, and is a component of tissues and organs such as skeleton, ligaments, and cornea. Collagen is the key component that maintains the modality of skin and tissues, and is the raw material to repair the injured tissues. The catabolism of collagen is regulated by hormonal modulation, and hormone metabolic disorders induced by thyroxine and reserpine might lead to disturbances of collagen metabolism. Definitely, the urine content of L-phenylalanyl-L-hydroxyproline decreased significantly in the rat model. 3-Methylguanine is amethylated purine base. Methylated purine bases are known to be present in normal urine and to change under pathological conditions, in particular in the development of leukemia, tumors, and immunodeficiency, by the altered turnover of nucleic acids typical of these diseases (Porcelli et al., 1997). Thyroxine and reserpine lead to dysfunctions of the whole body, so the content of 3-methylguanine was greatly increased in the rat model. 3-Methoxytyrosine is one of the main biochemical markers for aromatic L-amino acid decarboxylase deficiency, an inborn error of metabolism that affects serotonin and dopamine biosynthesis. Patients are usually identified during infancy due to developmental delay, hypotonia and extrapyramidal movements. Diagnosis is based on an abnormal neurotransmitter metabolite profile in cerebrospinal fluid (CSF) and reduced aromatic L-amino acid decarboxylase (AADC) activity in plasma. 3-Methoxytyrosine is elevated in CSF, plasma, and urine. After administration of thyroxine and reserpine, the urinary content of 3-methoxytyrosine increased significantly, which might be induced by disorders of substance and energy metabolism. 3-Methyldioxyindole and estrone were also detected in the negative mode, and the implications are as discussed previously.

All the results demonstrated that administration of thyroxine and reserpine disturbed the metabolic profile of healthy rats, and the 20 ions identified may be useful in differentiating metabolites for the clinical diagnosis of SYX. After administration of LW, SB, and SX, the metabolic profiles returned toward the baseline to different extents, and LW treated rats had the most similar metabolic profile to that of the control rats, indicating the greatest efficiency of the formula LW on the metabolomics level, which provides a novel method for evaluating the formulas of Chinese medicine.

11.5 CONCLUSIONS

In the present study, a metabolomic approach based on the UPLC-HDMS technique and chemometrics method was employed to demonstrate the urinary metabolic characteristics induced by thyroxine and reserpine. Pattern recognition techniques such as PCA applied to the chromatographic data clearly identified the different groups. Twenty differentiating metabolites were preliminary identified using the OPLS analysis. Based on the particular explanations of these ions, the relationship between differentiating metabolites and traditional bioinformatics was determined, which was also supported by blood biochemical examinations. These results suggest that metabolomics can connect the symptom pathogenesis of Chinese medicine with modern theory of medicine correctly, with differentiating metabolites as the apposite link. With the development of databases and analytical techniques, the "material bases" of Chinese medicine symptoms may become more clear, and may explain the scientific connotation of SYX.

References

Behr-Roussel, D., Rupin, A., Simonet, S., Bonhomme, E., Coumailleau, S., Cordi, A., Serkiz, B., Fabiani, J.N., Verbeuren, T.J., 2000. Effect of chronic treatment with the inducible nitric oxide synthase inhibitor *N*-iminoethyl-L-lysine or with L-arginine on progression of coronary and aortic atherosclerosis in hypercholesterolemic rabbits. Circulation 102 (9), 1033–1038.

Kim, I.Y., Lee, D.W., Lee, S.B., Kwak, I.S., 2014. The role of uric acid in kidney fibrosis: experimental evidences for the causal relationship. Biomed. Res. Int. 2014, 638732.

Kikuchi, T., Orita, Y., Ando, A., Mikami, H., Fujii, M., Okada, A., Abe, H., 1981. Liquid-chromatographic determination of guanidino compounds in plasma and erythrocyte of normal persons and uremic patients. Clin. Chem. 27 (11), 1899–1902.

Łebkowska, U., Małyszko, J., Brzósko, S., Łebkowski, W., Małyszko, J.S., Janica, J., Kowalewski, R., Gacko, M., Myśliwiec, M., Walecki, J., 2006. Renal artery resistance index, thyroid hormones, and thyroid volume in the early kidney transplants recipients. Transplant. Proc. 38 (1), 62–65.

Spindler, S.A., Sarkar, F.H., Sakr, W.A., Blackburn, M.L., Bull, A.W., LaGattuta, M., Reddy, R.G., 1997. Production of 13-hydroxyoctadecadienoic acid (13-HODE) by prostate tumors and cell lines. Biochem. Biophys. Res. Commun. 239 (3), 775–781.

Topp, H., Duden, R., Schöch, G., 1993. 5,6-Dihydrouridine: a marker ribonucleoside for determining whole body degradation rates of transfer RNA in man and rats. Clin. Chim. Acta 218 (1), 73–82.

Porcelli, B., Muraca, L.F., Frosi, B., Marinello, E., Vernillo, R., De Martino, A., Catinella, S., Traldi, P., 1997. Fast-atom bombardment mass spectrometry for mapping of endogenous methylated purine bases in urine extracts. Rapid Commun. Mass Spectrom. 11 (4), 398–404.

Wang, Z.R., 2005. Review on experimental studies on Yin-yang theory of traditional Chinese medicine. Zhonghua Yi Shi Za Zhi 35 (1), 22–24.

Wang, P., Sun, H., Lv, H., Sun, W., Yuan, Y., Han, Y., Wang, D., Zhang, A., Wang, X., 2010. Thyroxine and reserpine-induced changes in metabolic profiles of rat urine and the therapeutic effect of Liu Wei Di Huang Wan detected by UPLC-HDMS. J. Pharm. Biomed. Anal. 53 (3), 631–645.

Yuhong, H., Qian, L., Yu, L., Yingqiang, Z., Yanfen, L., Shujing, Y., Shufang, Q., Lanjun, S., Shuxuan, Z., Baohe, W., 2013. An n-of-1 trial service in clinical practice: testing the effectiveness of Liuwei Dihuang decoction for Kidney-Yin Deficiency Syndrome. Evid. Based Complement. Alternat. Med. 2013, 827915.

CHAPTER 12

Metabolic Profiling of Healthy Persons Treated with Liuweidihuang Wan

Xijun Wang, Haitao Lv, Hui Sun, Aihua Zhang

CONTENTS

12.1 INTRODUCTION

Liuweidihuang Wan (LW) (Fig. 12.1) one of the most important TCM formulae consists of six herbs including *Rehmannia glutinosa* Libosch, *Cornus officinalis* Sieb., *Dioscorea opposite* Thunb., *Alisma orientale* Juz, *Poria cocos* Wolf, and *Paeonia suffruticosa* Andrews, is widely used in eastern Asia (Wang et al., 2013). In China, there are hundreds of pharmaceutical companies that can produce LW. It is reported that LW and its constituents exhibit a wide bioactive action, such as improving the rheological characteristic of blood, modulating immunity, scavenging free radical and antidotal effect, antiosteoporosis, enhancing cognition, and ameliorating memory, etc. (Liu et al., 2013; Liang et al., 2014). It has long been used clinically in the treatment of many kinds of diseases with the indications of ShenYinXu (SYX) Syndrome, such as dizziness, tinnitus, weakness, and soreness of the loins and the limbs, emission and sweating, the effectiveness has been well recorded during long-term clinical practice.

Our previous studies have reported upon the pharmacodynamic constituents of LW, a hypothesis has been raised that only those constituents absorbed into the blood postdose could most probably have become bioactive constituents of LW (Wang et al., 2008). As a result, eleven of the constituents were detected and isolated from the rats plasma, furthermore the evaluation of invigorating the kidney, revealed that the bioactivities of morroniside, sweroside, and loganin are the most preferred. However, the mechanism of invigorating the kidneys of healthy persons is not yet fully understood. The main purpose of this work is the use of a metabolomic method to characterize biochemical changes as proofed by variations of the urinary metabolites of pre- and post-dosed LW for healthy persons, so as to reveal the regulation and improvement of metabolic pathways induced by LW. The overall experimental process is shown in the flowchart (Fig. 12.2).

Chinmedomics. http://dx.doi.org/10.1016/B978-0-12-803117-9.00012-3

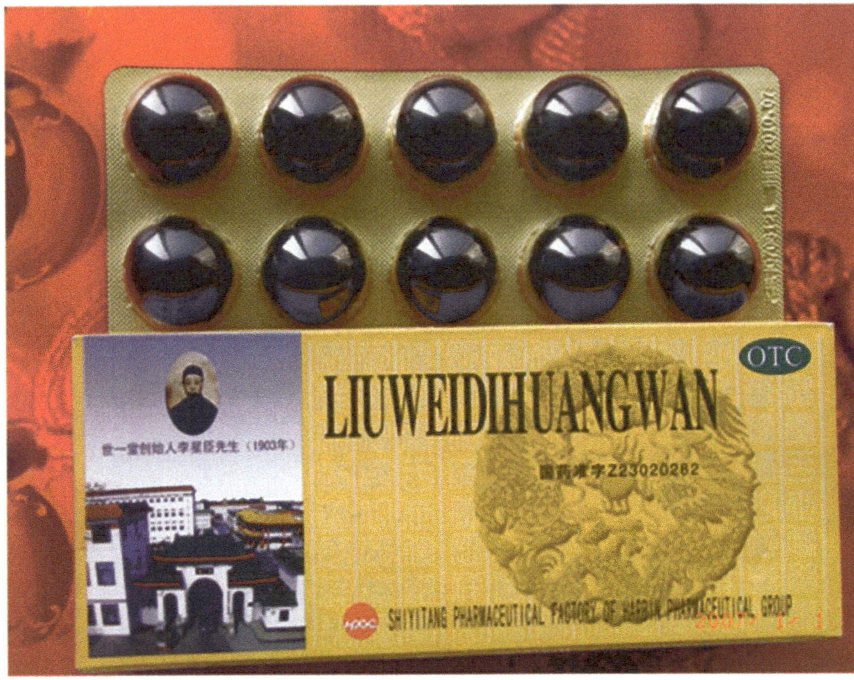

FIGURE 12.1 The product of Liuweidihuang Wan produced by Harbin Shiyitang Pharmaceutical Co. Ltd (Harbin, China).

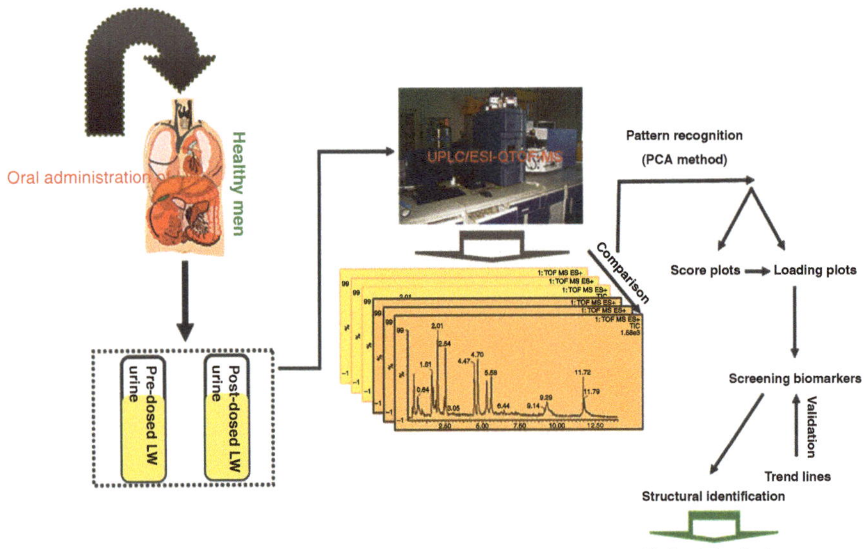

FIGURE 12.2 Overall experimental flowchart of metabolomics used in this study.

12.2 EXPERIMENTAL

12.2.1 Quality Control of LW

The patent product of LW was purchased from Harbin Shiyitang Pharmaceutical Co. Ltd (Harbin, China). High-performance liquid chromatography-diode array detection (HPLC-DAD) fingerprint combined with simultaneous determination of representative compounds have been developed for the quality control of LW. The chromatographic separation was performed on a Water Symmetry Shield™ RP_{18} column (3.9× 150 mm, 5 μm) in Waters 2695 (Milford Corp., MA, USA). Mobile phase was a gradient of acetonitrile to 0.15% phosphoric acid. The effluent was monitored on a DAD detector. Ten separated LW samples were prepared as follows: 1 g of powder of LW were accurately weighed and extracted with 5.0 mL of methanol for 15 min in an ultrasonic bath. The extraction was repeated two times. The extracted solutions were combined and concentrated to dryness, and the residue was accurately dissolved in 10.0 mL of a volumetric flask with methanol, filtered through a 0.45 μm filtration membrane, and then 10 μL of the clear solution that was injected into HPLC. The HPLC fingerprint of 10 separated LW samples was established based on the data 254 nm (Fig. 12.3). In fingerprint, 10 general chromatogram peaks were determined and marked; the peaks four and nine representing respectively loganin and paeonol were simultaneously determined based on the above fingerprint. As a result, the fingerprint resemblances are all better than

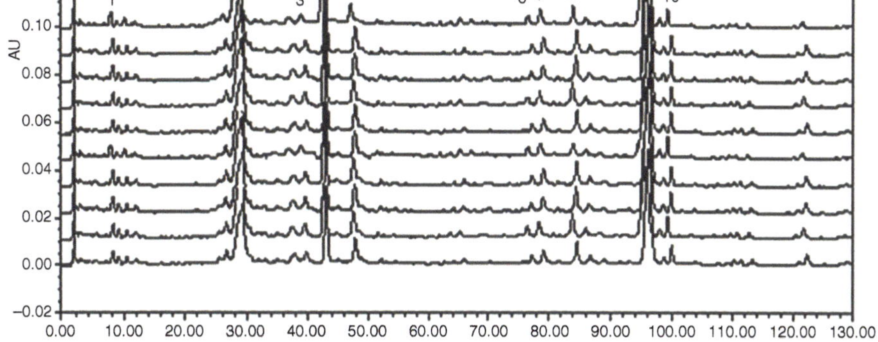

FIGURE 12.3 Representative fingerprint chromatogram of Liuweidihuang Wan.

0.9996, and LW contained 0.37 ± 0.01 mg/g of loganin, and 0.93 ± 0.03 mg/g of paeonol. It illustrates that the quality of LW is stable.

12.2.2 Collection and Preparation of Human Urine Samples

The studies had been approved by the Ethics Review Committee for Human Experimentation of Heilongjiang University of Chinese Medicine. Eight healthy male students (age range 25–30 years) were appointed as research subjects, due to the fact that LW was mainly applied to the male population. All participants gave informed written consent. The control morning urine samples (predosed LW urine samples) were collected from every participant between 5:00 a.m. and 7:00 a.m. before breakfast on day 1. Subsequently LW was orally administrated to all the healthy persons at a dose of 130 mg/kg of body weight between 8:00 a.m. and 9:00 a.m. for the first time, between 5:00 p.m. and 6:00 p.m. for the second time from day 1 to day 7. Individually dosed morning urine samples (postdosed LW urine samples) were collected from healthy persons between 5:00 a.m. and 7:00 a.m. before breakfast on day 9. Food (standard diet) and water (uniform labelled water) were available throughout the study, and the intake of alcohol and drugs was forbidden during the study. Samples were stored at −20°C prior to analysis; and UPLC/MS samples were prepared by centrifugation at 13,000 rpm (5°C).

12.2.3 Metabolomics Analysis Platform

Chromatographic analysis was performed using a Waters ACQUITY UPLC system using a 2.1 × 100 mm ACQUITYC_{18} column (Milford, MA, USA) at a flow rate of 0.4 mL/min and a 5 μL injection volume. The column eluent was directed to the mass spectrometer without split. All samples were eluted using a bilinear gradient, 0–10% B over 1.5 min, 10–20 % B from 1.5 min–4.0 min, 20–35 % B from 4.0 min–6.0 min, and then 35–90 % B, from 6.0 min–10.0 min (A: 0.1% formic acid in water; B: 0.1% formic acid in acetonitrile).

Mass spectrometry analysis was performed on a Micromass QTOF (Waters MS Technologies, Manchester, UK) with an electrospray ionization source (ESI) operating in positive and negative ionization mode. The desolvation gas rate was set to 550 L/h at a temperature of 300°C; the cone gas rate was set at 60 L/h, and the source temperature at 110°C. The capillary voltage and the cone voltage were set at 3200 V and 35 V in positive ion mode, and 2900 V and 60 V in negative ion mode, respectively. Data were collected in centroid mode from m/z 100 to m/z 900 with a lockspray frequency of 0.48 s, and data averaging over 10 scans. The raw data were analysed by the Micromass MarkerLynx applications manager Version 1.0 (Waters, UK). A list of the intensities of the peaks detected is then generated for the first sample, using the retention time (RT) and m/z data pairs as the identifier for each peak. The ion intensities for each peak detected are then normalized, within each sample, to the sum of the peak intensities in that sample.

12.3 RESULTS

12.3.1 UPLC-MS Analysis in Positive Ion Mode

A typical total ion chromatogram (TIC) obtained using UPLC/ESI-QTOF-MS of urine sample obtained from the pre- and postdosed LW for day 1 and day 9 is shown in Fig. 12.4A in positive ionization mode. These chromatograms reveal clear differences between the profiles for the pre- and postdosed LW men. In order to fully visualize and describe the similarities and differences amongst these complex data sets, the raw UPLC-MS data generated on samples from the pre- and postdosed LW men were subjected to chemometric analysis method. From the PCA scores plot (Fig. 12.4B) it is clear that a distinct clustering of pre- and postdosed LW male urine samples has been achieved. The corresponding loadings plot indicated the ions and their chromatographic retention times that were attributable to the clustering observed in the scores plot. The ions in the positive ionization data identified as contributing significantly to the observed statistical variation were determined to be *m/z* 310.23 (5.35 min), 286.23 (4.73 min), 293.18 (4.48 min) and 287.13 (2.54 min) (Fig. 12.4C) which were decreased in predosed LW men urine samples when compared to that in the postdosed LW men urine samples (Fig. 12.4D). The representative extracted ion chromatograms and mass spectra fragments of four ions are shown in Fig. 12.4E. These compounds were considered to be intrinsic ones and were detected from the predosed human urine samples. The structure of *m/z* 310.23 (5.35 min) was preliminarily identified as *N*-acetylneuraminic acid, and the structures of others molecules are currently unknown and are now in progress.

12.3.2 UPLC-MS Analysis in Negative Ion Mode

Figure 10.5A shows the representative TIC chromatograms for the negative ionizations mode UPLC/ESI-QTOF-MS analysis of pre- and postdosed LW human urine samples collected respectively at day 1 and day 9. Although some differences can be visually noted among the two sets of details illustrated in Fig. 12.5A, more subtle variations can be demonstrated using a pattern recognition approach, e.g., PCA. From the PCA scores plot (Fig. 12.5B) it is obvious that a distinct clustering human urine sample of pre- and postdosed LW has been obtained. The corresponding loadings plot showed the ions, and their chromatographic retention times that were attributable to the clustering observed in the scores plot. Among them, three dominant ions *m/z* 212.01 (4.74 min), 107.02 (5.63 min), and 191.03 (0.61 min), respectively (Fig. 12.5C) and the amount of three ions from predosed LW urine samples was statistically different from that of the human urine samples of the postdosed LW. The trend lines for the three ions are shown in Fig. 12.5D. These three compounds were also considered to be intrinsic compounds, and were detected from the predosed human urine samples. The existence ratio of *m/z* 212.01 in the postdosed

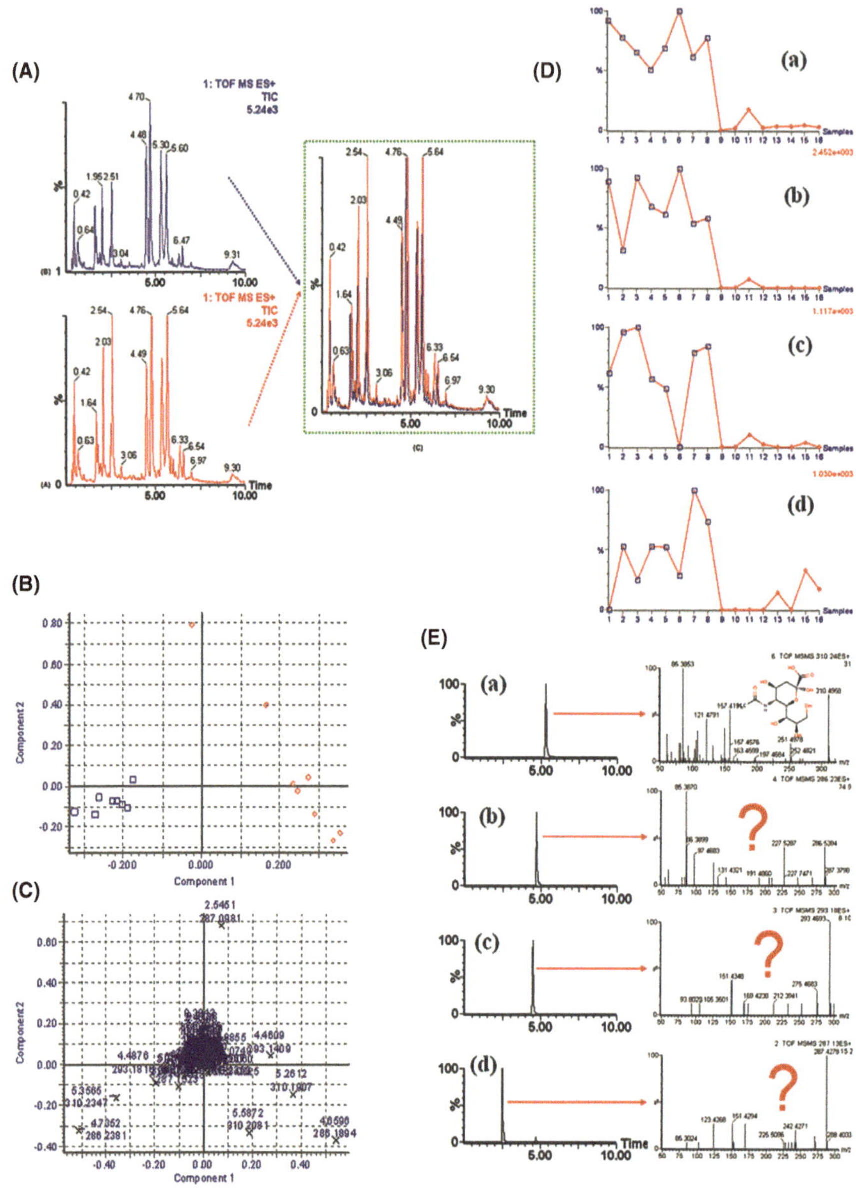

FIGURE 12.4 **UPLC-MS analysis in positive ion mode for urine samples of healthy persons.**
(A) Representive total ion current chromatograms of urine samples in positive ionization mode for post-dosed LW, predosed LW.
(B) Scores plot of PCA data obtained from pre and postdosed LW urine samples of healthy persons. Symbols: (□) Predosed LW urine samples; (◇) Postdosed LW urine samples.
(C) Loadings plot of PCA data obtained from pre and postdosed LW urine samples of healthy persons.
(D) Trends of the (a) *m/z* 310.23 (5.35 min), (b) 286.23 (4.73 min), (c) 293.18 (4.48 min) and (d) 287.13 (2.54 min) ions for pre and postdosed LW urine samples. Symbols: (□) Predosed LW urine samples; (◇) Postdosed LW urine samples.
(E) Extracted ion chromatograms and the corresponding mass spectra fragments of (a) *m/z* 310.23 (5.35 min), (b) 286.23 (4.73 min), (c) 293.18 (4.48 min), and (d) 287.13 (2.54 min) ions using UPLC/ESI-QTOF-MS in positive ion mode.

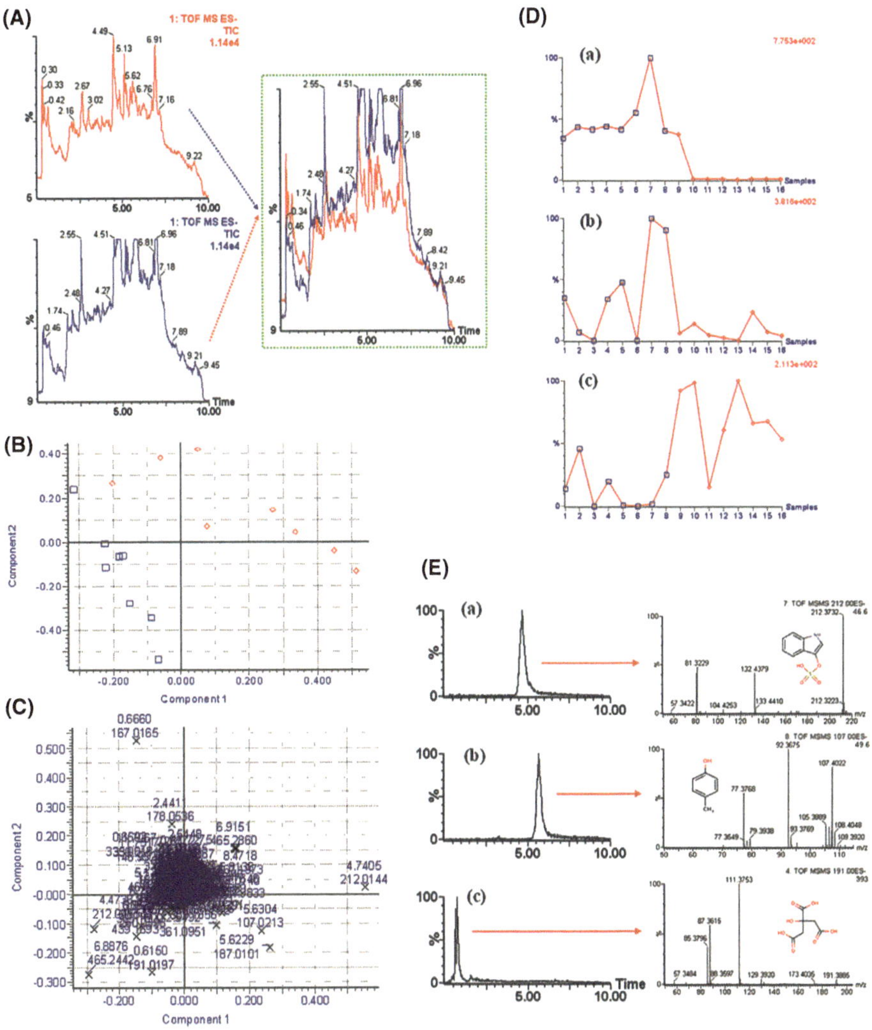

FIGURE 12.5 UPLC-MS analysis in negative ion mode for urine samples of healthy persons.
(A) Typical total ion current chromatograms of urine samples in negative ionization mode for postdosed LW, predosed LW.
(B) Scores plot of PCA data obtained from pre and postdosed LW urine samples of healthy persons. Symbols: (□) Predosed LW urine samples; (◇) Postdosed LW urine samples (B).
(C) Loadings plot of PCA data obtained from pre and postdosed LW urine samples of healthy persons.
(D) Trends of the (a) *m/z* 212.01 (4.74 min), (b) 107.02 (5.63 min), (c) 191.03 (0.61 min) ions for pre and postdosed LW in urine samples. Symbols: (□) Predosed LW urine samples; (◇) Postdosed LW urine samples.
(E) Extracted ion chromatograms and the corresponding mass spectra fragments of (a) *m/z* 212.01 (4.74 min), (b) 107.02 (5.63 min), (c) 191.03 (0.61 min) ions using UPLC/ESI-QTOF-MS in negative ion mode.

human urine samples at day 9 were approximately zerofold higher when compared to the predosed human urine samples, the content of *m/z* 107.02 in the predosed human urine samples was approximately fourfold higher than that in the postdosed human urine samples at day 9; however, the level of *m/z* 191.03 in the postdosed human urine samples at day 9 was about fivefold higher, compared to that in the predosed human urine samples. The representative extracted ion chromatograms and mass spectra fragments of the four ions are shown in Fig. 12.5E. The compounds were considered to be intrinsic ones and were detected from the predosed human urine samples. Currently, the structures of *m/z* 212.01 (4.74 min), 107.02 (5.63 min), and 191.03 (0.61 min) were preliminarily identified as indoxyl sulfate, *p*-cresol, and citric acid.

12.4 DISCUSSION

The aim of this study was to investigate whether the response to metabolic urinary profiles of health persons is influenced by LW administration, and elucidate the mechanism of nourishing the kidney from the metabolic levels of low molecular weight metabolites. The metabolomic analysis conducted in the present study reveals LW induced changes in the metabolic profiles of certain endogenous metabolites. In Figs 12.4D and 12.5D the pre- and postdosed LW displays a clear difference of biochemical trend. LW significantly up-regulated TCA cycles as an indicator of the elevation of urinary citric acid (Fig. 12.6A), impacted on the structures and the activities of gut microbiota resulting in the obviously excreted reduction of urinary indoxyl sulfate and *p*-cresol (Fig. 12.6B and C), and LW can also intervene in the sialic acid biosynthetic pathway as a result of inhibiting *N*-acetylneuraminic acid (Fig. 12.6D). In summary, preintervention with LW can cause changes in the energy metabolism pathway and gut microbiota metabolism, and their metabolic activities. Thus, these findings may reveal some underlying unknown mechanism of the action of LW.

The clear difference between pre- and postdosed LW in energy metabolism-related metabolites was revealed. Citric acid (citrate) is a key metabolite of the energy metabolism, which is formed in the tricarboxylic acid cycle, or that may be introduced with diet. We proposed that the tricarboxylic acid (TCA) cycle intermediates were significantly different between pre- and postdosed LW healthy persons with a regulatory variation of citrate. An up-regulated level of citrate in human urine may be an indication of more energy formation of healthy persons postdosing LW, when compared to that in predosing LW. Therefore, the elevated concentration of citrate in human urine samples may be the result of LW up-regulating the TCA cycle, furthermore inhibiting the precipitation of calcium salts in the kidneys, for preventing kidney degradation, which indicates a more robust immunity system to xenophobic intervention.

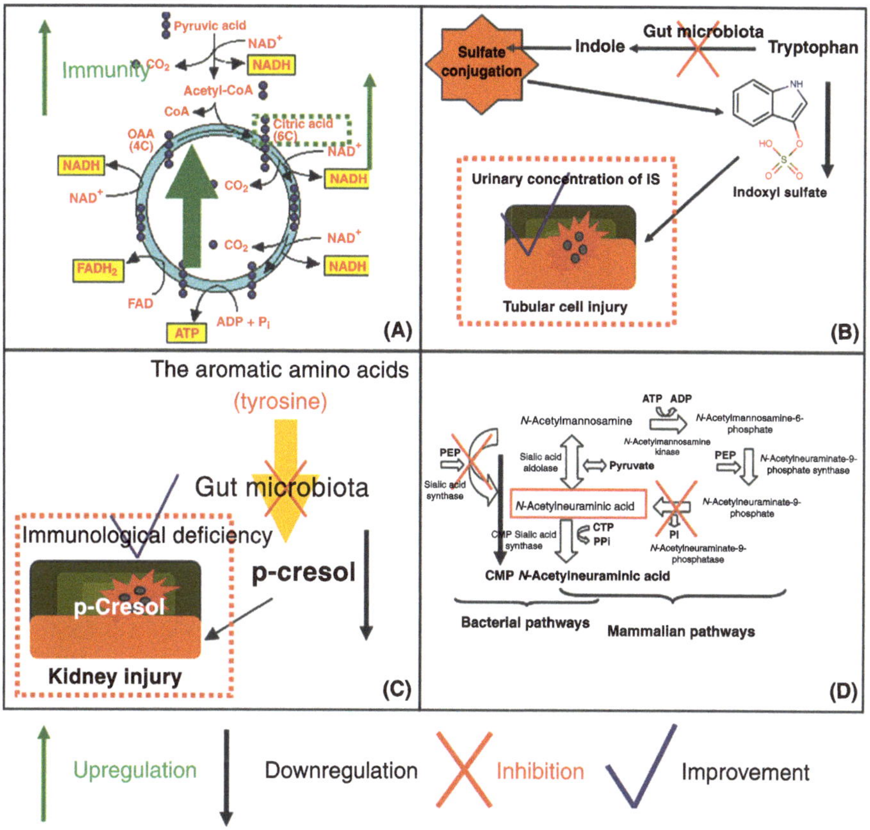

FIGURE 12.6 The variations of metabolic pathways in response to LW preintervention. LW significantly up-regulated TCA cycles as an indicator of the elevation of urinary citric acid: (A) impacted on structures and activities of gut microbiota resulting in the obviously excreted reduction of urinary indoxyl sulfate (B) and *p*-cresol (C); and LW can also intervene in the sialic acid biosynthetic pathway as a result of inhibiting *N*-acetylneuraminic acid (D).

LW may characterize the pharmacological effects of invigorating the kidney partly through up-regulating the immune system function.

In this study, preadministration of LW has significantly down-regulated the urinary level of indoxyl sulfate (IS), the result may again reveal that LW could prevent or treat kidney failures or toxicity, the mechanism of action may be two points: one is enforcing antioxidative systems and clearing free radicals, another is directly improving gut macrobiotic structures or its activities, so as to inhibit the production of IS. In the current work, the down-regulation of production of urinary *p*-cresol induced by LW administration suggests that LW might change the production environment of *p*-cresol, and indirectly result in the reduction of the urinary levels of *p*-cresol, effectively protecting the kidney organ. This study was conducted based on dietary standardization, the intake of high protein and high fatty food was strictly forbidden, so the influences of fasting, high protein, and high fat intake have been excluded from this study, whereas LW administration may directly, and indirectly regulate gut microbiota structures, or the activities of healthy persons to reduce production of *p*-cresol in the gut, furthermore kidney excreted levels of *p*-cresol were markedly reduced, so that the effects of nourishing and protecting kidneys were well demonstrated.

The term sialic acid indicates a general name for nine carbon acidic sugars that includes *N*-acetylneuraminic acid, *N*-glycolylneuraminic acid, and 2-keto-3-deoxy-D-glycero-D-galactonononic acid (a-D-KDN). The most main sialic acid (SA) is *N*-acetyl-neuraminic acid (NeuAc) in humans, and circulates in tissues and body fluids in two forms: as free sialic acid (FSA) and as conjugated sialic acid (CSA) (Mathé et al., 2014). Synthesis of SA begins with glucose, which undergoes several modifications to become activated uridine-diphosphate-*N*-acetylglucosamine. This central metabolite is converted to *N*-acetylmannosamine by an enzyme that is feedback inhibited by cytidine monophosphate-SA (Niethamer et al., 2012; Plumbridge and Vimr, 1999). Failure of this feedback mechanism results in overproduction and elevation of cytoplasmatic SA. Such over-production of free *N*-acetyl-neuraminic acid (NeuAc) will result in sialic acid storage diseases (SSDs). The sialic acid storage diseases (SSDs) are autosomal recessive neurodegenerative disorders (Mochel et al., 2009).

In this study, the distinct down-regulation of free *N*-acetylneuraminic acid in urine was illustrated in postdosed LW, which revealed that LW could treat SSDs; this effect was directly correlated to the conventional effect of LW. The LW was widely used for therapy "Wuchi and Wuruan" (WCWR) of infants in ancient China. According to the theories of TCM, the WCWR was mainly induced by kidney deficiency, which is characterized by inborn developmental delays, growth retardation, hypotonia, ataxia, and development delays in hearing, teething, and speaking, usually during the developing infants' period, and resulting in a slowly progressive adult form. It is found that the synthetic symptoms of WCWR of infants were obviously in accordance with that of SSDs, and infantile free sialic acid storage disease (ISSD). Thus, we can infer that the pathogenesis of WCWR in infants may be partly attributed to the development of sialic acid storage diseases in the infants period, hence LW showed effectively the therapeutic effect on WCWR of infants by mainly depending on its effect of anti-SSDs through antagonizing over production, and accumulation of *N*-acetylneuraminic acid. Simultaneously, we also pinpointed that the SSDs may be actually attributed to a renal deficiency in infants being in accordance with the pathogenesis of WCWR, whereas LW showed a potential for treating or preventing SSDs.

12.5 CONCLUSIONS

The differences in the urinary metabolite profiles of pre- and postdosed LW of healthy persons were identified with an UPLC/MS-based metabolomic strategy. Furthermore, a PCA model based on the MS-spectral profiles allowed characterization, and elucidation of pre- and postdosed LW of healthy persons. These results highlight the potential role of metabolomics as a novel approach for fundamental investigations of the action and the mechanism of LW formulae. In this study, metabolomics strategy offers a practical method for performing therapeutic assessments of TCM formulae based on metabolomics in the future.

References

Liang, X., Li, H., Li, S., 2014. A novel network pharmacology approach to analyse traditional herbal formulae: the Liu-Wei-Di-Huang pill as a case study. Mol. Biosyst. 10 (5), 1014–1022.

Liu, J.P., Feng, L., Zhang, M.H., Ma, D.Y., Wang, S.Y., Gu, J., Fu, Q., Qu, R., Ma, S.P., 2013. Neuroprotective effect of Liuwei Dihuang decoction on cognition deficits of diabetic encephalopathy in streptozotocin-induced diabetic rat. J. Ethnopharmacol. 150 (1), 371–381.

Mathé, E.A., Patterson, A.D., Haznadar, M., Manna, S.K., Krausz, K.W., Bowman, E.D., Shields, P.G., Idle, J.R., Smith, P.B., Anami, K., Kazandjian, D.G., Hatzakis, E., Gonzalez, F.J., Harris, C.C., 2014. Noninvasive urinary metabolomic profiling identifies diagnostic and prognostic markers in lung cancer. Cancer Res. 74 (12), 3259–3270.

Mochel, F., Sedel, F., Vanderver, A., Engelke, U.F., Barritault, J., Yang, B.Z., Kulkarni, B., Adams, D.R., Clot, F., Ding, J.H., Kaneski, C.R., Verheijen, F.W., Smits, B.W., Seguin, F., Brice, A., Vanier, M.T., Huizing, M., Schiffmann, R., Durr, A., Wevers, R.A., 2009. Cerebellar ataxia with elevated cerebrospinal free sialic acid (CAFSA). Brain 132 (Pt 3), 801–809.

Niethamer, T.K., Yardeni, T., Leoyklang, P., Ciccone, C., Astiz-Martinez, A., Jacobs, K., Dorward, H.M., Zerfas, P.M., Gahl, W.A., Huizing, M., 2012. Oral monosaccharide therapies to reverse renal and muscle hyposialylation in a mouse model of GNE myopathy. Mol. Genet. Metab. 107 (4), 748–755.

Plumbridge, J., Vimr, E., 1999. Convergent pathways for utilization of the amino sugars *N*-acetylglucosamine, *N*-acetylmannosamine, and *N*-acetylneuraminic acid by *Escherichia coli*. J. Bacteriol. 181 (1), 47–54.

Wang, X.J., Zhang, N., Sun, H., 2008. Comparative analysis of components absorbed into blood after oral administration of Liuwei Dihuang Wan and its related prescriptions. Zhongguo Zhong Yao Za Zhi 33 (15), 1881–1884.

Wang, P., Lv Ht, Zhang Ah, Sun, H., Yan Gl, Han, Y., Wu Xh, Wang, Xj., 2013. Improved ultra-performance liquid chromatography with electrospray ionization quadrupole-time-of-flight high-definition mass spectrometry method for the rapid analysis of the chemical constituents of a typical medical formula: Liuwei Dihuang Wan. J. Sep. Sci. 36 (21–22), 3511–3516.

CHAPTER 13

Metabolic Profiling and Potential Biomarkers Analysis of ShenYangXu Syndrome

Aihua Zhang, Qi Liu, Hui Sun, Xijun Wang

CONTENTS

13.1 INTRODUCTION

Traditional Chinese Medicine possesses a concept that "the kidney is the regulator of growth and development of the body." In the TCM overview, the kidney is in charge of vital body functions and its role is broader than just the anatomical kidney. It performs reproductive, urinary, endocrine, skeletal, blood and central nervous functions, while also storing primordial yin and yang (known as the inherited kidney yin and yang). Hence, the TCM kidney is the most important system of the body. It is the root of life activities; warms and promotes the functions of other organs, and is the origin of our congenital (inherited) foundation. Since the TCM kidney affects so many vital functions, ShenYangXu (SYX) Kidney-yang deficiency syndrome patients present with a wide range of symptoms, including: sore and/or pain in the lower back, cold and weak limbs, water retention in the legs, copious and pale urine, aversion to cold, poor appetite, spiritual fatigue, depression, sexual dysfunction, premature ejaculation, adrenal fatigue, hypothyroidism, and a weak pulse (Huang et al., 2013; Lu et al., 2011). Another TCM principle is viewed as follows: "Kidney yang motivates the power of human vitality, and controls the storage of vital essence." In short, kidney yang is the pilot light for our energy system; and the diming of this light represents SYX, and consequently results in the insufficient functioning of the reproductive, urinary, endocrine, skeletal, and nervous systems.

SYX was first documented in *Huangdi Neijing*, an early systematic monograph in TCM, and is characterized by warm dysfunction, and the metabolic disorder of body fluids, causing an aversion to cold, cold limbs, cold in the waist and back, aches in the waist and knees, tinnitus, impairment of hearing, and looseness of teeth (Shen et al., 2006). Modern medicine research indicates that the functional disorder with different degrees of hypothalamic-pituitary-target gland (adrenal, thyroid, and gonad) axis is the significant mechanism of SYX (Shen, 2004). A high dose of exogenous glucocorticoid injected rats is a classic method mimicking the syndrome of SYX (Tan et al., 2014) that leads

Chinmedomics. http://dx.doi.org/10.1016/B978-0-12-803117-9.00013-5

to a decrease of the steroid hormone from the adrenal cortex. More importantly, kidney yang works closely with the neural-endocrine-immune network thus connecting the immune response to the environment. The pertinent studies have shed light on the complex patterns of the transcriptomes of SYX patients (Shen et al., 2008). Therefore, the present study adopted the established metabolomic method based on UPLC-Q-TOF-HDMS in order to acquire the metabolic characteristics of SYX induced by a high dose of corticosterone in rats, and proposed related metabolic pathways. Our findings indicate that this approach may facilitate a further understanding of the metabolic basis of SYX, and will allow novel strategies for a rational therapy of this disease.

13.2 MATERIALS AND METHODS

13.2.1 Animal Handling

Male Wistar rats (weighing 220 ± 20 g) were supplied by the GLP Center of Heilongjiang University of Chinese Medicine (Harbin, China). All the rats were randomly divided into two groups of 15 rats each as follows: model group and control group. All animals were allowed to acclimatize in metabolism cages for 1 week prior to treatment. The model rats were injected subcutaneously corticosterone at a dose of 1 mg/kg once daily for 21 days. Rats in the normal control group were given olive oil alone under the same conditions.

13.2.2 Collection and Preparation of Urine Samples

Urinary samples for metabolomic study were collected. Urine was collected from metabolism cages at an ambient temperature throughout the whole procedure, and centrifuged at 13,000 rpm at 5 °C for 15 min, and the supernatants were stored frozen at −80 °C until metabolomic analysis. The quality control sample was used to optimize the condition of UPLC-Q-TOF-HDMS, as it contained the most information of whole urine samples.

13.2.3 Metabolic Profiling Platform

13.2.3.1 Chromatography

UPLC/MS analysis was carried out using a Waters ACQUITY ultra performance liquid chromatography system (Waters Corp., Milford, USA) controlled with Masslynx (V4.1, Waters Corp., Milford, USA). An aliquot of 6 μL of sample solution was injected onto an ACQUITY UPLC HSS T_3 column (100 × 2.1 mm i.d., 1.8 μm, Waters Corp., Milford, USA) at 40 °C, and the flow rate was 0.4 mL/min, and the injection volume was 3 μL. The optimal mobile phase consisted of a linear gradient system of: (A) 0.1% formic acid; and (B) 0.1% formic acid in water in acetonitrile; 0–1 min, 1% A; 1–2.5 min, 1–13% A; 2.5–6.5 min, 13–40% A; 6.5–8.0 min, 40–99% A; 8.0–10.5 min, 99% A; 10.5–11.0 min, 99–1% A; 11–13 min, 1 A. Every day, after the instrument was calibrated, the QC sample was first analyzed to test the stability of the instrument in order to ensure consistent performance of the system. All the samples were kept at 4 °C during the analysis.

13.2.3.2 Mass Spectrometry

The eluent was introduced into the high-definition mass spectrometer (Waters Corp., Milford, USA) analysis, and the optimal conditions of analysis were as follows: the source temperature was set at 110 °C, desolvation gas temperature was 450 ° C, cone gas flow was 50 h, desolvation gas flow was 600 L/h. In positive ion mode, the capillary voltage was 3.0 kV, the sampling cone voltage was 25 V, and extraction cone voltage was 3.0 V, desolvation gas flow was 600 L/h. In negative ion mode, the capillary voltage was 3.0 kV, the sampling cone voltage was 20 V, and extraction cone voltage was 4.0 V, desolvation gas flow was 700 L/h. The data acquisition rate was set to 0.4 s/scan, with a 0.1 s inter scan delay. Data were collected in centroid mode, and the mass range was set at *m*/*z* 50–1000 using an extended dynamic range. For accurate mass acquisition, a lock-mass of leucine enkephalin at a concentration of 0.2 ng/mL was used via a lock spray interface at a flow rate of 100 μL/min monitoring for positive ion mode ($[M + H]^+ = 556.2771$) and negative ion mode ($[M-H]^- = 554.2615$) to ensure accuracy during the MS analysis.

13.2.3.3 Data Processing

Each sample was represented by a total ion current (TIC) chromatogram. The LC-MS raw data were processed using the Micromass MarkerLynx, version 4.0 (Waters Corp., Milford, USA). And then the raw data were transformed into a single matrix containing aligned peaks with the same *m*/*z* retention time pair along with normalized peak intensities and sample name. The three-dimensional matrix, including retention time and *m*/*z* pairs (variable indices), sample names (observations), and normalized ion intensities (variables), was exported to an EZinfo software package for multivariate data analysis.

13.2.3.4 Multivariate Data Analysis

The multivariate data matrix was analyzed by EZinfo software 2.0 (Waters Corp., Milford, USA) to cope with highly multivariate, noisy, collinear, and possibly incomplete data. From the loading plots of PCA and OPLS-DA, various metabolites could be identified as being responsible for the separation, and were therefore viewed as potential biomarkers. Potential biomarkers of interest were extracted from the VIP-plot constructed following analysis with OPLS–DA, and the markers were chosen on the basis of their contribution to the correlation within the data set.

13.2.3.5 Biomarkers and Metabolic Pathway Identification

The MassFragment™ application manager (Waters Corp., Milford, USA) was used to facilitate the MS/MS fragment ion analysis process by way of chemically intelligent peak-matching algorithms. Furthermore, MassFragment conducts rapid putative ion identification, and putative biologically relevant analysis via incorporation of four major small molecule databases: ChemSpider, KEGG, HMDB, and Lipid Maps. The construction, interaction, and pathway analysis of potential biomarkers were performed with MetaboAnalyst software based on database source to identify the metabolic pathways. The possible biological roles were evaluated by the enrichment analysis.

13.2.4 Statistical Analysis

The PCA was used to uncover unknown trends in the treated groups. Statistically significant differences in mean values were tested by using two-tailed, two-sample Student's t-test, and $p < 0.05$ was considered statistically significant. Prior to multivariate analysis, the resultant data matrices from two analytical techniques were mean-centered and paretoscaled.

13.3 RESULTS

13.3.1 Urinary Metabolic Profiling by UPLC-MS

The raw LC–MS data from metabolic profiling were pretreated following the procedure described in the section of metabolic profiling platform. The typical UPLC/MS TIC of rat urine from the control group, SYX group, and JSP treatment group were shown in Fig. 13.1A and B in positive ionization mode. There were 8302 ion signals in positive mode and 7289 ion signals in negative mode detected in the UPLC-Q-TOF-HDMS data. It was hard to identify the ions that contributed to the discrimination of SYX and control groups from thousands of signals using traditional statistical methods. PLS-DA is a well-established supervised multivariate statistical analysis method that has been widely used in metabolomic studies. In this study, a PLS-DA method was established, and employed to identify biomarkers that were related to SYX development.

13.3.2 Multivariate Statistical Analysis of LC/MS Data

Subtle changes could be found using a pattern recognition approach, such as PCA and PLS-DA. PCA and PLS-DA are the two most popular pattern recognition methods to gain information for classification and to identify metabolites. Figure 11.2A and B showed the PCA score plot representing the distribution among control, and SYX group in two dimensions. Obvious separation between the control and SYX groups suggested that urine biochemical perturbation occurred significantly in the SYX group. The 3D plot of PCA showed metabolic profiles in the control, and the model group were separated clearly (Fig. 13.2C and D). The ions showed significant difference in abundance between the control, and treated animals, and contributed to the observed separation, and were selected from the respective VIP-plots as potential markers in

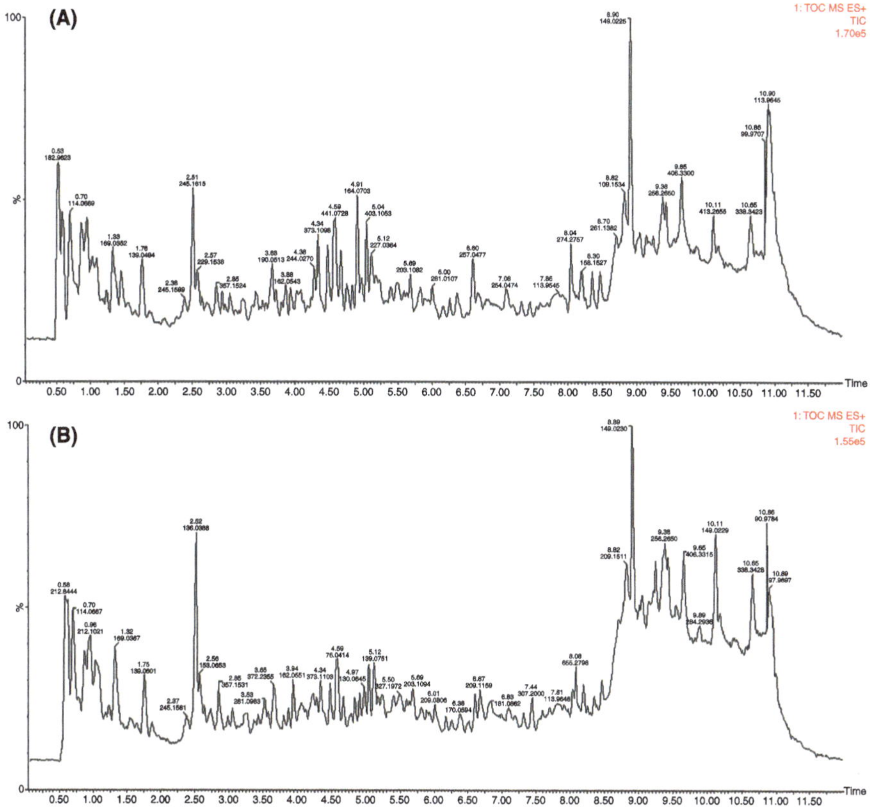

FIGURE 13.1 Typical total ion chromatograms (TIC) obtained from rat urine in the positive mode. Control group (A), model group (B).

positive and negative ion mode (Fig. 13.2E and F). To identify biomarkers in the SYX and the control groups, the value of the variable importance in the projection (VIP) of the PLS-DA model was obtained. The Student's t-test was used to reveal the significant differences of identified metabolites between the SYX and the control groups.

13.3.3 Identification of Biomarkers Related to SYX and Their Function

All collected samples were analyzed and low molecular weight metabolites were represented as the chromatographic peaks in the TIC chromatograms. The information including the retention time, the exact mass, and the ms/ms data were supplied by the robust LC-MS platform. The precise molecular mass was determined within a reasonable degree of measurement error using Q-TOF-MS, and the potential element composition, degree of fractional isotope

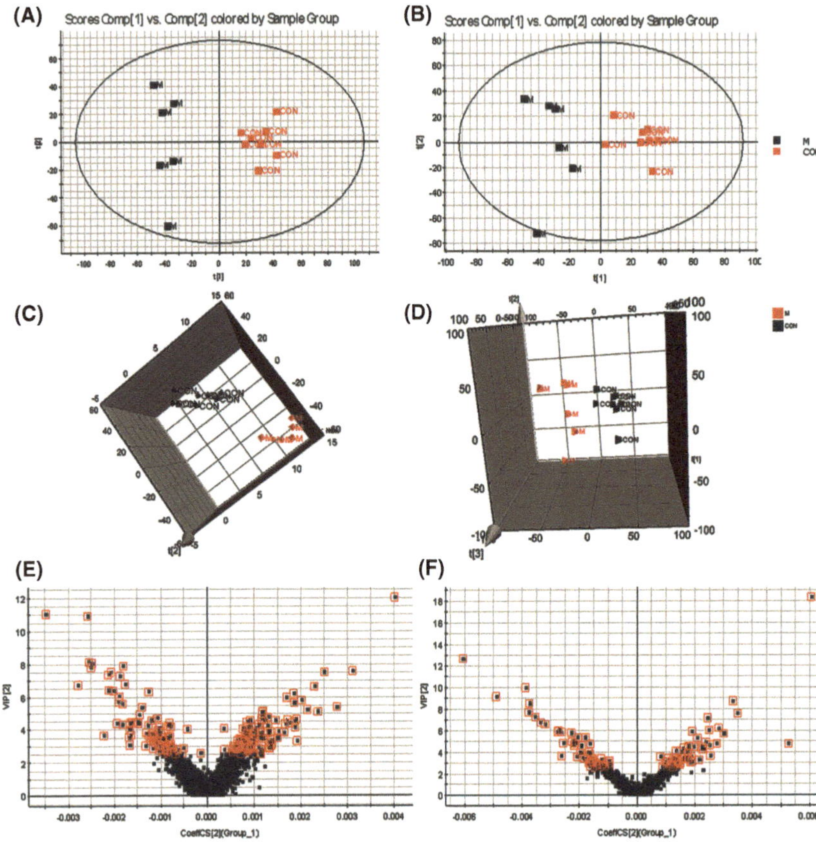

FIGURE 13.2 Metabolic profiling and potential biomarkers analysis.
PCA score plots for control (■) and model (■) group in positive mode (A); and negative mode (B); 3D plot of PCA of SYX in positive ion mode (C); and negative ion mode (D); VIP- plot of OPLS-DA for SYX vs controls in positive ion mode (E); and in negative ion mode (F).

abundance was also obtained. Furthermore, metabolite identification was conducted with high resolution MS, and MS/MS fragments, as well as database analyses. In accordance with the protocol, described previously, a number of potential biomarkers were listed in Table 13.1. In this study, a total of 60 variables (the value of VIP > 2.0, $p < 0.05$) contributed to the metabolic variation. These variables (40 in positive mode and 20 in negative mode) were predicted by comparing the MS, and MS/MS fragments with the metabolites found by searching in databases. As shown in Fig. 13.3, 28 of the 60 biomarkers identified were upregulated, and 32 of them were depressed in the SYX group. The related biomarker pathway was identified by searching the KEGG PATHWAY database. The detailed construction of the metabolism pathways with higher

Table 13.1 Identification and Change Trends of Marker Metabolites Related to SYX-Based Urine Metabolic Profiling

No.	Rt (min)	*m/z* Determined	*m/z* Calculated	Error (ppm)	Ion Form	Formula	Metabolite Name	VIP Value	Trend
1	6.52	255.0651	255.0651	0	$[M+H]^+$	$C_7H_{14}N_2O_6S$	5-L-Glutamyl-taurine	8.62363	↑
2	4.06	121.0657	121.0653	3.3	$[M+H]^+$	C_8H_8O	Phenylacetaldehyde	4.48357	↑
3	5.04	162.0551	162.0555	−2.5	$[M+H]^+$	$C_9H_7NO_2$	4,6-Dihydroxyquinoline	5.86728	↑
4	3.46	206.0449	206.0453	−1.9	$[M+H]^+$	$C_{10}H_7NO_4$	Xanthurenic acid	7.24889	↓
5	4.06	180.1017	180.1025	−4.4	$[M+H]^+$	$C_{10}H_{13}NO_2$	(R)-Salsolinol	2.83317	↑
6	1.44	190.1194	190.1192	1.1	$[M+H]^+$	$C_7H_{15}N_3O_3$	L-Homocitrulline	3.47929	↑
7	0.73	129.0660	129.0664	−3.1	$[M+H]^+$	$C_5H_8N_2O_2$	Hydrouracil	3.29001	↓
8	5.75	349.2378	349.2379	−0.3	$[M+H]^+$	$C_{21}H_{32}O_4$	11β,21-Dihydroxy-5β-pregnane-3,20-dione	2.8392	↑
9	4.21	126.0224	126.0225	−0.8	$[M+H]^+$	$C_2H_7NO_3S$	Taurine	4.40552	↑
10	0.87	229.1194	229.1188	2.6	$[M+H]^+$	$C_{10}H_{16}N_2O_4$	Prolylhydroxyproline	6.2257	↓
11	1.19	124.0397	124.0399	−1.6	$[M+H]^+$	$C_6H_5NO_2$	Niacin (Nicotinic acid)	3.89617	↓
12	0.87	176.1030	176.1035	−2.8	$[M+H]^+$	$C_6H_{13}N_3O_3$	Citrulline	4.05663	↓
13	3.68	190.0502	190.0504	−1.1	$[M+H]^+$	$C_{10}H_7NO_3$	Kynurenic acid	6.27684	↓
14	0.92	127.0505	127.0508	−2.4	$[M+H]^+$	$C_5H_6N_2O_2$	Thymine	7.86767	↑
15	0.92	259.0929	259.0930	−0.4	$[M+H]^+$	$C_{10}H_{14}N_2O_6$	Ribosylimidazoleacetic acid	5.34566	↑
16	1.26	112.0516	112.0511	4.5	$[M+H]^+$	$C_4H_5N_3O$	Cytosine	2.89274	↓
17	1.27	138.0923	138.0919	2.9	$[M+H]^+$	$C_8H_{11}NO$	Tyramine	3.08608	↑
18	5.15	215.0327	215.0321	2.8	$[M+H]^+$	$C_5H_{11}O_7P$	Deoxyribose 1-phosphate	3.55279	↓
19	2.51	174.1244	174.1243	0.6	$[M+H]^+$	$C_7H_{15}N_3O_2$	Apo-[3-methylcrotonoyl-CoA:carbon-dioxide ligase (ADP-forming)]	3.56549	↑

(Continued)

Table 13.1 Identification and Change Trends of Marker Metabolites Related to SYX-Based Urine Metabolic Profiling *(cont.)*

No.	Rt (min)	*m/z* Determined	*m/z* Calculated	Error (ppm)	Ion Form	Formula	Metabolite Name	VIP Value	Trend
20	1.15	204.1229	204.1236	−3.4	$[M + H]^+$	$C_9H_{17}NO_4$	Acetylcarnitine	3.25817	↓
21	2.93	197.0309	197.0311	−1.0	$[M + H]^+$	$C_6H_4N_4O_4$	Xanthine-8-carboxylate	2.87283	↑
22	5.12	302.1394	302.1392	0.7	$[M + H]^+$	$C_{17}H_{19}NO_4$	Noroxycodone	2.75347	↓
23	4.6	315.0594	315.0593	0.3	$[M + H]^+$	$C_8H_{15}N_2O_9P$	5′-Phosphoribosyl-N-formylglycinamide (FGAR)	2.50632	↓
24	4.99	130.0653	130.0657	−3.1	$[M + H]^+$	C_9H_7N	Quinoline	7.92895	↑
25	1.27	314.1236	314.1240	−1.3	$[M + H]^+$	$C_{14}H_{19}NO_7$	Tyramine glucuronide	5.50204	↑
26	5.04	270.0382	270.0379	1.1	$[M + H]^+$	$C_7H_{12}NO_8P$	*N*-Acetyl-L-glutamate 5-phosphate	5.33969	↓
27	6.4	170.0590	170.0582	4.7	$[M + H]^+$	$C_4H_{12}NO_4P$	Phosphorylcholine	6.66665	↑
28	1.78	139.0505	139.0508	−2.2	$[M + H]^+$	$C_6H_6N_2O_2$	Nicotinamide *N*-oxide	4.60596	↓
29	0.77	143.0825	143.0821	2.8	$[M + H]^+$	$C_6H_{10}N_2O_2$	Ectoine	4.60012	↓
30	2.3	271.1397	241.1406	−3.3	$[M + H]^+$	$C_{11}H_{18}N_4O_4$	2-[3-Carboxy-3-(methylammonio)propyl]-L-histidine	4.22563	↑
31	1.03	141.0660	141.0664	−2.8	$[M + H]^+$	$C_6H_8N_2O_2$	1,4-Methylimidazoleacetic acid	3.15977	↓
32	0.94	194.0923	194.0930	−3.6	$[M + H]^+$	$C_9H_{11}N_3O_2$	4-(Nitrosoamino)-1-(3-pyridinyl)-1-butanone	2.95164	↓
33	1.04	190.0168	190.0174	−3.2	$[M + H]^+$	$C_6H_7NO_4S$	Lanthionine ketimine	3.8959	↑
34	0.73	138.0551	138.0555	−2.9	$[M + H]^+$	$C_7H_7NO_2$	Anthranilic acid	2.99686	↑
35	2.58	145.0502	145.0501	0.7	$[M + H]^+$	$C_6H_8O_4$	3-Hexenedioic acid	2.82142	↓
36	5.14	399.1465	399.1451	3.5	$[M + H]^+$	$C_{15}H_{22}N_6O_5S$	S-Adenosylmethionine	2.70582	↓
37	2.93	345.0771	345.0786	−4.3	$[M + H]^+$	$C_{12}H_{17}N_4O_4PS$	Thiamine monophosphate	2.68747	↓
38	1.08	86.0603	86.0606	−3.5	$[M + H]^+$	C_4H_7NO	2-Pyrrolidinone	2.60115	↓

39	3.44	178.0531	178.0538	−3.9	$[M + H]^+$	$C_6H_{11}NO_3S$	*N*-Formylmethionine	2.55577	↓
40	4.23	181.0868	181.0865	1.7	$[M + H]^+$	$C_{10}H_{12}O_3$	3-Methoxybenzenepropanoic acid	2.51977	↓
41	2.94	144.0655	144.0661	−4.2	$[M-H]^-$	$C_6H_{11}NO_3$	Allysine	4.88062	↓
42	3.91	228.0879	228.0872	3.1	$[M-HCOO]^-$	$C_9H_{13}NO_3$	Epinephrine	2.70176	↓
43	1.12	187.0715	187.0719	−2.1	$[M-H]^-$	$C_7H_{12}N_2O_4$	Nα-Acetyl-L-glutamine	5.30931	↓
44	3.09	225.0865	225.0875	−4.4	$[M-HCOO]^-$	$C_9H_{12}N_2O_2$	5-Hydroxykynurenamine	2.55661	↓
45	3.59	280.0821	280.0821	0	$[M-H]^-$	$C_{13}H_{15}NO_6$	4-Hydroxyphenylacetylglutamine	3.92032	↑
46	5.64	248.0934	248.0923	4.4	$[M-HCOO]^-$	$C_{12}H_{13}NO_2$	3-Indolebutyric acid	3.4496	↑
47	7.41	365.2314	365.2328	−3.8	$[M-H]^-$	$C_{21}H_{34}O_5$	5a-Tetrahydrocortisol	3.28775	↑
48	1.39	96.9596	96.9596	0	$[M-H]^-$	H_2O_4S	Sulfuric acid	5.79567	↑
49	3.74	144.0451	144.0449	1.4	$[M-H]^-$	C_9H_7NO	3-Methyleneoxindole	4.23378	↓
50	2.78	264.0518	264.0508	3.8	$[M-HCOO]^-$	$C_{11}H_9NO_4$	5-Hydroxyindolepyruvate	4.42036	↑
51	2.95	74.0243	74.0242	1.4	$[M-H]^-$	$C_2H_5NO_2$	Glycine	3.5757	↓
52	6.75	239.0920	239.0919	0.4	$[M-H]^-$	$C_{12}H_{16}O_5$	3-Carboxy-4-methyl-5-propyl-2-furanpropanoic acid	3.02591	↑
53	3.87	242.0435	242.0430	2.1	$[M-HCOO]^-$	$C_5H_{12}NO_5P$	2-Amino-5-phosphopentanoic acid	6.01274	↑
54	4.94	159.1015	159.1021	−3.8	$[M-H]^-$	$C_8H_{16}O_3$	7-Hydroxyoctanoic acid	2.91417	↓
55	2.83	142.0504	142.0504	0	$[M-H]^-$	$C_6H_9NO_3$	Vinylacetylglycine	2.81308	↓
56	5.47	236.0934	236.0923	4.7	$[M-HCOO]^-$	$C_{11}H_{13}NO_2$	5-Methoxytryptophol	3.42729	↑
57	6.6	187.1326	187.1334	−4.3	$[M-H]^-$	$C_{10}H_{20}O_3$	3-Hydroxycapric acid	4.66398	↓
58	5.1	147.0403	147.0406	−2.0	$[M-HCOO]^-$	$C_3H_6N_2O_2$	*N*-Formiminoglycine	2.53224	↑
59	1.59	238.0351	238.0352	−0.4	$[M-HCOO]^-$	$C_9H_7NO_4$	L-Dopachrome	2.8595	↑
60	3.86	338.0817	338.0876	−1.5	$[M-H]^-$	$C_{15}H_{17}NO_8$	6-Hydroxy-5-methoxyindole glucuronide	3.62112	↓

Notes: '↑' and '↓' represent the compounds that were up- and down regulated in the SYX group, compared with the control group.

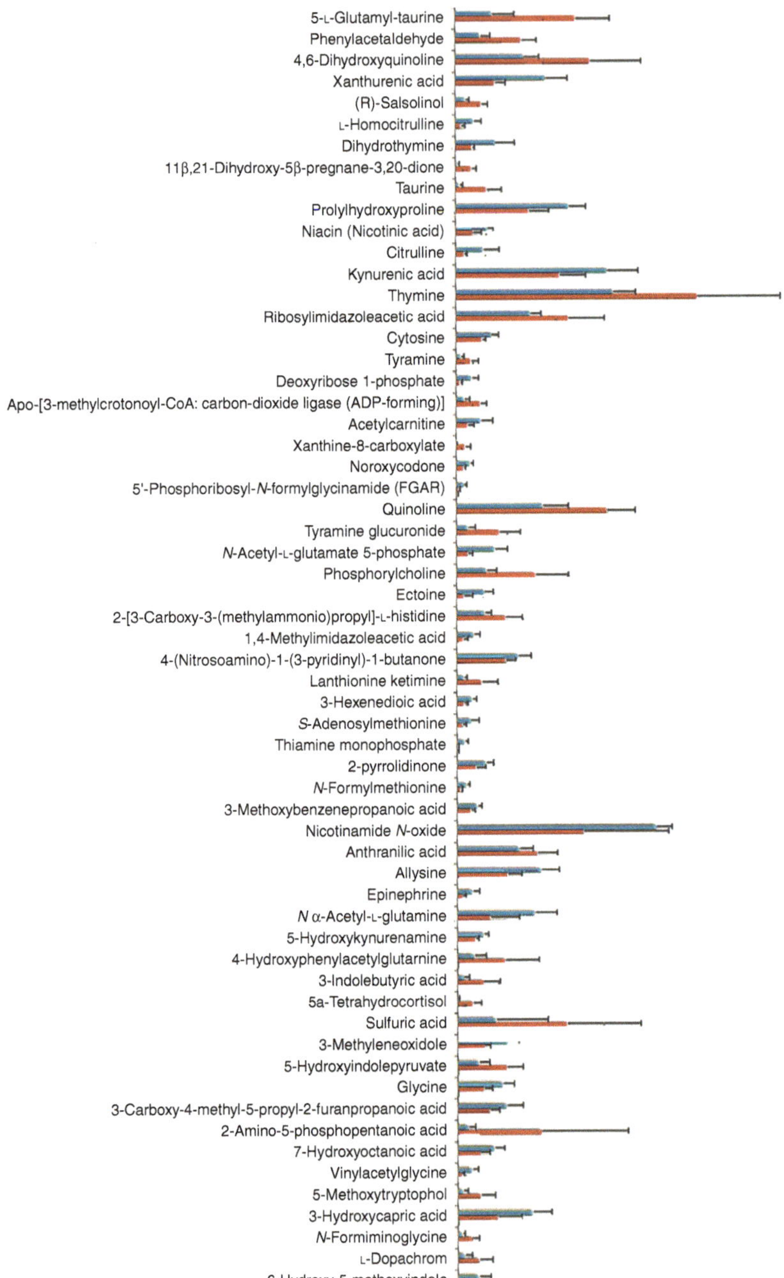

FIGURE 13.3 Changes in the relative intensity of target metabolites identified by UPLC/MS. The corresponding markers represented in Table 11.1. A two-tailed, parametric t-test was used to determine the significance of the change in relative intensity for each metabolite. Bars represent the mean relative metabolite concentration and standard deviations. Compared with model group $^*p < 0.05$.

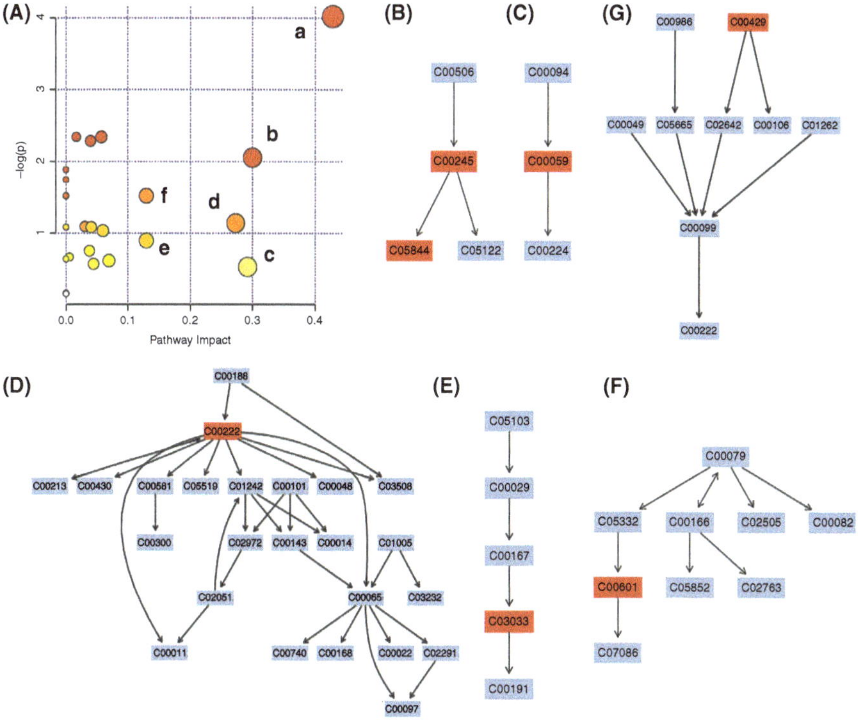

FIGURE 13.4 Metabolic pathway analysis.

Summary of pathway analysis with MetPA (A); taurine and hypotaurine metabolism (B); sulfur metabolism (C); glycine, serine, and threonine metabolism (D); pentose and glucuronate interconversions (E); phenylalanine metabolism (F); β-alanine metabolism (G).

scores was shown in Fig. 13.4. The results suggested that these target pathways showed the marked perturbations over the time-course of SYX, and could contribute to the development of SYX. As shown in Fig. 13.4 and Table 13.2; the six main pathological processes including taurine and hypotaurine metabolism, sulfur metabolism, glycine, serine and threonine metabolism, pentose and glucuronate interconversions, phenylalanine metabolism, and β-alanine metabolism were involved in SYX development, which indicated that dysfunction of multipathways were involved in the pathological process of SYX. In these biomarkers, 5-L-glutamyl-taurine and taurine were related to taurine and hypotaurine metabolism, and sulfuric acid is related to sulfur metabolism. Biomarkers including 5-L-glutamyl-taurine and taurine belong to the taurine and hypotaurine metabolism; which indicated that SYX was accompanied by inflammation. Furthermore, sulfuric acid indicated that the balance of sulfur metabolism was broken during SYX formation. An overview of the metabolic pathway alterations associated with the SYX group is shown in Fig. 13.5.

Table 13.2 Result From Ingenuity Pathway Analysis With MetPA

Name	Total	Expected	Hits	Raw p	Impact
Taurine and hypotaurine metabolism	8	0.2168	2	0.0181	0.4286
Tryptophan metabolism	41	1.1113	3	0.0966	0.0166
Pyrimidine metabolism	41	1.1113	3	0.0966	0.0573
Tyrosine metabolism	42	1.1384	3	0.1021	0.0397
Sulfur metabolism	5	0.1355	1	0.1285	0.3000
Cyanoamino acid metabolism	6	0.1626	1	0.1523	0.0000
Thiamine metabolism	7	0.1897	1	0.1753	0.0000
Methane metabolism	9	0.2439	1	0.2197	0.0000
Nitrogen metabolism	9	0.2439	1	0.2197	0.0000
Phenylalanine metabolism	9	0.2439	1	0.2197	0.1296
Pentose and glucuronate interconversions	14	0.3795	1	0.3206	0.2727
Arginine and proline metabolism	44	1.1926	2	0.3366	0.0304
Histidine metabolism	15	0.4066	1	0.3392	0.0000
Pantothenate and CoA biosynthesis	15	0.4066	1	0.3392	0.0408
Primary bile acid biosynthesis	46	1.2468	2	0.3567	0.0595
β-Alanine metabolism	19	0.5150	1	0.4088	0.1296
Starch and sucrose metabolism	23	0.6234	1	0.4712	0.0378
Glutathione metabolism	26	0.7047	1	0.5137	0.0057
Porphyrin and chlorophyll metabolism	27	0.7318	1	0.5272	0.0000
Cysteine and methionine metabolism	28	0.7589	1	0.5402	0.0694
Glycerophospholipid metabolism	30	0.8131	1	0.5653	0.0444
Glycine, serine, and threonine metabolism	32	0.8673	1	0.5891	0.2920
Aminoacyl-tRNA biosynthesis	67	1.8160	1	0.8484	0.0000
Purine metabolism	68	1.8431	1	0.8527	0.0000

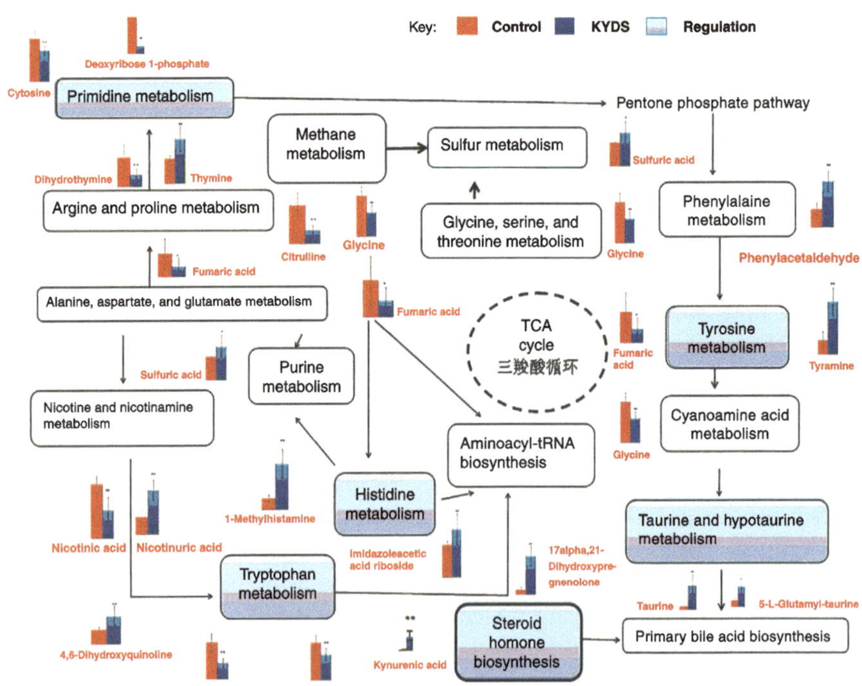

FIGURE 13.5 An overview of the metabolic pathway alterations associated with SYX.

13.4 DISCUSSION

The theory of *Yin-Yang* is a conceptual framework used for observing and analyzing the material world in ancient China. This philosophical approach has also been adopted for the studies of TCM syndrome, and the guidance processes, clinical diagnosis, treatments, and prevention strategies for thousands of years (Wang et al., 2011, 2012). In order to certify, and clarify the efficacy of therapies in traditional medicine, the work of establishing proper and optimal animal models plays an essential role in elucidating the pathogenesis of different syndromes in different traditional medicine theories, and exploring TCM to better prevent and treat SYX. Over the past half century, more than 20 experimental models with a variable range of clinical manifestations, similar to those observed in the SYX human have been developed. Injecting animals with high doses of exogenous glucocorticoid is one of classical methods of establishing SYX. This animal model mimicked the pathological state of suppression of hypothalamic-pituitaryadrenal (HPA) axis to some extent in the SYX human, and contributed greatly to important advances in the current understanding of the underlying mechanisms of SYX, as well as treatments. While there has been significant research on the physiology of SYX, few studies have been conducted on its metabolic background, largely due to the lagged development of appropriate methodology for the extreme complexity of the metabolic basis of TCM syndromes.

The kidneys in TCM are viewed as the root of life activities, and represent the origin of our congenital or inherited foundation. They regulate reproductive, urinary, endocrine, skeletal, blood, and central nervous functions while also storing primordial yin, and primordial yang (known as the inherited kidney yin and yang). Therefore, a congenital deficiency is one of the critical causes of SYX. For a further understanding of the metabolic alternation of SYX in Chinese medicine, a new UHPLC–MS metabolomics method with PCA was developed for the metabolic profiling analysis of rat urine. Metabolomics represents a paradigm shift in metabolic research away from methods that focus on a limited number of single pathways, to an approach that attempts to capture the whole range of metabolic networks. Fortunately, metabolomics may offer the possibility of identifying marker metabolites, and the pathways activated in SYX. This study has explored globally the metabolomics characters of SYX.

In this study, we explored novel biomarkers to increase the mechanism understanding of SYX, using a metabolomic approach, and the metabolite network of SYX was also predicted via the ingenuity pathway methods. The levels of 60 metabolites exhibited the marked changes compared with that of the control rats. Urinary excretion of these metabolites was also shown to have high specificity, and sensitivity as markers. The pathway analysis showed that potential markers were identified from taurine and hypotaurine metabolism, sulfur metabolism, glycine, serine, threonine metabolism, pentose, glucuronate interconversions, phenylalanine metabolism, and β-alanine metabolism – that changed specifically in the setting of SYX. Network reconstruction has led to the integration of metabolites associated with the causes of perturbation in multiple pathways. Our results also show that the combination of high-resolution metabolomics analytical tools with pattern recognition techniques provided a new methodological cue for dissecting the underlying efficacies and mechanisms of TCM syndrome.

13.5 CONCLUSIONS

In this study, metabolomics combined with pattern recognition approaches were integrated to discover differentiating metabolites, and to elucidate the metabolic characteristics of the SYX at the global metabolomics levels. Sixty metabolites were identified as potential biomarkers in SYX. Taurine and hypotaurine metabolism, sulfur metabolism, glycine, serine, and threonine metabolism, pentose and glucuronate interconversions, phenylalanine metabolism, and β-alanine metabolism were the main perturbed pathways in this pathological process. This study suggests that the LC-MS-based metabolomic strategy is a powerful approach, when used for the elucidation of the mechanisms of TCM, and opens up new perspectives to using the metabolomics platform to resolve special TCM issues.

References

Huang, D., Yang, J., Lu, X., Deng, Y., Xiong, Z., Li, F., 2013. An integrated plasma and urinary metabonomic study using UHPLC-MS: intervention effects of Epimedium koreanum on 'Kidney-Yang Deficiency Syndrome' rats. J. Pharm. Biomed. Anal. 76, 200–206.

Lu, X., Xiong, Z., Li, J., Zheng, S., Huo, T., Li, F., 2011. Metabonomic study on 'Kidney-Yang Deficiency syndrome' and intervention effects of Rhizoma Drynariae extracts in rats using ultra performance liquid chromatography coupled with mass spectrometry. Talanta 83 (3), 700–708.

Shen, Z.Y., 2004. Study on the regulation of HPAT axis molecular network in patients with aging-physiologic shen-deficiency syndrome. Zhongguo Zhong Xi Yi Jie He Za Zhi 24 (9), 841–843.

Shen, Z.Y., Chen, Y., Huang, J.H., 2006. Syndrome differentiation through drug effects in mapping the two regulatory pathways of gene networks in Shen deficiency syndrome. Zhongguo Zhong Xi Yi Jie He Za Zhi 26 (6), 521–525.

Shen, Z.Y., Zhang, X.M., Lin, W., 2008. Establishment of a quantitative mathematical model of Shen-deficiency syndrome based on whole-genome transcriptional profiles. Zhongguo Zhong Xi Yi Jie He Za Zhi 28 (2), 131–134.

Tan, Y., Liu, X., Lu, C., He, X., Li, J., Xiao, C., Jiang, M., Yang, J., Zhou, K., Zhang, Z., Zhang, W., Lu, A., 2014. Metabolic profiling reveals therapeutic biomarkers of processed Aconitum carmichaeli Debx in treating hydrocortisone induced kidney-yang deficiency syndrome rats. J. Ethnopharmacol. 152 (3), 585–593.

Wang, X., Sun, H., Zhang, A., Sun, W., Wang, P., Wang, Z., 2011. Potential role of metabolomics approaches in the area of traditional Chinese medicine: as pillars of the bridge between Chinese and Western medicine. J. Pharm. Biomed. Anal. 55 (5), 859–868.

Wang, X., Zhang, A., Sun, H., 2012. Future perspectives of Chinese medical formulae: chinmedomics as an effector. OMICS 16 (7–8), 414–421.

CHAPTER 14

Metabolic Evaluation of ShenQiWan Nourishing ShenYangXu Syndrome

Xiaohang Zhou, Aihua Zhang, Hui Sun, Xijun Wang

14.1 INTRODUCTION

Traditional Chinese Medicine (TCM) has been playing a very important role in health protection and disease control for thousands of years. The most common clinical practice of TCM is herb combination called formula. Formulae consist of several types of medicinal herbs. There are more than 100,000 formulae in use for disease therapy in China, and most of them have not yet been characterized for the development of new drugs. The key issue is that the actual value of TCM has not been fully recognized due to an insufficiency of scientific approaches. In recent years, TCM research has greatly accelerated with the advancement of analytical technologies and methodologies. Fortunately, significant advances with metabolomic techniques hold promise for overcoming previous limitations. Metabonomic research strategy is efficaciously coincident with the integrity and systemic features of TCM, and is increasingly used in the evaluation of TCM, and seeks to elucidate the therapeutic properties, and the metabolism of prescription, and the involved metabolic pathways. It will greatly facilitate the further elucidation of the action mechanisms of TCM.

ShenYangXu (SYX) Kidney-yang deficiency syndrome (KYDS) is one of the comment syndrome patterns in TCM. SYX is a complex kidney disorder involving many diseases such as nephrotic syndrome and chronic renal failure. It amounts to 10.3% of the national burden of disease in China, and is predicted to rise to 18% by 2020. Currently, a new focus on the pursuit of TCM is being sought for the treatment of SYX, and the therapeutic effects have also been verified by a host of clinical studies. ShenQiWan (SQW, JSP) a well-known formulae for invigorating SYX that has been extant for thousands of years in Asia, was first recorded in *Synopsis of Golden Chamber*, consisting of *Radix Rehmanniae Preparata, Fructus Macrocarpii, Rhizoma Dioscoreae Oppositae, Poria, Rhizoma Alismatis, Cortex Moutan Radicis, Achyranthes bidentata*, and *Plantago asiatica*. It may yield multitargets, and retain the balance of metabolic pathways in disease treatment, since it possesses many bioactive components that may affect various pathways. Nevertheless, due to its complex constituents, it is not easy to explore the action mechanism of SQW using traditional methods, therefore, little is known about

CONTENTS

Chinmedomics. http://dx.doi.org/10.1016/B978-0-12-803117-9.00014-7

the change of the whole metabolites in SYX treated with SQW. Although SQW has been used to improve the treatment effect of SYX patients in the clinical practice, however, there has been a sparcity of literature reporting about the detailed therapeutic mechanisms. In light of this, we performed a metabolomic approach to evaluate the intervention effects of SQW on SYX.

14.2 MATERIALS AND METHODS

14.2.1 Materials and Reagents

SQW were purchased from the Beijing Shiyitang Drug Store (Harbin, China). All other reagents were of analytical grade. Corticosterone was produced by Sigma Ltd. Corticotropin releasing hormone (CRH), corticosterone (CORT), adrenocorticotropic hormone (ACTH) kit were provided by Beijing Huaying Biological Ltd; thyroid stimulating hormone (TSH), T3, T4, luteinizing hormone (LH), testosterone (T), estradiol (E2) kit were provided by Beijing North Biotechnology Institute; cAMP kit, cGMP kit, gonadotropin-releasing hormone (GnRH), thyrotropin-releasing hormone (TRH) kit were provided by R&D. Acetonitrile, HPLC grade, was obtained from Merck (Darmstadt, Germany); methanol (HPLC grade) was purchased from Fisher Scientific Corp. (Loughborough, UK); leucine enkephalin was purchased from Sigma–Aldrich (St. Louis, MO, USA). All other reagents were of analytical grade.

14.2.2 Animal Handling

All the male Wistar rats (weighing 220 ± 20 g) were randomly divided into three groups of eight rats each as follows: model group, SQW-treated group, and control group. All animals were allowed to acclimatize in metabolism cages for 1 week prior to treatment. After 1 week they were administered orally SQW at a dose of 3.389 g/kg once daily for the following 8 days, all rats were injected subcutaneously corticosterone at a dose of 1 mg/kg once daily for 21 days, and then administered orally SQW. Mice in the normal control group were given saline alone under the same conditions.

14.2.3 Urine Collection and Preparation

Urinary samples for metabolomic study were collected. Urine was collected from metabolism cages at ambient temperature throughout the whole procedure and centrifuged at 13,000 rpm at 5°C for 15 min, and the supernatants were stored frozen at −80°C < until metabolomic analysis. The quality control sample was used to optimize the condition of UPLC-MS, as it contained the most information from the whole urine samples.

14.2.4 Metabolic Profiling Platform

14.2.4.1 Chromatography Analysis

Chromatography analysis was carried out using ultra performance liquid chromatography system (Waters Corp., Milford, USA) controlled with Masslynx (V4.1,

Waters Corp., Milford, USA). An aliquot of 6 μL of sample solution was injected onto an ACQUITY UPLC HSS T_3 column (100 × 2.1mm i.d., 1.8 μm, Waters Corp., Milford, USA) at 40°C and the flow rate was 0.4 mL/min, and the injection volume was 3 μL. The optimal mobile phase consisted of a linear gradient system of: (A) 0.1% formic acid; and (B) 0.1% formic acid in water in acetonitrile: 0–1 min, 1% A; 1–2.5 min, 1–13% A; 2.5–6.5 min, 13–40% A; 6.5–8.0 min, 40–99% A; 8.0–10.5 min, 99% A; 10.5–11.0 min, 99–1% A; 11–13 min, 1 A.

14.2.4.2 Mass Spectrometry Condition

The source temperature was set at 110°C, desolvation gas temperature was 450°C, cone gas flow was 50 h, desolvation gas flow was 600 L/h. In positive ion mode, the capillary voltage was 3.0 kV, the sampling cone voltage was 25 V, and extraction cone voltage was 3.0 V, desolvation gas flow was 600 L/h. In negative ion mode, the capillary voltage was 3.0 kV, the sampling cone voltage was 20 V, and extraction cone voltage was 4.0 V, desolvation gas flow was 700 L/h. For accurate mass acquisition, a lock-mass of leucine enkephalin at a concentration of 0.2 ng/mL was used via a lock spray interface at a flowrate of 100 μL/min monitoring for positive ion mode ($[M + H]^+ = 556.2771$) and negative ion mode ($[M-H]^- = 554.2615$) to ensure accuracy during the MS analysis.

14.2.5 Statistical Analysis

The LC-MS raw data was exported to EZinfo software 2.0 (Waters Corp., Milford, USA) for multivariate data analysis. The PCA was used to uncover unknown trends in the treated groups. The pathway analysis of potential biomarkers was performed with MetaboAnalyst based on database source. Statistically significant differences in mean values were tested by using two-tailed, two-sample Student's t-test, and $p < 0.05$ was considered statistically significant. Prior to multivariate analysis, the resultant data matrices from two analytical techniques were mean centered and pareto scaled.

14.3 RESULTS

14.3.1 Biochemical Analysis and Histopathological Observations

The subcutaneous injection of corticosterone for 3 weeks effectively induced SYX in rats, as evidenced by the markedly biochemical indexes. The biochemistry parameters of the control, the SYX, SQW groups were summarized in the supplementary Table 14.1. A significant elevation of plasma CORT and E2 were observed in SYX rats compared with controls over the entire time course, whereas the levels of CRH, ACTH, 17-OHCS, TSH, T, cAMP, cGMP were decreased. Based on the previous results, the SYX model was considered to be successfully established. All these results indicated that the rats presented the typical pathological features of SYX. Meanwhile, disease severity was further verified by the H&E staining of the hypothalamic, pituitary, adrenal, and

Table 14.1 Blood Biochemical SYX Characteristics Results Post-SQW Treatment

Groups	CRH (ng/dL)	ACTH (g/mL)	CORT (ng/mL)	17-OHCS (μmol/L)	TRH (μmol/L)	TSH (nIU/mL)	T3 (ng/dL)
Control	70.4 ± 12.3	23.79 ± 3.02	271.0 ± 7.47	77.26 ± 29.43	35.6 ± 5.72	33.00 ± 24.5	12.20 ± 2.95
SYX	57.4 ± 6.1**	20.89 ± 0.71*	292.3 ± 10.0**	53.45 ± 6.6**	33.9 ± 13.07	19.69 ± 9.4**	8.27 ± 1.89**
SQW	55.80 ± 4.60	20.47 ± 1.92	275.81 ± 5.42#	76.66 ± 16.01##	25.30 ± 7.96	49.88 ± 33.2#	14.28 ± 3.41##

Groups	T4 (ng/mL)	GnRH (ng/mL)	FSH (μIU/dL)	LH (mIU/dL)	T (pg/mL)	E_2 (nmol/dL)	cAMP (ng/mL)	cGMP (pg/mL)
Control	26.2 ± 4.1	9.35 ± 2.16	16.60 ± 4.89	29.2 ± 8.5	108.96 ± 75.19	10.13 ± 5.97	27.78 ± 11.89	23.06 ± 8.93
SYX	17.39 ± 5.09**	6.59 ± 3.67	11.53 ± 5.01	28.9 ± 4.5	16.25 ± 5.6**	20.05 ± 15.75	13.68 ± 2.37**	14.92 ± 5.67*
SQW	26.50 ± 5.58##	4.53 ± 3.20	4.85 ± 2.53##	20.70 ± 3.40##	134.8 ± 87.90##	37.53 ± 14.72	25.77 ± 9.75##	25.22 ± 2.05##

*Significant difference from control at $p < 0.05$.
**Significant difference from control at $p < 0.001$.
#Significant difference from model at $p < 0.05$.
##Significant difference from model at $p < 0.001$.

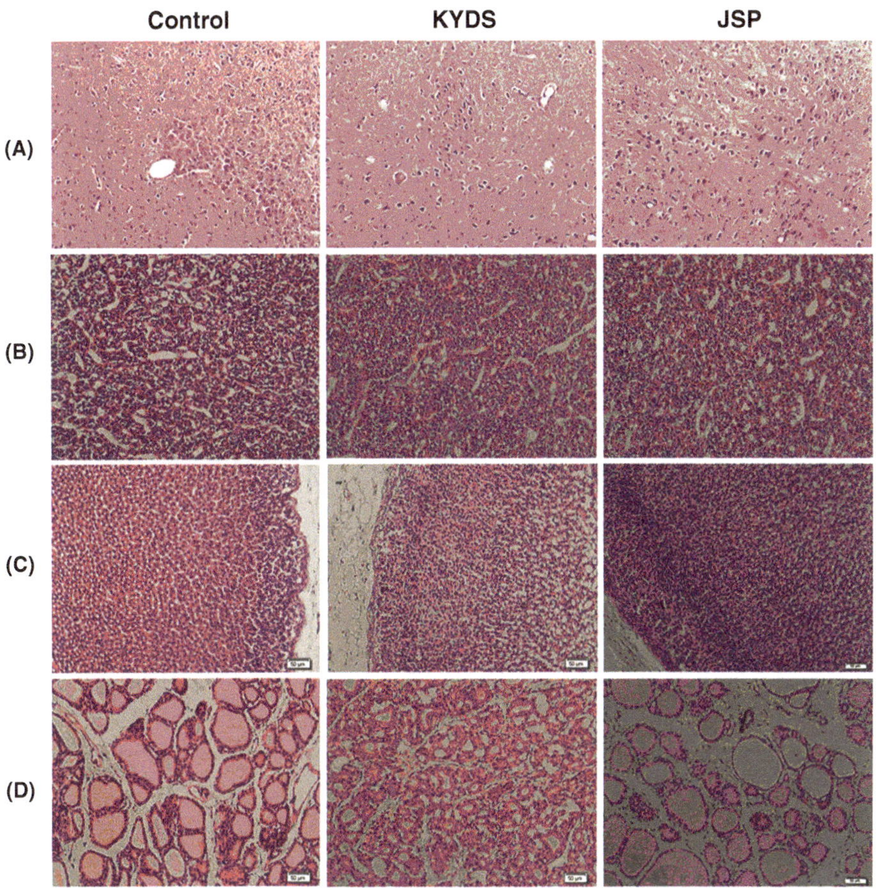

FIGURE 14.1 **H&E staining for histological evaluation.**
Typical photographs of hypothalamic: (A) pituitary; (B) adrenal; (C) thyroid; (D) sections stained with H&E. (Magnification 200×.)

thyroid sections. As demonstrated in Fig. 14.1, the histopathological examination of tissue for the control group, the SYX, and the SQW-treated group was clearly depicted. This staining showed that SYX rats exhibited the typical pathological features of HPG, HPA, and HPT axes. Administration with SQW could effectively ameliorate these symptoms, as demonstrated by the marked biochemical index compared with the SYX rats.

14.3.2 Urinary Metabolite Profiling by UPLC-MS

Typical UPLC/MS TIC of rat urine from the control group, SYX group, and the SQW treatment group are shown in Fig. 14.2A–C, in positive ionization mode. There were 8302 ion signals in positive mode and 7289 ion signals

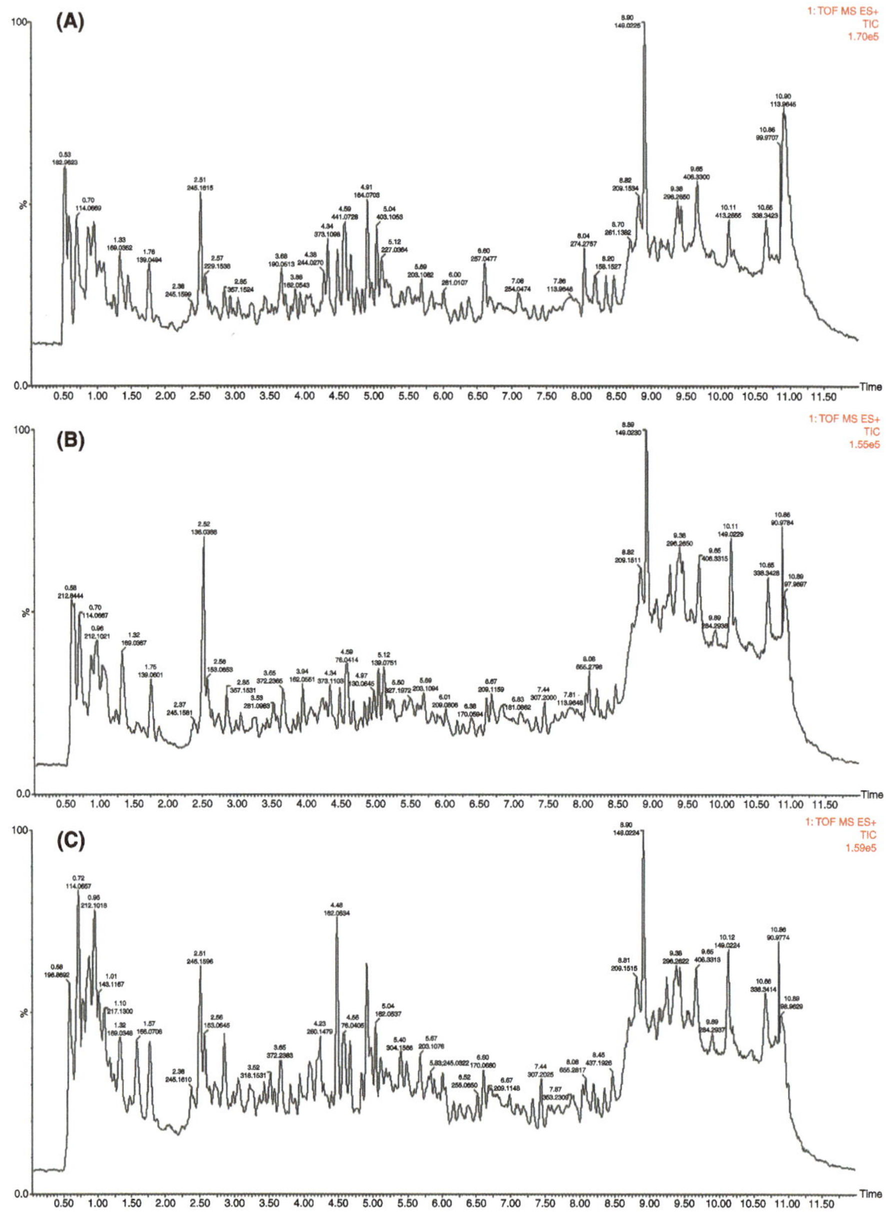

FIGURE 14.2 Typical total ion chromatograms (TIC) obtained from rat urine in the positive mode. (A) Control group; (B) model group; (C) SQW-treated group.

in negative mode detected in the UPLC-Q-TOF-HDMS data. It was difficult, using traditional statistic methods, to identify the ions that contributed to the discrimination of SYX and control groups; with thousands of signals involved.

14.3.3 Effects of SQW on SYX Based on Metabolite Profile

Using the presented UPLC-MS method, the 60 potential biomarkers from UPLC-MS have been found, it is reasonable to take them as the potential drug targets for further investigating the intervening mechanisms of SQW to SYX. Therefore, the levels of the biomarkers from UPLC-MS on day 21 were introduced as variables to PCA, respectively, performed on control, SYX, and SQW-treated groups. The PCA score plot showed that the control, SYX, and SQW treatment groups were separated clearly (Fig. 14.3) and the SQW treatment group was closer to the control group than the SYX group, which suggested that SQW could reverse the pathological process of SYX. By comparing the identified biomarker level in the control, SYX, and SQW treated groups, 31 of the identified biomarkers are completely reversed by SQW (Fig. 14.4). The relative mean height intensity of different metabolites was graphed in Fig. 14.4. All of the reversed biomarkers belonged to the taurine and hypotaurine metabolism, primidine metabolism, tyrosine metabolism, tryptophan metabolism, histidine metabolism, and steroid homone biosynthesis metabolism pathways (Fig. 14.5). Monitoring changes of these metabolites may predict the development of SYX. Interestingly, according to the parametric t-test, we found that the relative concentration of these metabolites could be reversed after taking the SQW. Compared with the alterations of SYX-related metabolites, most of them were reset to a normal state after SQW administration.

14.4 DISCUSSION

Traditional Chinese Medicine is a unique medical system that assisted the ancient Chinese in dealing with disease, and has since contributed to the development of important therapeutics to combat diseases worldwide. However, the targets and modes of action of TCM formulae are seldom researched, thus hindering the drug development process. Metabolomics approaches have provided a vast array of significant information data relevant to drug discovery. It seeks to provide a comprehensive profile of all the metabolites present in a biological sample. This property agrees with the holistic thinking of TCM, and plays an increasingly important role in explaining the scientific meaning of evidence-based TCM, and may pave a new way to evaluate the efficacy of TCM. SYX, recorded first in *Neijing*, an early Chinese systematic and theoretical monograph. As a formula combining eight herbal medicines SQW is effective in treating SYX. Nevertheless, its holistic mechanism still requires defining.

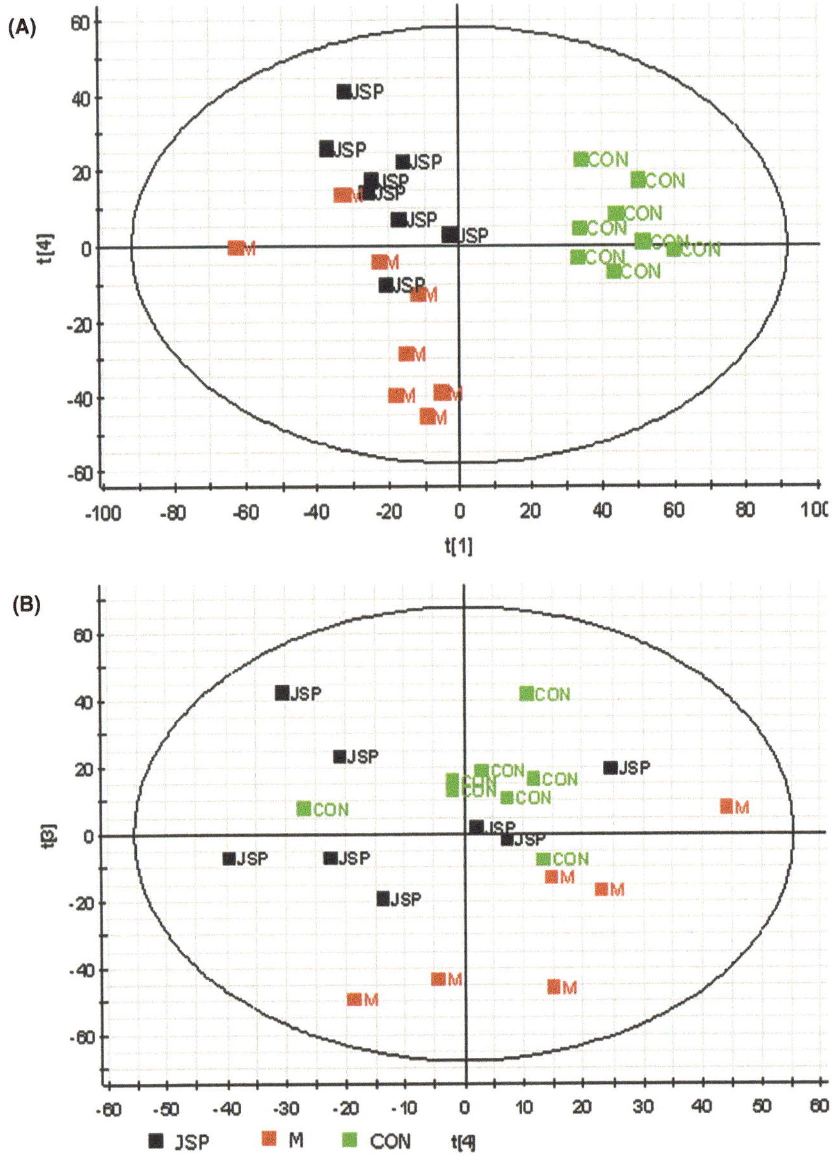

FIGURE 14.3 PCA score plots for the SYX after SQW treatment in positive mode (A); and negative mode (B); (■) Control group, (■) model group, (■) SQW (JSP) group.

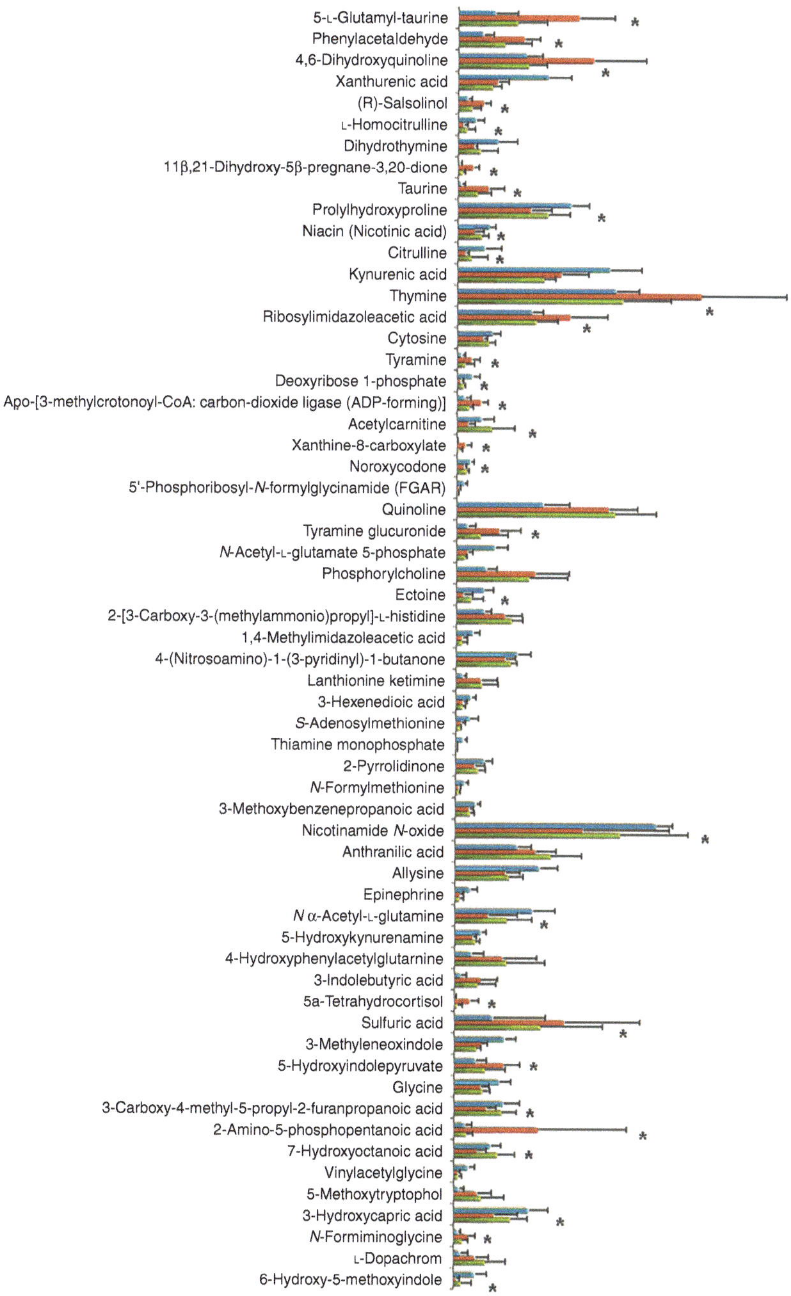

FIGURE 14.4 Changes in the relative intensity of target metabolites as identified by UPLC/MS. A two-tailed, parametric t-test was used to determine the significance of the change in relative intensity for each metabolite. Bars represent the mean relative metabolite concentration and standard deviations. Compared with model group $^{*}p < 0.05$.

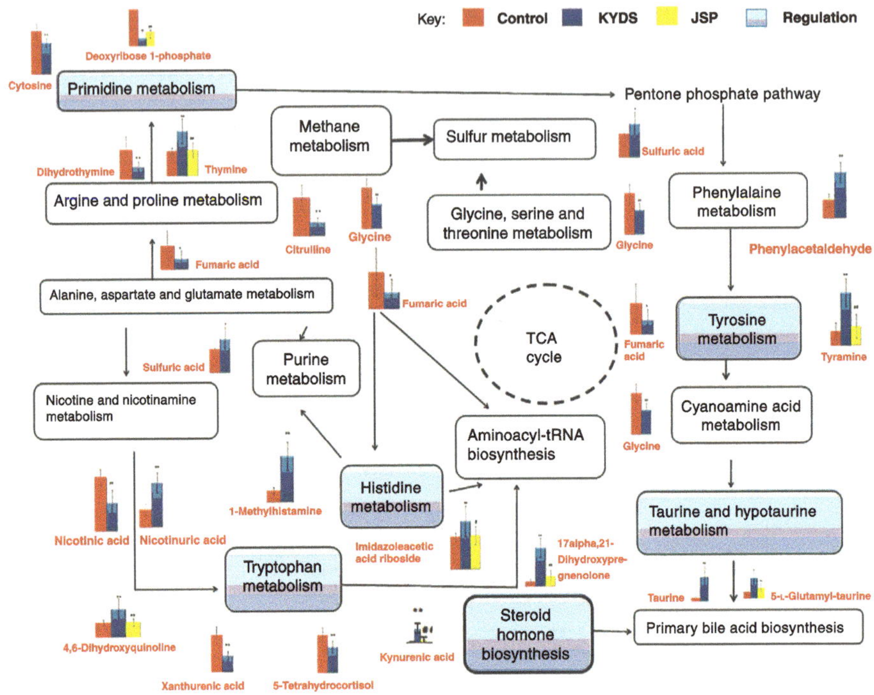

FIGURE 14.5 An overview of the metabolic pathway alterations associated with SQW-treated on SYX group.

The biological parameters suggest an intervention effect for SQW in the regulation of the hypothalamic-pituitary-target gland axis at multiple potential sites. The therapeutic effect of SQW on the treatment of SYX can be demonstrated not only from biochemical analysis and histopathological observations but also in metabolomic study. Next, we aimed to explore novel biomarkers to increase the understanding of SYX, and investigate the therapeutic mechanism of SQW using the metabolomic approach. With PCA analysis, a clear separation of model and control group was achieved. Some potential biomarkers, like 5-L-glutamyl-taurine, xanthurenic acid, L-homocitrulline, etc., have been found. In the potential biomarkers, 5-L-glutamyl-taurine, and taurine is related to taurine and hypotaurine metabolism; sulfuric acid is related to sulfur metabolism. Biomarkers including 5-L-glutamyl-taurine and taurine belong to taurine and hypotaurine metabolism, which indicated that SYX was accompanied by inflammation. Furthermore, sulfuric acid indicated that the balance of sulfur metabolism was broken during SYX formation. Next, in order to examine whether SQW was able to influence the metabolite levels of SYX: the relative intensities of the metabolites together in three groups (as displayed in

Fig. 14.4) – showed that postadministration of SQW the levels of these metabolites had tendencies to the control group.

The rats were administered orally SQW once daily for the following 8 days, and then injected with corticosterone once daily for 21 days to simulate SYX. Combined with the results of the biological parameters assay, these changes in urinary metabolites suggest that the disorders of metabolism pathways are related to SYX, and the potential effect of SQW. Urinary excretion of these metabolites was also shown to have both high specificity and sensitivity as markers. Metabolic trajectory analysis has illustrated that SQW can be regulated back toward the levels, to that of the control group, demonstrating that SQW could have therapeutic efficacy. After treatment with SQW, the relative content of these metabolites were effectively regulated, which suggests that the therapeutic effects of SQW on SYX may involve regulating disturbances to the metabolism of primidine metabolism, histidine metabolism, steroid homone biosynthesis, tyrosine metabolism, tryptophan metabolism, taurine and hypotaurine metabolism, etc. We had built the metabolomic features of SQW to protect against SYX. A total of 31 metabolites may correlate to the regulation of SQW treatment on SYX. Combining urine metabolomic results with the pharmacological assay, we can obtain a greater knowledge of the therapeutic effects brought about through regulating the disturbed metabolism pathways. Our work also confirms that the metabolomics platforms are able to dissect the underlying efficacies and mechanisms of TCM.

14.5 CONCLUSIONS

In this study, the urine metabolomics approach has successfully established biomarker exploration for SYX, and the mechanism studies of SQW. This is the first study on the intervention effects of SQW in SYX rats unravelled using UPLC/Q-TOF MS metabolomics, and multivariate data analysis combined with the biochemical tests. As a result, 31 metabolites are completely reversed by SQW through regulating the disturbed metabolism pathways, such as primidine metabolism, histidine metabolism, steroid homone biosynthesis, tyrosine metabolism, tryptophan metabolism, taurine and hypotaurine metabolism, etc., to their normal state. This is helpful in further understanding the therapeutic mechanism of SQW against SYX for clinical use.

CHAPTER 15

Metabolic Profiling and Biomarkers Analysis of XinQiXu Syndrome

Xijun Wang, Hui Sun, Aihua Zhang

CONTENTS

15.1 INTRODUCTION

XinQiXu (XQX) Syndrome is a manifestation of myocardial ischemia and is a leading cause of death and disability worldwide (Wang et al., 1998). XQX is critical both for diagnosing coronary heart disease, and for selecting and evaluating the response to therapeutic interventions. Currently, myocardial ischemia is diagnosed through an ECG ST-segment, and the T-wave changes testing (Ekerstad et al., 2011; Zhai et al., 2011). This approach, however, is often unsatisfactory because of the fact that the transient nature of ECG changes is not suitable for detecting serious XQX. The existing evidence also demonstrates that XQX is associated with significantly higher healthcare and productivity costs, studies in this area have a number of limitations (Calvert et al., 2011). Furthermore, the diagnosis and prognosis of patients with XQX remains very poor because the molecular mechanism underlying XQX is not fully understood. In this context, there is an urgent need to identify the novel biomarkers of XQX. Fortunately, significant advances with metabolomic techniques hold promise for overcoming these limitations. Metabolomics offers a global analysis of low-molecular-weight metabolite in biological samples, attempts to capture global changes, and physiological status in biochemical networks and pathways, in order to elucidate the sites of perturbations. Thus, UHPLC/TOF-MS-based metabolomics was conducted to identify possible biomarkers, and metabolism pathways on XQX (Wang et al., 2013a).

15.2 MATERIALS AND METHODS

15.2.1 Animal Handling and Sample Preparation

All the rats were randomly divided into two groups of sixteen rats each as follows: control and model groups. All animals were allowed to acclimatize in metabolism cages for 1 week prior to treatment. The rat model was performed as described previously (Liang et al., 2011). Urinary samples for metabolomic

Chinmedomics. http://dx.doi.org/10.1016/B978-0-12-803117-9.00015-9

study, serum samples for biochemical measurement, and heart samples for histopathology were collected. The serum was used for biochemical indicators detection according to the instructions of the commercial kits. The heart was cut for pathological section diagnosis and electron microscopic observation. Urine was collected from metabolism cages at ambient temperature throughout the whole procedure, and centrifuged at 13,000 rpm at 5°C for 15 min, and the supernatants were stored frozen at −20°C until metabolomic analysis.

15.2.2 Metabolic Profiling Platform

15.2.2.1 Chromatography

Chromatographic analysis was performed in a Waters ACQUITY UHPLC system controlled with Masslynx (V4.1, Waters Corp., Milford, USA). An aliquot of 6 μL of sample solution was injected onto an ACQUITY UHPLC HSS T3 column (50 × 2.1 mm, 1.8 μm, Waters Corp., Milford, USA) at 45°C, and the flow rate was 0.5 mL/min. The optimal mobile phase consisted of a linear gradient system of: (A) 0.1% formic acid in water; (B) 0.1% formic acid in acetonitrile, 0–1 min, 1% B; 1–2.5 min, 10% B; 2.5–6.5 min, 20% B; 6.5–8.5 min, 40% B; 8.5–11.5 min, 99% B. In addition, the QC sample was used to optimize the condition of UHPLC-Q-TOF-HDMS.

15.2.2.2 Mass Spectrometry

The eluent was introduced into the high-definition mass spectrometer (Waters Corp., Milford, USA) analysis, and the optimal conditions of analysis were as follows: the source temperature was set at 110°C, desolvation gas temperature was 300°C, cone gas flow was 50 h, desolvation gas flow was 600 L/h. In positive ion mode, the capillary voltage was 3.4 kV, the sampling cone voltage was 30 V, and extraction cone voltage was 3.0 V. In negative ion mode, the capillary voltage was 3.0 kV, the sampling cone voltage was 35 V, and extraction cone voltage was 4.0 V. The data acquisition rate was set to 0.4 s/scan, with a 0.1 s inter scan delay. Data were collected in centroid mode from 100–1000 Da.

15.2.2.3 Multivariate Data Analysis and Metabolic Pathway

The intensity of each ion was normalized with respect to the total ion count to generate a data matrix that consisted of the retention time, the m/z value, and the normalized peak area. The multivariate data analysis (PCA and OPLS-DA) was analyzed by EZinfo software 2.0 (Waters Corp., Milford, USA). Potential markers of interest were extracted from the S-plots constructed following analysis with OPLS-DA, and markers were chosen based on their contribution to the variation, and correlation within the data set. This information was then submitted for database searching, either inhouse or using the online database.

The pathway analysis of potential biomarkers was performed by MetPA software based on a biochemical database.

15.3 RESULTS

15.3.1 Biochemical Analysis and Histopathological Observations

To explore the XQX animal model, the biochemical analysis was performed, and the level of superoxide dismutase (SOD), maleic dialdehyde (MDA), cyclic adenosine monophosphate (cAMP), cyclic guanosine monophosphate (cGMP), atrial natriuretic peptide (ANP), and troponins I (Tn-I) in serum were compared with that of the control rats. The biochemistry parameters of the control, and model groups were summarized in Table 15.1. The SOD, cAMP, cGMP, MDA, and ANP values of the model group showed significant difference compared to that of the control group ($p < 0.05$). Hematoxylin–eosin staining was further used to survey the histopathological changes in the heart tissue. As demonstrated in Fig. 15.1, the histopathological examination of cardiac muscle tissue was clearly depicted in myonecrotic areas. From the results of the pathological sections, there was atrophy of the myocardial fiber, dilation of blood vessels among the myocardial fiber, and dilation of myocardial fiber – in the model group compared with the control group.

15.3.2 Changes in Myocardial Ultrastructure

The ultrastructures of the surrounding areas of the myocardial tissues were also examined. Figure 15.1 presents the representative electron micrographs of the myocardial tissues. The disorder of the cardiac muscle fiber, the edema of the cell matrix, and the swelling of the mitochondria were clearly observed in the model group, compared with the control group. Besides, some of the cristae in the mitochondria were dissolved. And the substance on the intercalated disk reduced slightly. While in the model group, the myocomma was disordered, and the light and dark band of the myocomma was not clear. There is no significant change in the structure of the intercalated disk, and the cristae in the mitochondria. A dramatic injury in cardiac muscle cells, as indicated by the disrupted myofibrils, and the ruptured mitochondria.

Table 15.1 Biochemistry Results in Different Groups (mean ± SD, $n = 8$)

Group	SOD (U/mL)	MDA (nmol/mL)	cAMP (pmol/mL)	cGMP (pmol/mL)	ANP (ng/mL)	cTn1
Control	129.53 ± 5.87	2.25 ± 0.29	16.97 ± 4.35	14.05 ± 2.36	0.27 ± 0.06	6.62 ± 0.33
Model	107.14 ± 12.88*	3.06 ± 0.54*	20.88 ± 2.05*	17.45 ± 2.90*	0.42 ± 0.04*	7.54 ± 1.75

**Significant difference from control group at* $p < 0.05$

(A) (B)

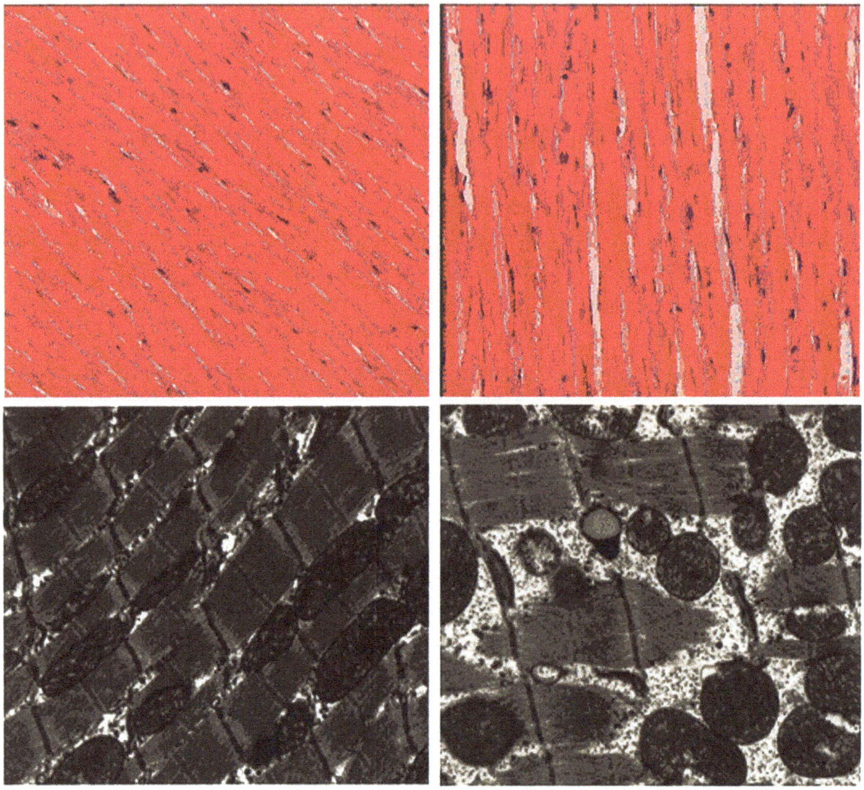

(C) (D)

FIGURE 15.1

Representative photomicrographs of cardiac tissue sections stained with H&E: (A) control group; (B) model group; changes of myocardial ultrastructure; (C) control group; (D) model group.

15.3.3 Metabolic Study

15.3.3.1 Analysis of Metabolic Profiling

All the data containing the retention time, peak intensity, and exact mass were imported in the Masslynx™ software for multiple statistical analyses. Typically, the OPLS-DA was built to find individual differences, and the potential biomarkers that can really reflect the differences between the model and the control groups (Fig. 15.2A and D). The loading plot of OPLS-DA showed metabolic profiles in the control and model groups were separated clearly (Fig. 15.2B and E). The ions showed significant difference in abundance between the control and the treated animals, and contributed to the observed separation and were selected from the respective S-plots as potential markers in positive and negative ion mode (Fig. 15.2C and F).

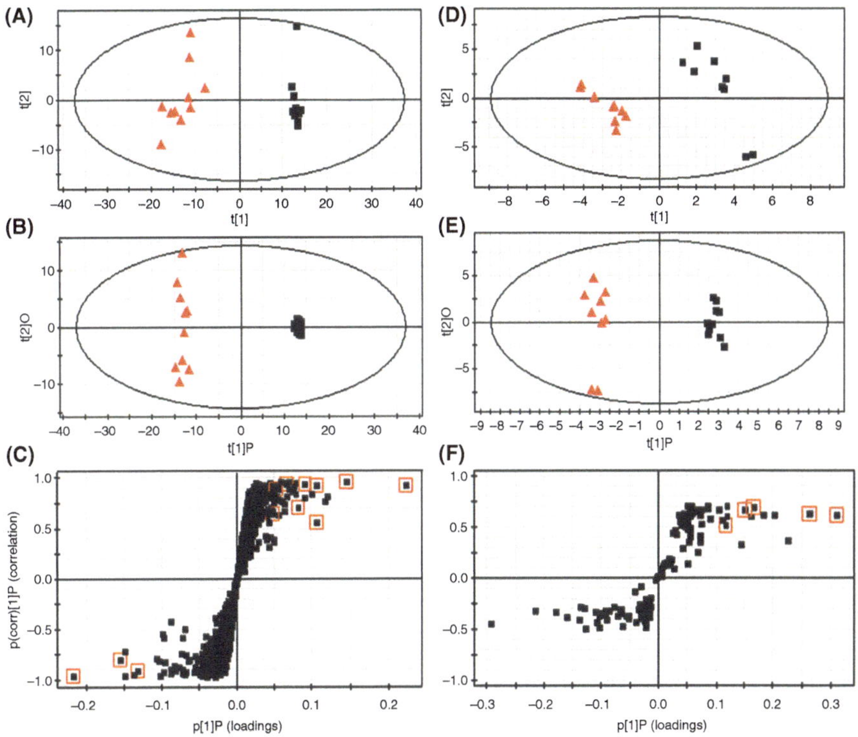

FIGURE 15.2

OPLS-DA model result for control (■ and model group (▲ in positive ion mode)) (A). Loading plot of OPLS-DA of XQX in positive ion mode. (B) S-plot of OPLS-DA of XQX in positive ion mode (C). OPLS-DA model result for control (■ and model group (▲ in negative ion mode)) (D). Loading plot of OPLS-DA of XQX in negative ion mode (E). S-plot of OPLS-DA of XQX in negative ion mode (F).

15.3.3.2 Biological Pathway and Function Analysis

Metabolite identification was conducted with high-resolution MS and MS/MS fragments, as well as database analyses. The potential biomarkers of significant contribution were characterizated as 12 in positive ion mode, and 5 in negative ion mode (Table 15.2 with corresponding retention time, *m/z*, ion mode, and related trends). The relative mean height intensity of different metabolites was graphed in Fig. 15.3. Metabolic pathway analysis revealed that potential biomarkers were identified from glycolysis or gluconeogenesis, biosynthesis of unsaturated fatty acids, fatty acid biosynthesis, and purine metabolism that changed specifically in the setting of XQX. Of the 17 distinct metabolites identified from these pathways, many were in the various progress stages of XQX. The detailed construction of the glycolysis or gluconeogenesis metabolism pathways with higher scores where shown in Fig. 15.4. Results suggested that

Table 15.2 Identification and Trends of Potential Biomarkers Related with XQX vs. the Control Group in Positive and Negative ESI Mode

No.	t_R (min)	Ion (*m/z*)	ESI Mode	Metabolites	Content Variance
1	0.86	270.1342	+	3-Oxohexadecanoic acid	↓
2	8.96	407.2763	+	7-Ketodeoxycholic acid	↑
3	9.30	409.2711	+	3b,7b,12a-Trihydroxy-5b-cholanoic acid	↑
4	1.07	267.1317	+	Glyceric acid 1,3-biphosphate	↓
5	8,52	409.2711	+	3a,7b,12b-Trihydroxy-5b-cholanoic acid	↑
6	2.36	297.1417	+	13(S)-HODE	↓
7	3.61	162.0517	+	L-Carnitine	↓
8	0.62	301.1496	+	3-Hydroxyoctadecanoic acid	↓
9	0.50	169.0336	+	Uric acid	↓
10	0.44	361.1711	+	Tetracosatetraenoic acid	↓
11	2.65	311.1588	+	12(13)Ep-9-KODE	↓
12	1.84	285.1542	+	Stearic acid	↓
13	3.82	295.1200	−	9(S)-HODE	↓
14	8.71	329.2280	−	9,10,13-TriHOME	↓
15	4.47	295.1278	−	9-HODE	↓
16	9.16	329.2255	−	9,12,13-TriHOME	↓
17	0.58	222.9897	−	3-Hydroxykynurenine	↓

Notes: ↑, content increased; ↓, content decreased.

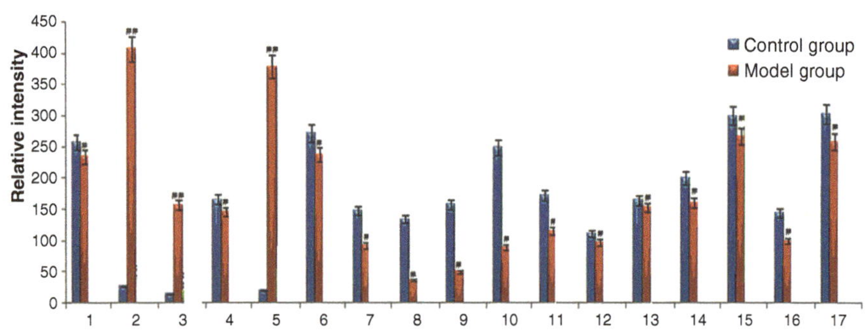

FIGURE 15.3 Changes in the relative intensity of target metabolites identified by UHPLC/MS. The corresponding markers are represented in Table 15.2. A two-tailed, parametric t-test was used to determine the significance of the change in the relative intensity for each metabolite. Bars represent the mean relative metabolite concentration, and the standard deviations. Compared with the model group $^{*}p < 0.05$; $^{**}p < 0.01$.

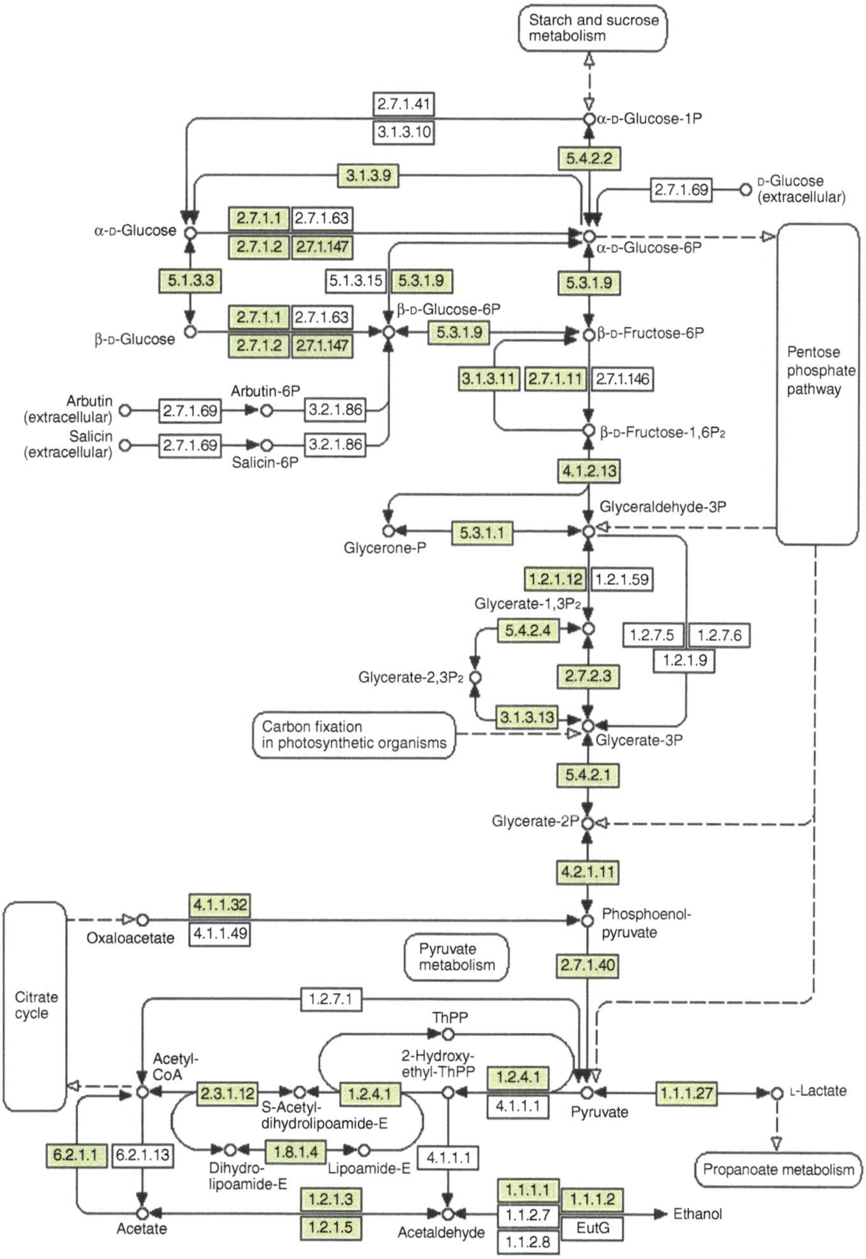

FIGURE 15.4 Construction of the glycolysis or gluconeogenesis metabolism pathways in rats. The map was generated using the reference map by KEGG (www.genome.jp/kegg/). (Green boxes indicate enzymatic activities with putative cases of analogy in rats.)

these target pathways showed the marked perturbations over the time-course of XQX, and could contribute to the development of XQX.

15.4 DISCUSSION

Coronary artery disease is one of the leading causes of death worldwide. Despite extensive studies, however, its metabolic mechanism associated with XQX remains unclear. New biomarkers are needed to augment the information to illuminate disease mechanisms. Metabolomics represents a paradigm shift in the metabolic research away from method, that focuses on a limited number of single pathways to approach, that attempts to capture the complexity of metabolic networks (Zhang et al., 2013). Investigators have begun to incorporate the metabolomics approach to the identification of disease states. Additionally, the high-throughput metabolomics makes it ideal to perform biomarker screens for diseases or drug efficacy (Wang et al., 2013b). In this study, we aimed to explore novel biomarkers to increase the understanding of the mechanism of XQX using a metabolomic approach, and the metabolite network of XQX was also predicted via ingenuity pathway methods.

The levels of 17 metabolites exhibited the marked changes compared with that of the control rats. Urinary excretion of these metabolites was also shown to have both high specificity and sensitivity as markers. Glyceric acid 1,3-biphosphate or 1,3-bisphosphogylcerate primarily exists as a metabolic intermediate in glycolysis during respiration. They have been recognized as regulatory signal implicated in the control of metabolism, oxygen affinity of red cells, and other cellular functions. 3-oxo-hexadecanoic acid and 3-hydroxyoctadecanoic acid are intermediates in fatty acid biosynthesis. 9(S)-HODE, 9-HODE, and 13-HODE are endogenous fatty acid peroxisomal proliferator-activated receptor-gamma (PPARgamma) agonists, synthesized in the body from linoleic acid. They can activate PPARgamma in human endothelial cells increasing plasminogen activator inhibitor type-1 expression. 9,10,13-TriHOME, 9,12,13-triHOME are trihydroxyoctadecenoic acid metabolites of linoleic acid, one of the major fatty acids found in lipids. Vascular tissue converts various polyunsaturated fatty acids to monohydroxy and trihydroxy metabolites derived from hydroperoxides that may be involved in regulating prostaglandin synthesis. L-Carnitine is an essential factor in fatty acid metabolism in mammals. It is most important known metabolic function is to transport long chain fatty acids into the mitochondria of muscle cells, including those in the heart, for their utilizations in metabolism through β-oxidation. Hearts undergoing severe arrhythmia quickly deplete their stores of carnitine. Carnitine supplementation has improved some patients who have angina secondary to coronary artery disease. L-Carnitine supplementation also has a protective role on exhaustive exercize-induced oxidative stress. In this study, the concentrations of the ions described above were significantly decreased in the model group, than in those of control group ($p < 0.05$). It suggested that XQX syndrome caused oxidative stress, and lipid peroxidation that further caused disorder of the metabolism, and biosynthesis of fatty acid (including linoleic acid metabolism). The metabolism disorder brought insufficient energy supply to cardiac muscle, and induced physiological effects, e.g., the decrease in the contractility of cardiac myocytes, arrhythmia, and myocardial ischemia. Bile acids are physiological detergents that facilitate excretion, absorption, and transport of fats and sterols, including cholesterol, in the intestine and liver. 7-ketodeoxycholic acid, 3b,7b,12a-trihydroxy-5b-cholanoic acid, 3a,7a,12b-trihydroxy-5b-cholanoic acid are all bile acids. In this study, the concentrations of the bile acids were significantly increased in the model group than in the control group. The concentrations of uric acid in the urine were significantly decreased in the model group, than in the control group, this may induce the elevation of serum uric acid, and furthermore, increase the risk for cardiovascular diseases. The significant decrease of 3-hydroxyanthranilic in the model group compared with the control group was speculated to be the lessened intake of tryptophan, one of the essential amino acids.

Metabolomics for the screening of biomarker patterns, and the elucidation of biochemical processes during the postgenomic era has increased contemporaneously with the progress in global systems biology. It will expand into a deeper understanding of the disease pathophysiology, which could lead to the discovery of new targets, and the identification of biomarkers in a clinical setting. Application of the metabolomic technologies for the study of XQX will increase our understanding of the pathophysiological processes involved, and this should help us to identify potential biomarkers to develop new therapeutic strategies. System analysis of metabolic networks that are a central paradigm in biology will help us in identifying new biomarkers which will generate a more indepth understanding of the XQX mechanism, and thus provide better guidance for drug efficacy. Additionally, we suggest that the application of the metabolite-profiling tool can provide an efficient means for early diagnosis of XQX, and provide a practical method for monitoring therapeutic intervention. Indeed, future metabolomic studies in human populations with XQX will be needed to validate the biomarkers found in the animal model.

15.5 CONCLUSIONS

Our study highlights the importance of the power of metabolomics to elucidate the metabolic characters of XQX at the global metabolome level. Analyzing the topology of the network, we have detected 17 potential biomarkers, and predicted the major metabolites network of XQX using the validated pattern recognition methods, and computational systems analysis. These biomarkers that represent molecular targets for XQX disease patterns were identified, and related to the acutely perturbed pathways of: glycolysis or gluconeogenesis, biosynthesis of unsaturated fatty acids, fatty acid biosynthesis, and purine metabolism. Taken together, our findings provide important evidence that the metabolomic technologies can be used to identify XQX biomarkers, which is a crucial step towards improving current diagnosis, developing new presymptomatic treatments, and assessing drug efficacy.

References

Calvert, J.W., Condit, M.E., Aragón, J.P., Nicholson, C.K., Moody, B.F., Hood, R.L., Sindler, A.L., Gundewar, S., Seals, D.R., Barouch, L.A., Lefer, D.J., 2011. Exercise protects against myocardial ischemia-reperfusion injury via stimulation of β(3)-adrenergic receptors and increased nitric oxide signaling: role of nitrite and nitrosothiols. Circ. Res. 108 (12), 1448–1458.

Ekerstad, N., Swahn, E., Janzon, M., Alfredsson, J., Löfmark, R., Lindenberger, M., Carlsson, P., 2011. Frailty is independently associated with short-term outcomes for elderly patients with non-ST-segment elevation myocardial infarction. Circulation 124 (22), 2397–2404.

Liang, X., Chen, X., Liang, Q., Zhang, H., Hu, P., Wang, Y., Luo, G., 2011. Metabonomic study of Chinese medicine Shuanglong formula as an effective treatment for myocardial infarction in rats. J. Proteome Res. 10 (2), 790–799.

Wang, S., Zhao, M., Lu, X., 1998. Clinical study on relationship among Heart-Qi Deficiency, left ventricular function and myocardial ischemia in patients with coronary heart disease. Zhongguo Zhong Xi Yi Jie He Za Zhi 18 (8), 457–460.

Wang, X., Wang, Q., Zhang, A., Zhang, F., Zhang, H., Sun, H., Cao, H., Zhang, H., 2013a. Metabolomics study of intervention effects of Wen-Xin-Formula using ultra high-performance liquid chromatography/mass spectrometry coupled with pattern recognition approach. J. Pharm. Biomed. Anal. 74, 22–30.

Wang, X.J., Yan, G.L., Zhang, A.H., Sun, H., Piao, C.Y., Li, W.Y., Sun, C., Wu, X.H., Li, X.H., Chen, Y., 2013b. Metabolomics and proteomics approaches to characterize and assess proteins of bear bile powder for hepatitis C virus. Chin. J. Nat. Med. 11 (6), 653–665.

Zhai, P., Sciarretta, S., Galeotti, J., Volpe, M., Sadoshima, J., 2011. Differential roles of GSK-3β during myocardial ischemia and ischemia/reperfusion. Circ. Res. 109 (5), 502–511.

Zhang, A.H., Sun, H., Han, Y., Yan, G.L., Yuan, Y., Song, G.C., Yuan, X.X., Xie, N., Wang, X.J., 2013. Ultraperformance liquid chromatography-mass spectrometry based comprehensive metabolomics combined with pattern recognition and network analysis methods for characterization of metabolites and metabolic pathways from biological data sets. Anal. Chem. 85 (15), 7606–7612.

CHAPTER 16

Active Constituents Screening Based on Correlation Analysis Between Marker Metabolites and the Absorbed Constituents in WenXin Formulae

Xijun Wang, Hui Sun, Aihua Zhang

CONTENTS

16.1 INTRODUCTION

Traditional Chinese Medicine (TCM) has its own unique medical system with the significant characteristics of pursuing an overall therapeutic effect with a multitarget treatment (Zhang et al., 2013). It consists of multiple medicinal plants (called "formulae") that can regulate balance, and homeostasis of the body in a holistic fashion, and exert synergistic effects via the interaction of individual metabolites in the formulae (Wang et al., 2008). Due to the complexity of the chemical compositions, and many endogenous intervening substances, it is a challenge to identify the chemical constituents, and metabolites related to TCM. Fortunately, the serum pharmacochemistry method has been proven to assist in the elucidation of the effective constituents, and metabolites of TCM (Wang, 2006; Dong et al., 2012). In recent years, many analytical methods have been successfully applied in the analysis of TCM formulae, and liquid chromatography coupled with tandem mass spectrometry (LC-MS) was widely applied for the analysis of herbal constituents *in vivo* (Wang et al., 2013a; Sun et al., 2013; Yin et al., 2013). Especially, UPLC has, with minor particles (sub 1.8 μm), greatly improved the sensitivity, resolution, and speed, and integration with MS, to form a high-performance platform for TCM. The MetaboLynx™ data processing approach is concerned with the analysis and interpretation of complex data variables (Sun et al., 2013).

WenXin Formulae (WXF) is one of the most famous traditional formulae, and has been efficiently and widely used to treat XQX Syndrome for thousands of years in Asia. It is composed of eight herbs: *Ginseng Radix et Rhizoma, Cinnamomi Ramulus, Allii Macrostemonis Bulbus, Trichosanthis Fructus, Pinelliae Rhizoma Praeparatum Cum Alumine, Paeoniae Radix Rubra, Ophiopogonis Radix*, and *Coptidis Rhizoma* (Wang et al., 2013b). All the herbs are included in the *Chinese Pharmacopoeia* (version 2010). Despite the popular usage of WXF, few reports are available on the systematic analysis of its chemical constituents, and little

Chinmedomics. http://dx.doi.org/10.1016/B978-0-12-803117-9.00016-0

CONTENTS

information is known about the changes of metabolism postoral administration of WXF. Although WXF has been used to improve cardiovascular patients in clinical practice, however, there is a sparcity of literature that has reported about the detailed therapeutic mechanisms. It is of significance to systematically analyze the chemical constituents, and *in vivo* metabolites in WXF formulae so as to interpret its material basis for the pharmacological effects. Therefore, the present study was undertaken to investigate the chemical constituents, and the metabolites *in vivo* (Cao et al., 2014) and to further screen the absorbed bioactive components in WXF.

16.2 EXPERIMENTAL

16.2.1 Chemicals and Materials

The HPLC grade methanol was purchased from Merck (Darmstadt, Germany). Water was produced by a Milli-Q Ultra-pure water system (Millipore, Billerica, USA). Leucine enkephalin was purchased from Sigma–Aldrich (MO, USA). Formic acid was purchased from Tianjin Kermel Chemical Reagent Co., Ltd (Tianjin, China). The plant materials: *Ginseng Radix et Rhizoma, Cinnamomi Ramulus, Allii Macrostemonis Bulbus, Trichosanthis Fructus, Pinelliae Rhizoma Praeparatum Cum Alumine, Paeoniae Radix Rubra, Ophiopogonis Radix, Coptidis Rhizoma* were purchased from the Harbin Tongrentang Drug Store (Harbin, China). All crude drugs were authenticated by pharmacognosist, Professor Xijun Wang from the Department of Pharmacognosy of Heilongjiang University of Chinese Medicine. Voucher specimens were deposited at the authors' laboratory.

16.2.2 Preparation of WXF Samples for LC/MS Analysis

According to the original composition and preparation method of WXF as prescribed in writing in the *Synopsis of the Golden Chamber,* WXF was prepared in the following procedure. Weigh 125 g of the powder of WXF, suspend in 1000 mL of 70% (v/v) methanol, and extract backflow for three times, each for 1 h. The extracted solution was collected by vacuum suction filtration, and finally the solution was freeze-dried into a powder. The solution was filtered through 0.22 μm membranes (pore size) prior to use, and a 5 μL aliquot was injected for analysis.

16.2.3 Animal Experiments

16.2.3.1 Constituents *In Vivo* Study

Male Wistar rats (weighing 220 ± 20 g) were purchased from the Laboratory Animal Center of Heilongjiang University of Traditional Chinese Medicine. The animals were housed in an animal room (24 ± 2°C, 60 ± 5% relative humidity) for 1 week. A 12 h dark/light cycle was set, and they were given water, and fed normal food *ad libitum*. Prior to drug administration, the experimental animals were deprived of food for 24 h once more, and were free to access water. The freeze-dried WXF powder was dissolved in distilled water, then the mixture was grinded adequately, and sonicated in an ultrasonic bath for 30 min to prepare the decoction (0.2 g/mL). Rats were orally administered with WXF extract (1 mL per 100 g body weight). The control group was orally administrated with an equivalent volume of distilled water. After 1 h the rats were anesthetized by an intraperitoneal injection of 1% pentobarbital sodium (0.15 mL per 100 g body weight).

16.2.3.2 Metabolites *In Vivo* Study

The male Wistar rats were randomly divided into four groups of 16 rats each as follows: control group, model group, low-dose WXF group, and high-dose WXF group. All rats were allowed to acclimatize in metabolism cages for 1 week prior to treatment. Urinary samples for metabolomic study, serum samples for biochemical measurement, and heart samples for histopathology were collected. The serum was used for biochemical indicators detection according to the instructions of the commercial kits. The heart was cut for pathological section diagnosis, and electron microscopic observation. Urine was collected from metabolism cages at ambient temperature throughout the whole procedure and centrifuged at 13,000 rpm at 5 °C for 15 min, and the supernatants were stored frozen at −20 °C until metabolomic analysis. The quality control sample was used to optimize the condition of UPLC-MS, as it contained most of the information of the whole urine samples.

16.2.4 Preparation of Serum Samples

The blood samples were collected from hepatic portal vein at 1 h postadministration (n = 6 for each group) and the rats were sacrificed. Then, the serum was separated immediately by centrifuging at 4000 rpm for 10 min at 4 °C. Forty micron liter of phosphoric acid was added to 2.0 mL of the above supernatant, and ultrasonicated for 1 min, and then vortexed for 30 s. The mixed solution was applied to preactivated OASIS HLB solid phase extraction C_{18} columns (Waters Corp., USA). Before that, the column was washed with 3 mL of methanol, and 3 mL of water, successively. Then, 100% methanol and elutes were collected and dried under N_2 at room temperature. The residues were redissolved in 100 μL methanol, centrifuged at 4000 rpm for 15 min at 4 °C. The sample was filtered through a 0.22 μm membrane filter unit before analysis, and a 4 μL aliquot was injected for UPLC/MS analysis.

16.2.5 Constituents Analysis *In Vivo*

16.2.5.1 Chromatography Analysis

Chromatographic analysis was performed in a Waters Acquity™ ultra performance LC systems (Waters Corp., Milford, USA) controlled with Masslynx (v4.1). Separation was performed on an ACQUITY UPLC HSS T3 Column (2.1×50 mm, 1.8 μm, Waters Corp., Milford, USA) held at 45 °C, injection volume was 4 μL, and the flow rate was 0.5 mL/min. The optimal mobile phase consisted of A (HCOOH:H_2O = 0.1:100, v/v) and B (HCOOH:methanol = 0.1:100, v/v). The linear elution gradient program was used as follows: 0–3 min, 1–15 % B; 3–10 min, 15–25% B; 10–15 min, 25–40% B; 1.5–20 min, 40–50% B; 20–26 min, 50–99% B.

16.2.5.2 Mass Spectrometry Experiment

UPLC was interfaced with a Waters Synapt™ High Definition TOF Mass system (Waters Corp., Milford, USA) equipped with an electrospray ion source operating in either positive or negative ESI mode. The optimal conditions of analysis were as follows: ESI^+ mode, capillary voltage of 2.8 kV, sampling cone voltage was 440 V, extraction cone voltage was 4.0 V; ESI^- mode, capillary voltage of 2.8 kV, sampling cone voltage was 35.0 V, and extraction cone voltage was 4.0 V. The temperature was set at 110 °C, desolvation gas temperature was 350 °C, and desolvation gas flow was 600 L/h. The mass spectrometer was calibrated using a solution of sodium formate before experiment. The full-scan MS data were produced across the mass range of 50–1500 Da with collision energy of 10 eV.

16.2.5.3 Data Processing

By comprehensively searching databases, such as Chemspider, PubMed, Pubchem compound, an inhouse molecular formula database was established by use of Masslynx™ (v4.1) software MassFragment which includes the compound name, molecular formula, chemical structure. For further confirmation of the structure, and the source of the metabolites, the raw data were processed by a Metabolynx (v 4.1), which employs a comprehensive list of potential biotransformation reactions with the elemental compositions of each possible metabolite, and generates a sequence of extraction ion chromatograms (XICs). Comparing the XICs of the samples postadministration, and control samples allows for the identification of the potential drug metabolites. Elemental composition for expected, and unexpected metabolite peaks was created using an exhaustive list of potential formulas of the metabolites dependant on the ion mode.

16.2.6 Metabolic Profiling Analysis Platform

Chromatographic analysis was performed in a Waters ACQUITY UPLC system controlled with Masslynx (v4.1), Waters Corp., Milford, USA). The optimal mobile phase consisted of a linear gradient system of: (A) 0.1% formic acid in water; (B) 0.1% formic acid in acetonitrile, 0–1 min, 1% B; 1–2.5 min, 10% B; 2.5–6.5 min, 20% B; 6.5–8.5 min, 40% B; 8.5–11.5 min, 99% B. The eluent was introduced into the high-definition mass spectrometer (Waters Corp., Milford, USA) analysis, and the optimal conditions of analysis were as follows: the source temperature was set at 110°C, desolvation gas temperature was 300°C, cone gas flow was 50 h, and desolvation gas flow was 600 L/h. In positive ion mode, the capillary voltage was 3.4 kV, the sampling cone voltage was 30 V, and extraction cone voltage was 3.0 V. In negative ion mode, the capillary voltage was 3.0 kV, the sampling cone voltage was 35 V, and the extraction cone voltage was 4.0 V. The multivariate data matrix was analyzed by EZinfo (software 2.0) (Waters Corp., Milford, USA). The metabolite identities were confirmed by comparison of the mass spectra, and the chromatographic retention times with the available reference standards.

16.3 RESULTS AND DISCUSSION

16.3.1 Optimization of LC and MS Conditions

The LC chromatograms with good separation, different mobile phase compositions were screened, and it was found that methanol was the most suitable eluting solvent system, providing low background noise, and good chromatographic peak formation. In consideration of sensitivity and resolution: the ultimate flow rate was optimized at 0.5 mL/min. Furthermore, the LC column temperature was set at 45°C to reduce the column pressure resulting from a higher flow rate. To acquire maximum sensitivity for most constituents, MS parameters, such as source temperature, desolvation temperature, desolvation gas flow, and capillary voltage were optimized by flow injection analysis. The experiments were performed: desolvation temperature (300, 350, 375°C), source temperature (100, 110, 120°C), capillary voltage (2.6, 2.7, 2.8, 2.9 kV), desolvation gas flow (550, 600, 650 L/h), and the peak area was taken as the criteria for optimization. The optimum conditions were decided as follows: desolvation temperature 350°C, source temperature 110°C, desolvation gas (N_2) 600 L/h, capillary voltage 2.8 kV for positive ion ESI mode and negative ion ESI mode. When formic acid was added to the mobile phase; however, higher and narrower peaks were achieved. Using the optimal UPLC-MS conditions described earlier, most of the main peaks could be well separated by optimum gradient elution.

16.3.2 Characterization of Chemical Constituents From WXF

To characterize the chemical constituents of WXF, a UPLC-ESI-Q-TOF-MS method was established. All compounds' structures were characterized, or tentatively characterized, based on their chromatographic and spectrometric data. The component herb from which each compound was derived was confirmed by individually analyzing the herbs with the same UPLC-ESI-Q-TOF-MS method. Different components could be determined by their specific molecular ions and fragment ions, along with their chromatographic retention times, other compounds in WXF were preliminarily identified. Ultimately, a total

Table 16.1 UPLC/MS Identification of Chemical – Constituents in Rat Plasma Postoral Administration of WXF

No.	T_R (min)	Identified Compound	Negative Ion (*m/z*) Exp.	Negative Ion (*m/z*) Error (mDa)	Positive Ion (*m/z*) Exp.	Positive Ion (*m/z*) Error (mDa)	Formula	MW (Da)
1	2.2	Syringic acid-4-β-D-glucopyranoside	359.0967	−1.4	383.0964	−0.7	$C_{15}H_{20}O_{10}$	360.1059
2	2.49	Protocatechuic acid-3-glucoside	315.0693	−2.3	–	–	$C_{13}H_{16}O_9$	316.0794
3	3.91	Oxypaeoniflorin	495.1461	−4.1	519.1488	−1	$C_{23}H_{28}O_{12}$	496.1580
4	4.82	Magnoflorine	–	–	342.1728	−2.9	$C_{20}H_{24}NO_4$	342.1699
5	5.00	Albiflorin	479.1554 525.1528	0.1	503.1483	4.6	$C_{23}H_{28}O_{11}$	480.1631
6	5.89	Paeoniflorin	479.1551 525.1532	−0.2	503.154	−1.1	$C_{23}H_{28}O_{11}$	480.1631
7	7.66	Linalool 3,6-oxide-6-O-β-D-xylopyranosyl-β- D-glucopyranoside	463.2204	2.5	487.2149	0.6	$C_{21}H_{36}O_{11}$	464.2257
8	9.02	Berberrubine	–	–	322.1021	5.2	$C_{19}H_{16}NO_4$	322.1073
9	10.10	Galloylpaeoniflorin isomer	631.1638	−2.5	–	–	$C_{30}H_{32}O_{15}$	632.1741
10	10.7	Coptisine	–	–	320.0889	2.8	$C_{19}H_{14}NO_4$	320.0917
11	11.00	Epiberberine	–	–	336.1188	4.2	$C_{20}H_{18}NO_4$	336.1230
12	11.50	Jatrorrhizine	–	–	338.1387	−0.1	$C_{20}H_{20}NO_4$	338.1386
13	12.00	Berberine	–	–	336.1200	3	$C_{20}H_{18}NO_4$	336.1230
14	12.40	Palmatine	–	–	352.1518	2.5	$C_{21}H_{22}NO_4$	352.1543
15	13.10	3-Methyl ellagic acid	315.0112	−2.9	317.0262	3.5	$C_{15}H_8O_8$	316.0219
16*	14.00	Not identificated	546.2992	1.6	–	–	$C_{37}H_{41}NO_3$	547.3086
17	14.40	3,3′-Di-O-methylellagic acid	329.0271	−2.6	331.042	3.3	$C_{16}H_{10}O_8$	330.0375
18	15.10	Ginsenoside Rg1	845.4908	3.7	823.4835	−1.5	$C_{42}H_{72}O_{14}$	800.4922
19	15.90	Not identificated	295.1304	−3	–	–	$C_{18}H_{16}O_4$	296.1412
20	16.40	Not identificated	295.1362	2.8	–	–	$C_{18}H_{16}O_4$	296.1412
21	17.00	9(S),12(S),13(S)-Trihydroxy-10(E),15(Z)-octadecadienoic acid	327.2136	−3.5	–	–	$C_{18}H_{32}O_5$	328.2249
22	21.60	Ginsenoside Rb1	1107.5955	0.4	1131.5971	−4.4	$C_{54}H_{92}O_{23}$	1108.6029

*Note: *metabolites*

of 22 compounds (Tables 16.1 and 16.2) were identified, or tentatively characterized, among them, there were 21 constituents, derived from prototypes, including ginsenosides, isoquinoline alkaloids, paeoniflorin monoterpenoid glycosides, and unsaturated fatty acids, etc. Representative BPI chromatogram of WXF in both the positive and negative ion modes by UPLC-MS was included in Figs 16.1 and 16.2.

Table 16.2 MS/MS Data of the Compounds Detected in Rat Plasma Postoral Administration of WXF by UPLC-MS

No.	T_R (min)	Identified Compound	MS/MS Data
1	2.2	Syringic acid-4-β-D-glucopyranoside	359[M-H]$^-$;197[M-H-(Glc-H_2O)]$^-$;179[M-H-(Glc-H_2O)- H_2O]$^-$; 135[M-H-(Glc-H_2O)- H_2O-COO]$^-$;
2	2.49	Protocatechuic acid-3-glucoside	315[M-H]$^-$;153[M-H-(Glc-H_2O)]$^-$;109[M-H-(Glc-H_2O)-COO]$^-$;
3	3.91	Oxypaeoniflorin	495[M-H]$^-$;465[M-H-CH_2O]$^-$;333[M-H-(Glc-H_2O)]$^-$; 281[M-H-$C_9H_{10}O_6$]$^-$;
4	4.56	Magnoflorine	342[M]$^+$;312[M-2CH_3]$^+$;297[M-3CH_3]$^+$; 265[M-CH_3-$C_3H_{12}N$]$^+$
5	5.00	Albiflorin	479[M-H]$^-$;449[M-H-CH_2O]$^-$;357[M-H-$C_7H_6O_2$]$^-$; 327[M-H-CH_2O-$C_7H_6O_2$]$^-$;
6	5.89	Paeoniflorin	525[M+HCOO]$^-$;479[M-H]$^-$;449[M-H-CH_2O]$^-$; 357[M-H-$C_7H_6O_2$]$^-$;327[[M-H-CH_2O-$C_7H_6O_2$]$^-$;
7	7.66	Linalool 3,6-oxide 6-O-β-D-xylopyranosyl-β-D-glucopyranoside	463[M-H]$^-$;331[M-H-(Xyl-H_2O)]$^-$; 169[M-H-(Xyl-H_2O)-(Glc-H_2O)]$^-$;
8	9.02	Berberrubine	322[M]$^+$;307[M-CH_3]$^+$;306[M-CH_3-H]$^+$; 278[M-CH_3-H-CO]$^+$;
9	10.10	Galloylpaeoniflorin isomer	631[M-H]$^-$;613[M-H-H_2O]$^-$;509[M-H-$C_7H_6O_2$]$^-$; 491[M-H-H_2O-$C_7H_6O_2$]$^-$;463[M-H-H_2O-$C_7H_6O_2$-CO]$^-$
10	10.7	Coptisine	320[M]$^+$;318[M-2H]$^+$;292[M-CO]$^+$;290[M-2H-CO]$^+$; 262[M-2H-2CO]$^+$;
11	11.00	Epiberberine	336[M]$^+$;321[M-CH_3]$^+$;320[M-CH_3-H]$^+$;318[M-CH_3-H-2H]$^+$; 292[M-CH_3-H-CO]$^+$;290[M-CH_3-H-2H-CO]$^+$
12	11.50	Jatrorrhizine	338[M]$^+$;323[M-CH_3]$^+$;322[M-CH_3-H]$^+$;320[M-CH_3-H-2H]$^+$; 294[M-CH_3-H-CO]$^+$;292[M-CH_3-H-2H-CO]$^+$
13	12.00	Berberine	336[M]$^+$;321[M-CH_3]$^+$;320[M-CH_3-H]$^+$;306[M-2CH_3]$^+$; 292[M-CH_3-H-CO]$^+$;278[M-2CH_3-CO]$^+$;
14	12.40	Palmatine	352[M]$^+$;337[M-CH_3]$^+$;336[M-CH_3-H]$^+$; 334[M-CH_3-H-2H]$^+$;319[M-CH_3-3H-CH_3]$^+$; 318[M-CH_3-3H-CH_4]$^+$;308[M-CH_3-H-CO]$^+$;
15	13.10	3-Methyl ellagic acid	315[M-H]$^-$;300[M-H-CH_3]$^-$;271[M-H-COO]$^-$;243[M-H-COO-CO]$^-$
16*	14.00	Not identificated	546[M-H]$^-$;533[M-H-CH_3]$^-$; 515[M-H-H_2O]$^-$; 485[M-H-H_2O-2CH_3]$^-$
17	14.40	3,3′-Di-O-methylellagic acid	329[M-H]$^-$;314[M-H-CH_3]$^-$;299[M-H-2CH_3]$^-$;285[M-H-COO]$^-$;
18	15.10	Ginsenoside Rg1	845[M+HCOO]$^-$;799[M-H]$^-$;637[M-H-(Glc-H_2O)]$^-$; 475[M-H-2(Glc-H_2O)]$^-$
19	15.90	Not identificated	295[M-H]$^-$;277[M-H-H_2O]$^-$;233[M-H-H_2O-COO]$^-$; 205[M-H-H_2O-COO-CO]$^-$;
20	16.40	Not identificated	295[M-H]$^-$; 277[M-H-H_2O]$^-$;233[M-H-H_2O-COO]$^-$; 205[M-H-H_2O-COO-CO]$^-$;
21	17.00	9(S),12(S),13(S)-Trihydroxy-10(E), 15(Z)-octadecadienoic acid	327[M-H]$^-$;312[M-H-CH_3]$^-$;295[M-H-CH_3-OH]$^-$; 283[M-H-COO]$^-$
22	21.60	Ginsenoside Rb1	1107[M-H]$^-$;945[M-H-(Glc-H_2O)]$^-$;783[M-H-2(Glc-H_2O)]$^-$; 621[M-H-3(Glc-H_2O)]$^-$459[M-H-4(Glc-H_2O)]$^-$

*Note: *metabolites*

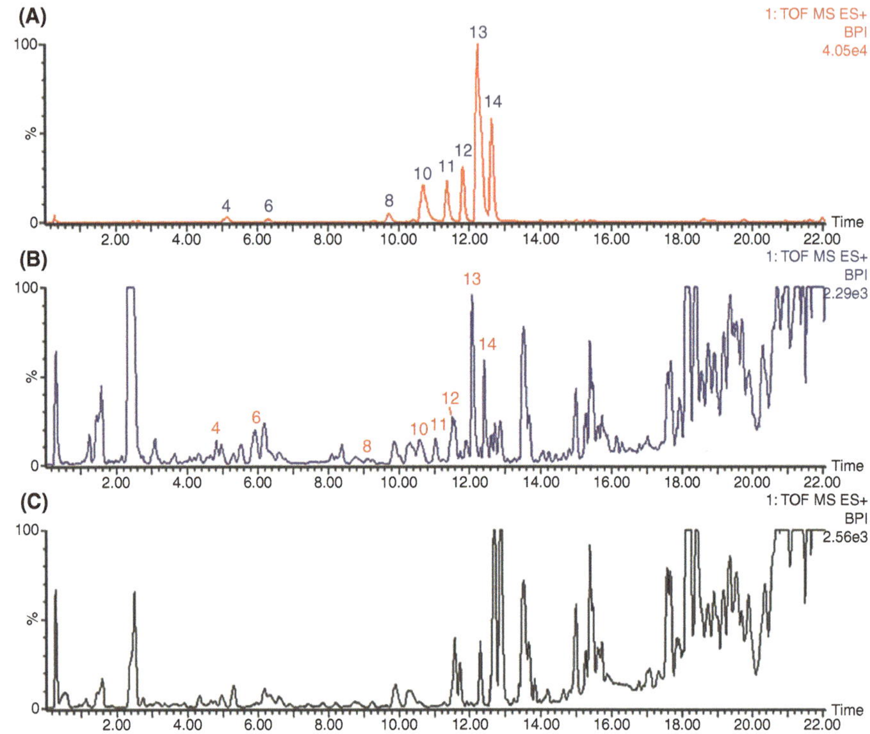

FIGURE 16.1

UPLC-HDMS chromatograms of WXF: (A) dosed rat serum; (B) controlled rat serum; and (C) in positive ion mode.

16.3.3 Analysis of the Prototype Constituents and Metabolites of WXF

The prototype constituents were identified by comparing their chromatographic, and MS/MS data with those of the authentic compounds, and the full scan analysis of WXF. Their element compositions were also calculated on the basis of mass accuracy (ppm) and i-FIT(Norm). Combining the mass accuracy less than 5.0 ppm and the i-FIT(Norm) value less than 1.0, the elemental composition can be unequivocally confirmed. The ions were then subjected to tandem MS/MS analysis. By scrutinizing their elemental compositions, and MS/MS fragmentations as well as retrieving the reference literature, in total, 10 constituents were discovered in rats' serum postdosing WXF, enabling further *in vivo* elucidation of the metabolites in the WXF (Figs 16.3 and 16.4). Among them, five prototype constituents were assigned for their identity, and the other five constituents were inferred as metabolites, and the detailed information was showed in Table 16.3. Peaks b, c, d, e, and f were glucuronide metabolites of the original berberis alkaloids; peak g, h, and j were unsaturated fatty acids;

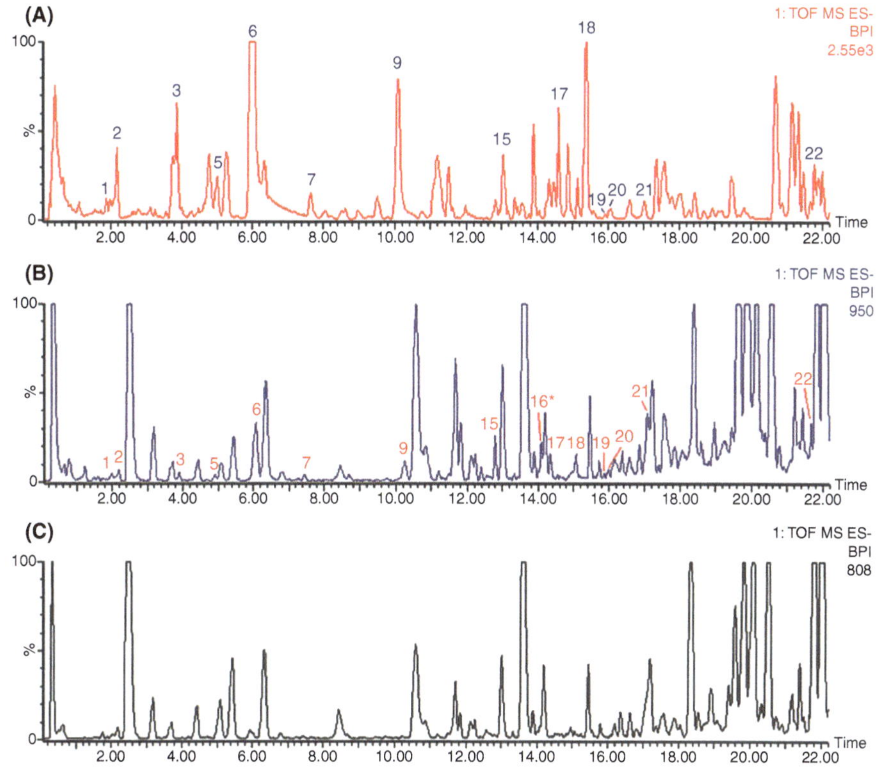

FIGURE 16.2
UPLC-HDMS chromatograms of WXF: (A) dosed rat serum; (B) controlled rat serum; and (C) in negative ion mode.

peak i was ginsenoside Rg1 isomers, and their MS ion information was shown in Table 16.3.

WXF have been used in clinic widely, however, its bioactive ingredients, and metabolites are not well understood, and, of course, this seriously restricts its further clinical application. To rapidly explore the potential chemical markers in rats postoral administration of WXF, an UPLC-MS coupled with MetaboLynx data processing method was proposed, and validated. To establish an integrated method for screening, and analyzing the effective components of TCM; serum pharmacochemistry was proposed, which based on the hypothesis that active constituents were the components absorbed into the blood, and this has provided significant advantages when quickly screening *in vivo* active compounds in medical plants. In the present study, serum pharmacochemistry based on the UPLC-ESI-QTOF/MS of the chromatographic fingerprint method was performed to screen for the bioactive compounds in WXF. By comparative analysis of the chemical profiles of the WXF extracts, control serum, and dosed

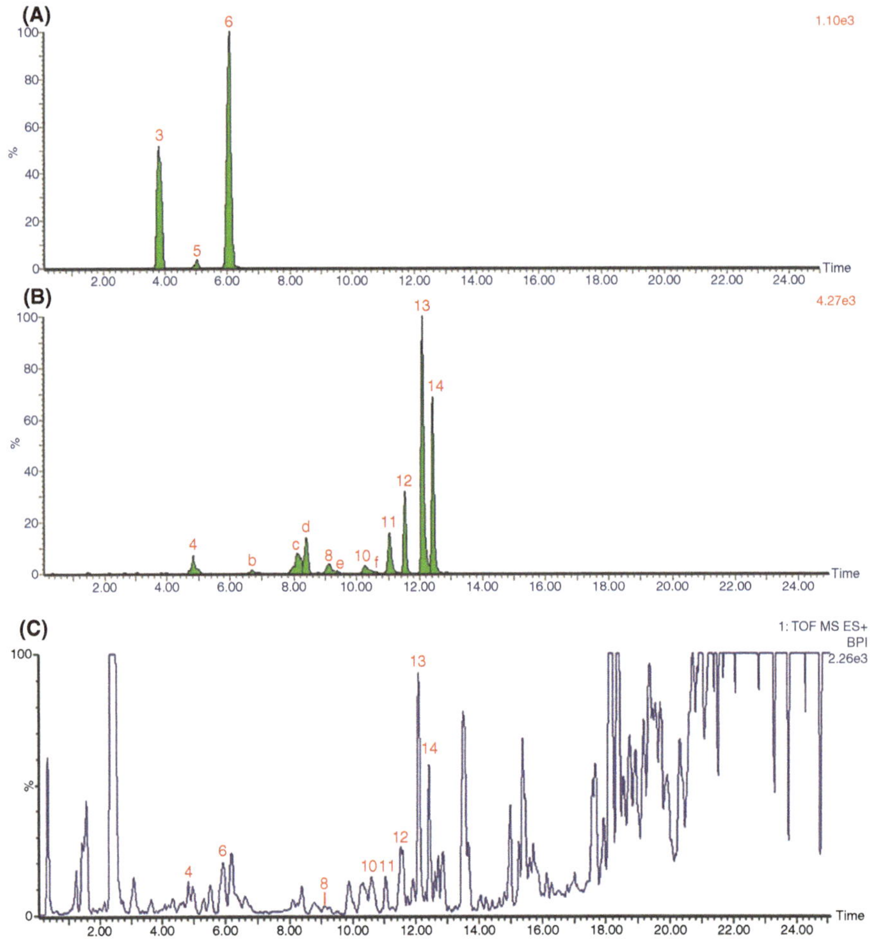

FIGURE 16.3 The BPI of WXF with Metabolynx in positive ion mode.
The UPLC-HDMS chromatogram of dosed rat plasma postadministration of WXF in the positive ion mode, and the extracted ions chromatograms of different compounds types with MetaboLynx. (A) Monoterpene glycosides; (B) protoberberine alkalidis; and (C) dosed rat plasma.

serum, the potential bioactive compounds in WXF may be discovered. The potential bioactive compounds were then identified based on their MS spectra. After the samples were analyzed using UPLC-MS, the data sets of retention time (RT)-m/z pairs, ion intensities, and sample codes were further processed with the MetaboLynx data processing tool which provides an effective and efficient approach in reducing matrix ions from biological samples, for the screening of drug metabolites in complex biological matrices. LC-MS coupled with MetaboLynx significantly improves the possibilities of extracting rich information among different chromatographic profiles.

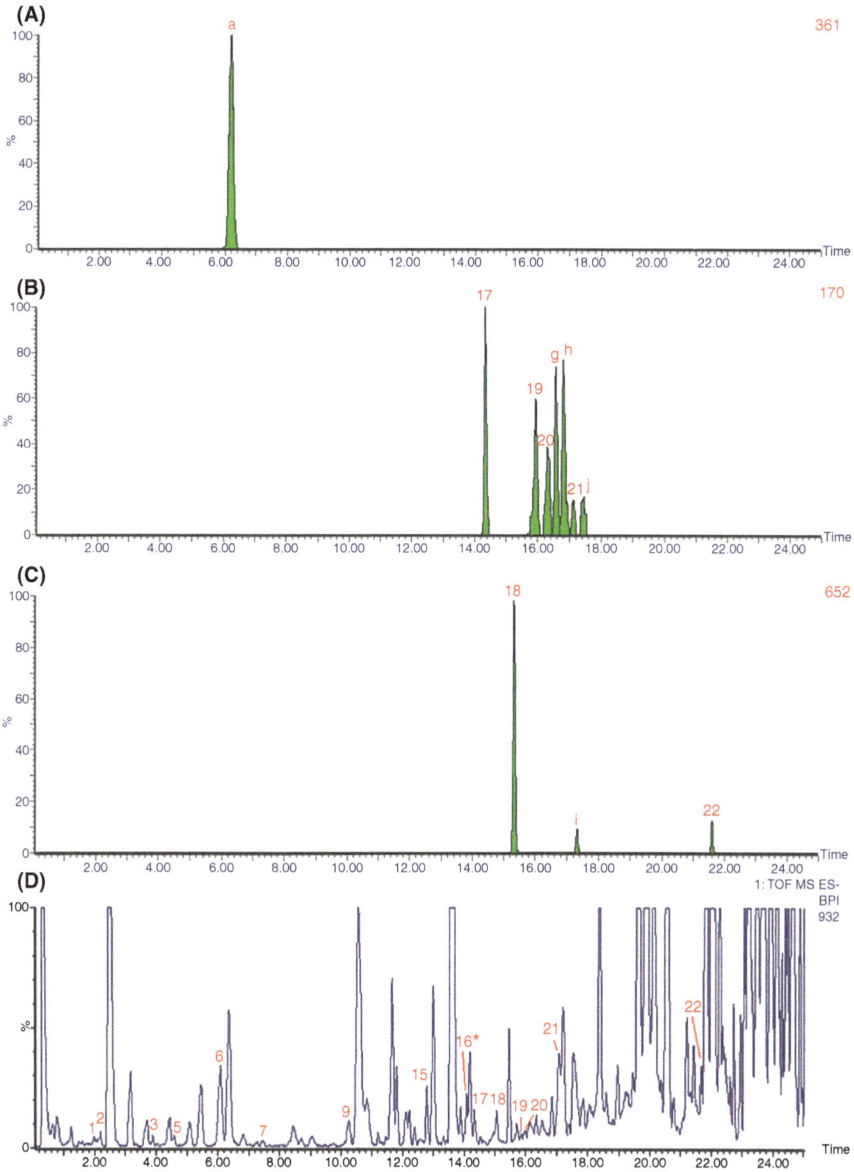

FIGURE 16.4 **UPLC-HDMS chromatogram of dosed rat plasma postoral administration of WXF in the negative ion mode, and the extracted ions chromatograms of different compounds types with MetaboLynx tool.**

(A) Feruloylquinic acids; (B) unsaturated fatty acids; (C) ginsenosides; (D) dosed rat plasma.

Table 16.3 Characterization of Metabolites From WXF by UPLC-MS Coupled With Masslynx Tool

No.	T_R (min)	Identified Compounds	Negative Ion (m/z) Exp.	Negative Ion (m/z) Error (mDa)	Positive Ion (m/z) Exp.	Positive Ion (m/z) Error (mDa)	Formula	MW (Da)
a	6.34	5-O-Feruloylquinic acid/3-O-Feruloylquinic acid	367.1001	−3	391.0996	1.1	$C_{17}H_{20}O_9$	368.1109
b	6.68	Berberrubine-9-O-β-D-glucuronide	–	–	498.1432	3.2	$C_{25}H_{24}NO_{10}$	498.1400
c	8.10	Jatrorrhizine-3-O-β-D-glucuronide	–	–	514.1722	0.9	$C_{26}H_{28}NO_{10}$	514.1713
d	8.30	Demethyleneberberine-2-O-β-D-glucuronide	–	–	500.1585	2.9	$C_{25}H_{26}NO_{10}$	500.1556
e	9.34	Demethylberberrubine-2-O-β-D-glucuronide	–	–	484.1204	−4	$C_{24}H_{22}NO_{10}$	484.1244
f	10.06	Demethyleneberberine	–	–	324.1233	−0.3	$C_{19}H_{18}NO_4$	324.1236
g	16.60	Unknown	295.1322	−1.2	–	–	$C_{18}H_{16}O_4$	296.1412
h	16.60	9(S),10(S),11(R)-Trihydroxy-12(Z),15(Z)-octadecadienoic acid	327.2145	−2.6	–	–	$C_{18}H_{32}O_5$	328.2249
i	17.40	Ginsenoside Rf	845.4907	0.3	823.4845	−2.5	$C_{42}H_{72}O_{14}$	800.4922
j	17.50	9(S),10(S),13(S)-Trihydroxy-11(E),15(Z)-octadecadienoic acid	327.2189	1.8	–	–	$C_{18}H_{32}O_5$	328.2249

Based on the fingerprinting technique and LC-MS, a method for rapid screening, and analysis of the multiple absorbed bioactive components, and metabolites postoral administration of WXF was developed. Postoral administration of WXF: 32 chemicals were detected in plasma, including 26 prototypes, and six biotansformed constituents. The results obtained from a comprehensive comparative analysis of the fingerprints of the WXF, and its metabolic fingerprints in rat biological samples indicated that 32 components in the WXF were absorbed into the rat's body. In addition, 10 components were found in the metabolic fingerprints, which suggested that they might be metabolites of some components in the WXF. The screening results of bioactive components in WXF will be helpful for the further research of its action mechanism. Both methods and results of this study could be valuable for further metabolic study of WXF.

16.3.4 Metabolomics Analysis

16.3.4.1 LC-MS Analysis of Metabolic Profiling

All the data containing the retention time, peak intensity, and the exact mass were imported in the Masslynx software for multiple statistical analyses. Trajectory analysis of the PCA score plots (3D) for the myocardial ischemia rats after WXF treatment showed clear segregation (Fig. 16.5). Systems analysis of the

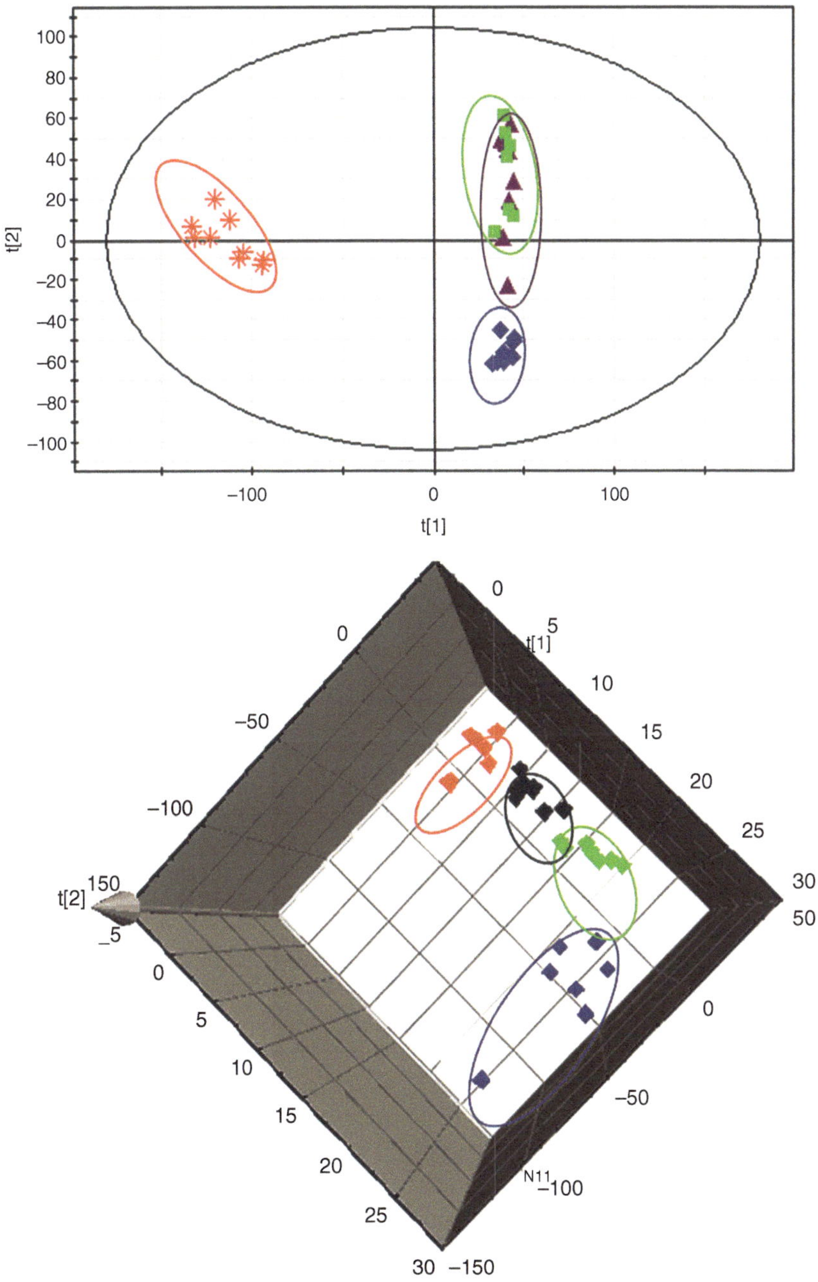

FIGURE 16.5 PCA score plots for the XQX after WXF treatment.

Control group (◆), model group (✳), high-dose WXF (▲), low-dose WXF (■). 3D plots of PCA for the XQX after WXF treatment (B). Control group (■), model group (■), high-dose WXF (■), low-dose WXF (■).

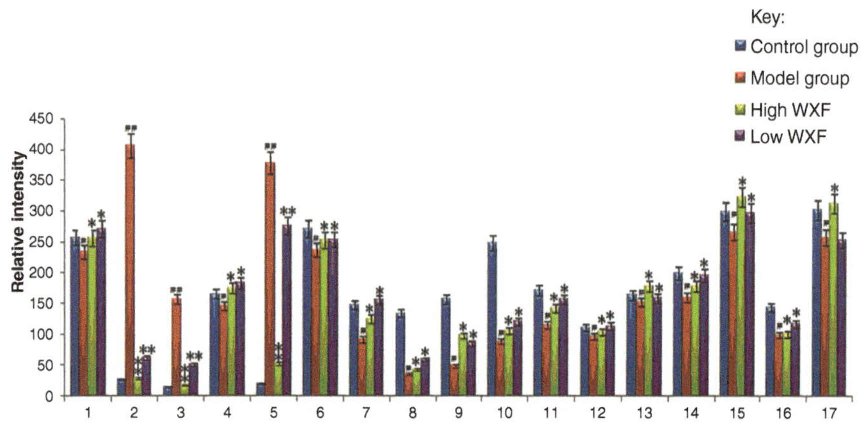

FIGURE 16.6 Changes in the relative intensity of target metabolites identified by UPLC/MS. The corresponding markers are represented in Table 15.2. A two-tailed, parametric t-test was used to determine the significance of the change in relative intensity for each metabolite. Bars represent the mean relative metabolite concentration and standard deviations. Compared with model group $^{*}p < 0.05$; $^{**}p < 0.01$; compared with model group $^{\#}p < 0.05$; $^{\#\#}p < 0.01$.

metabolomic alterations of the biosamples was then performed with MetaboAnalyst's tools. Acquired data was subjected to computational systems analysis to further investigate the effects of WXF on the metabolite profiles of XQX. The potential biomarkers of significant contribution for XQX were characterizated in the Chapter 15.

16.3.4.2 Changes of Relative Intensity of Biomarkers

According to the protocol detailed above, 17 endogenous metabolites contributing to the separation of the model group, and control group were detected in the urine samples. The relative mean height intensity of different metabolites was graphed in Fig. 16.6. Monitoring changes of these metabolites may predict the development of XQX. Additionally, the relative concentrations of the 12 endogenous metabolites were significantly affected by WXF treatment. Interestingly, according to the parametric t-test, we found that the relative concentration of these metabolites could be reversed after taking the WXF. Compared with the alterations of XQX-related metabolites, most of them were reset to a normal state post-WXF administration. Using metabolomic trajectory analysis it has been found that WXF can be regulated back toward baseline levels, as to that of the control group, demonstrating that WXF could have therapeutic efficacy. Of note, serum biochemistry and histopathological assay, also confirm that WXF exhibited therapeutic efficacies on XQX (data not shown). The results showed that WXF presented protective effects on XQX by reversing potential biomarkers to control levels, especially for the 17 metabolites in the pathway of glycolysis or gluconeogenesis, biosynthesis of unsaturated fatty acids, fatty acid biosynthesis, and purine metabolic pathways.

16.3.5 Correlation Analysis Between Marker Metabolites and Absorbed Constituents

To discover and screen the constituents, or metabolites absorbed into blood postoral administration of TCM formulae, furthermore, a correlation model between the marker metabolites, and their chemical composition was established. Six active constituents (albiflorin, paeoniflorin, coptisine, berberine, palmatine, ginsenoside Rg1) were evaluated, and screened using the plotting of correlation between marker metabolites, and serum constituents (PCMS) in WXF (graphed in Fig. 16.7). Currently, the activity of these compounds is ongoing verification by cell biological experiments at cell level. The works could provide key information for the identification, and the elucidation of bioactive constituents, and understanding the metabolism as well as the pharmacological actions for WXF. The present study was successfully applied for the rapid discovery of potentially bioactive components, and metabolites from WXF.

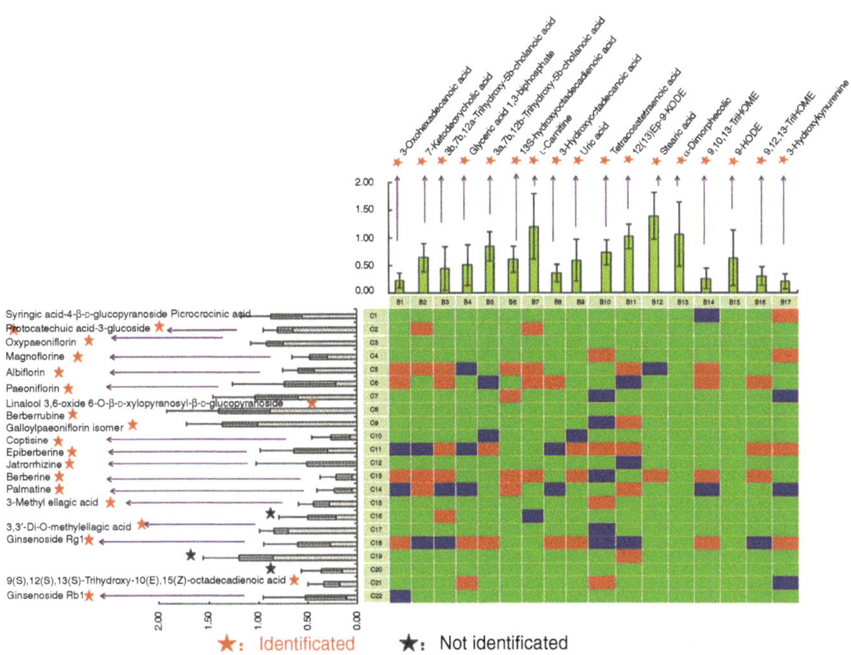

FIGURE 16.7 Plotting of the correlation between marker metabolites and serum constituents in WXF.

16.4 CONCLUSIONS

In this study, using the MetaboLynx approach and the UPLC-ESI-Q-TOF-MS technique, 32 components including 26 prototype components from WXF, and six metabolites were simultaneously observed, and provided helpful chemical information for further pharmacology research, and enhanced the speed and targeting analysis of bioactive constituents. Of note, this identification and structural elucidation of the chemical constituents may provide useful information for further mechanism studies of WXF. The therapeutic effects of WXF were further elucidated at the global metabolomics levels. WXF could provide satisfactory effects on XQX through regulating the multiple perturbed pathways to their normal state. More importantly, for the first time, albiflorin, paeoniflorin, coptisine, berberine, palmatine, and ginsenoside Rg1 were screened by the PCMS method in WXF. The PCMS strategy yielded a series of potential bioactive components in serum postoral administration of other herbal medicines, which are both useful for the drug discovery of compounds from TCM in a high-throughput manner. Taken together, our findings provide a type of validated rapid, and higher throughput methodology that would be useful for both the screening, and the identification of the constituents of herbal medicine.

References

Cao, H., Zhang, A., Zhang, F.M., Wang, Q.Q., Zhang, H., Song, Y.H., Zhou, Y., Sun, H., Yan, G.L., Han, Y., Wang, X., 2014. Ultra-performance liquid chromatography tandem mass spectrometry combined with automated MetaboLynx analysis approach to screen the bioactive components and their metabolites in Wen-Xin-Formula. Biomed. Chromatogr. 28 (12), 1774–1781.

Dong, W., Wang, P., Meng, X., Sun, H., Zhang, A., Wang, W., Dong, H., Wang, X., 2012. Ultra-performance liquid chromatography-high-definition mass spectrometry analysis of constituents in the root of Radix Stemonae and those absorbed in blood after oral administration of the extract of the crude drug. Phytochem. Anal. 23 (6), 657–667.

Sun, H., Wu, F., Zhang, A., Wei, W., Han, Y., Wang, X., 2013. Profiling and identification of the absorbed constituents and metabolites of schisandra lignans by ultra-performance liquid chromatography coupled to mass spectrometry. Biomed. Chromatogr. 27 (11), 1511–1519.

Wang, X.J., 2006. Progress and future developing of the serum pharmacochemistry of traditional Chinese medicine. Zhongguo Zhong Yao Za Zhi 31 (10), 789–792, 835.

Wang, L., Zhou, G.B., Liu, P., Song, J.H., Liang, Y., Yan, X.J., Xu, F., Wang, B.S., Mao, J.H., Shen, Z.X., Chen, S.J., Chen, Z., 2008. Dissection of mechanisms of Chinese medicinal formula Realgar-Indigo naturalis as an effective treatment for promyelocytic leukemia. Proc. Natl. Acad. Sci. USA 105 (12), 4826–4831.

Wang, H., Sun, H., Zhang, A., Li, Y., Wang, L., Shi, H., Dizou, X.L., Wang, X., 2013a. Rapid identification and comparative analysis of the chemical constituents and metabolites of Phellodendri amurensis cortex and Zhibai dihuang pill by ultra-performance liquid chromatography with quadrupole TOF-MS. J. Sep. Sci. 36 (24), 3874–3882.

Wang, X., Wang, Q., Zhang, A., Zhang, F., Zhang, H., Sun, H., Cao, H., Zhang, H., 2013b. Metabolomics study of intervention effects of Wen-Xin-Formula using ultra high-performance liquid chromatography/mass spectrometry coupled with pattern recognition approach. J. Pharm. Biomed. Anal. 74, 22–30.

Yin, Q., Wang, P., Zhang, A., Sun, H., Wu, X., Wang, X., 2013. Ultra-performance LC-ESI/quadrupole-TOF MS for rapid analysis of chemical constituents of Shaoyao–Gancao decoction. J. Sep. Sci. 36 (7), 1238–1246.

Zhang, A.H., Sun, H., Yan, G.L., Han, Y., Wang, X.J., 2013. Establishment of three-dimensional integrated serum pharmacochemistry-pharmacokinetics (pharmacodynamics)-systems biology for research of traditional Chinese medicine prescription and its application in Yinchenhao Tang. Zhongguo Zhong Yao Za Zhi 38 (21), 3786–3789.

CHAPTER 17

Targeted Synergism Effects of the Combined Active Constituents of Yinchenhao Tang

Xijun Wang, Aihua Zhang, Hui Sun, Wenjun Sun

17.1 INTRODUCTION

Currently, a paradigm shift is occurring in that there is a new focus on agents that modulate simultaneously multiple targets rather than working at the level of single protein molecules (Drews, 2000). Multiple target approaches have recently been employed to design medications that are used to treat atherosclerosis, cancer, depression, psychosis, and neurodegenerative diseases (van der Greef et al., 2007). During the past few years, the pharmaceutical industry has seen a shift from the "one disease, one target, one drug" and "one drug fits all" approaches, to the pursuit of combination therapies that include more than one active ingredient (Zheng et al., 2005; Yohannes et al., 2010). Due to the complexity of medicine, treatment protocols should be carefully designed, and formulae must be carefully developed to successfully fight the disease. A growing body of evidence has demonstrated that treating illnesses, such as human acute promyelocytic leukemia (Wang and Chen, 2008; Zhang et al., 2009), cancer (Zhang et al., 2010a), HIV (Perelson et al., 1997), chronic hepatitis C (Gao et al., 2010), and diabetic nephropathy (Zhang et al., 2008) with treatment regimens that use multiple drugs for combination therapy, and related mechanisms usually amplifies the therapeutic efficacy of each agent, yielding maximum therapeutic efficacy with minimal adverse effects (Pibiri et al., 2008). Such developments represent a triumph for modern medicine, and provide fertile ground for modern drug development (Lee et al., 2011; Xiao et al., 2005).

Interestingly, formulae consist of several kinds of crude drugs that originate from medicinal plants, animals or minerals; according to the rules of TCM theory, the famous formulae include four elements: the monarch, the minister, the assistant, and the servant. These elements work together harmoniously to achieve an ideal therapeutic outcome. One represents the principal component, and is called the monarch drug in TCM, whereas the others serve as adjuvant components that facilitate the delivery of the principal component to the disease site within the body (Keith et al., 2005). The therapeutic efficacy of TCM is usually attributed

CONTENTS

(Continued)

Chinmedomics. http://dx.doi.org/10.1016/B978-0-12-803117-9.00017-2

CONTENTS

to the synergistic properties, its capacity for minimizing adverse reactions, or its improved therapeutic efficacy. Synergism is a core principle of TCM or ethnopharmacology, and plays an essential role in improving the clinical efficacy of TCM (Wang et al., 2008a). Providing a scientific explanation for this type of synergy would certainly promote the reasonable, and effective application of TCM, and help to encourage rational approaches to safely combine the healthcare systems from various cultures. Multidrug combinations are increasingly important in modern medicine (Zhang et al., 2010b; Liao et al., 2010; Tang et al., 2014). However, the precise mechanisms through which formulae function manifests itself are poorly understood, and must be addressed using a molecular approach.

Yinchenhao Tang (YCHT) was recorded in *Shanghanlun*, a classic resource on TCM written by Zhongjing Zhang (150–215 A.D.), and is one of the most famous Chinese herbal formulae. YCHT consists of *Artemisia annua* L. (the monarch herb), *Gardenia jasminoides* Ellis (the minister herb), and *Rheum Palmatum* L. (the assistant and servant herb), and has been used for more than a thousand years to treat jaundice and liver disorders. Pharmacological studies and clinical practice have shown that it can be used clinically to treat cholestasis, hepatitis C, primary biliary cirrhosis, liver fibrosis, and cholestatic liver diseases. Our previous study showed 45 compounds in YCHT, and 21 compounds *in vivo* (Fig. 17.1A) were identified by the serum pharmacochemistry of TCM (Wang et al., 2008b). The compatibility of the different compounds in YCHT, and the effect on the absorption of these 21 compounds were investigated (Wang et al., 2011). Three potentially active constituents (6,7-dimethylesculetin (D), geniposide (G) and rhein (R)) were screened by the plotting of correlation between marker metabolites, and serum constituents (PCMS) in YCHT (Fig. 17.1B). Their chemical structures were shown in Fig. 17.2.

It has been proposed that DGR-based combination treatment for hepatic injury (HI) produces a synergistic effect. However, although much is known regarding the interactions between D, G, and R at pharmacokinetic sites (Zhang et al., 2011) there is little knowledge regarding the compound's synergistic properties. Understanding the synergistic effects of DGR represents an even greater challenge because multilayered regulation may be involved, with the three compounds having overlapping but distinct target properties. To gain insight into the complex biochemical mechanisms that underlie this effective HI therapy, we conducted an investigation incorporating advanced technologies using metabolomic, proteomic, and biochemical analyses throughout the treatment process for HI (Wang et al., 2013). In analyzing the formulae design in TCM, here we use the treatment of HI with DGR as a working model. D, G, and R – which are derived from *Artemisia annua* L., *Gardenia jasminoids* Ellis, and *Rheum Palmatum* L., respectively – were used as the active compounds in YCHT, and the efficacy and mechanisms of the DGR combination as used to treat HI were tested both *in vivo* and *in vitro*.

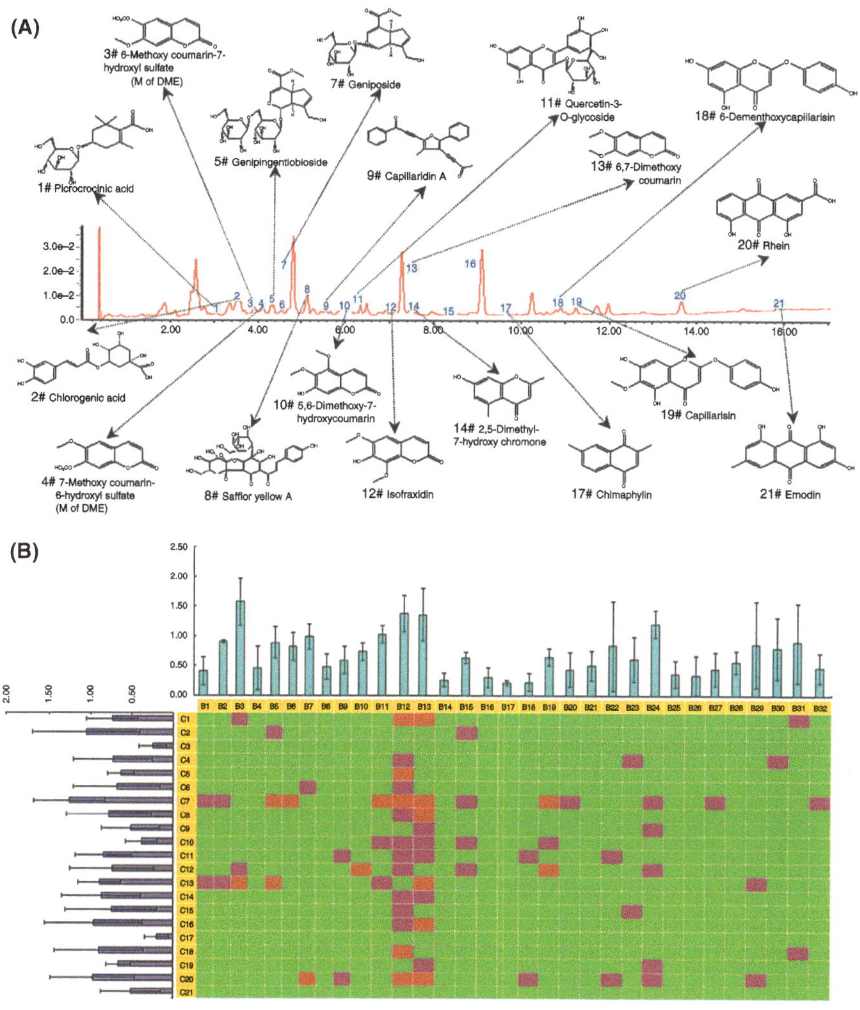

FIGURE 17.1 The fingerprint chromatogram of compounds: (A) *in vivo* postoral administration of YCHT; (B) correlation analysis of compounds *in vivo*, and marker metabolites.

17.2 MATERIALS AND METHODS

17.2.1 Reagents

Acetonitrile (HPLC grade) was purchased from Merck (Darmstadt, Germany); methanol (HPLC grade) was purchased from Fisher (USA). Distilled water was purchased from Watson's Food & Beverage Co., Ltd. (Guangzhou, China), and formic acid (HPLC grade) was purchased from the Beijing Reagent Company (Beijing, China). Leucine enkephalin was purchased from Sigma-Aldrich (MO, UK), and carbon tetrachloride (CCl_4) was purchased from the Chemicals

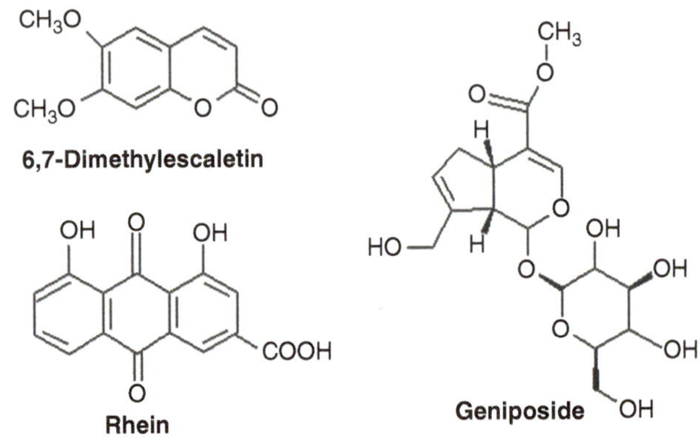

FIGURE 17.2 Chemical structures of 6,7-dimethylesculetin, geniposide, and rhein.

Factory (Tianjin, PRC). Glycerol was supplied by the Chemicals Factory. Other chemicals, except as noted, were analytical grade. D, G, and R were isolated within our laboratory, and identified by spectral analyses, primarily NMR and MS. After identification it was further purified by HPLC to yield an authorized compound with purity of at least 99%. Freeze-dried YCHT powder was produced by our laboratory. The assay kits were purchased from the Nanjing Jiancheng Biotech Company (Nanjing, China). The other reagents that were used in 2-DE were purchased from Bio-Rad.

17.2.2 Animals

Male Wistar rats were bred and maintained in a specific pathogen-free environment. The animals were allowed to acclimatize in metabolic cages for 1 week prior to treatment. The animals were randomly assigned to 10 groups of six rats each as follows: control, HI, YCHT, D, G, R, DG, DR, GR, and DGR groups. The rats in the control group were given distilled water for 8 consecutive days, and anesthetized via an intraperitoneal injection of 1% pentobarbital sodium (0.15 mL/100 g body weight) on day 9. The rats in the HI group were orally administrated 20% CCl_4 (1 mL/kg body weight) for 8 consecutive days (at 6:00 p.m.), and on day 9, the rats were anesthetized as in the control group. Simultaneously, other groups were administrated CCl_4 and various treatments for 9 consecutive days, then anesthetized as in the control group. The rats were orally given 1 mL/100 g each, and the doses of the three components used in this study were administered in accordance with the *Shanghanlun*. Blood was collected from the hepatic portal vein, and plasma was separated via centrifugation at 4500 rpm for 5 min at 4°C, flash frozen in liquid nitrogen, and stored at −80°C until the liver function tests, and metabolomics and proteomics analyses were performed. The plasma ALT, AST, ALP, GSH-PX, and SOD activities, and MDA, TG, GT, CHOL, TP, D-BIL, and T-BIL contents were measured according to the manufacturer's instructions. The rat livers were removed immediately after plasma collection and stored at −70°C until analysis. Urine was collected daily (at 6:00 a.m.) from the metabolic cages at ambient temperature throughout the entire procedure, and centrifuged at 10,000 rpm at 4°C for 5 min; the supernatants were then stored frozen at −20°C for subsequent metabolomic analysis.

17.2.3 Biochemical Analysis

We collected plasma samples in heparinized tubes, kept them on ice for 1 h and centrifuged them at 4500 rpm for 15 min at 4°C. We quantified the levels of plasma ALT, AST, ALP, GSH-PX, and SOD activity, and MDA, TG, GT, CHOL, TP, D-BIL, and T-BIL content using assay kits according to the manufacturer's instructions. The rat livers were removed immediately after plasma collection, and stored at −80°C until analysis.

17.2.4 Histology, Immunohistochemistry and TUNEL Assay

The livers were fixed in 4% neutral buffered formaldehyde at 4°C, and embedded in paraffin. The liver tissue was stained with H&E for histopathological analysis. Immunohistochemistry was performed using antibodies against Fas and BCL-2. TUNEL staining was performed to detect, and quantify apoptotic cells using the *in situ* cell death detection kit. The sections were viewed, and photographed using standard fluorescent microscopic techniques.

17.2.5 Metabolomics Analysis Platform

Urine and serum were collected for the UPLC-HDMS analysis. For the reversed-phase UPLC analysis, the ACQUITY UPLC BEH C_{18} column (50 × 2.1 mm i.d., 1.7 μm (Waters Corp., Milford, MA, USA) was used. The column temperature was maintained at 35°C, the flow rate during the mobile phase was 0.50 mL/min, and the injection volume was fixed at 2.0μL. Mobile phase A involved the use of 0.1% formic acid in acetonitrile, whereas mobile phase B involved the use of 0.1% formic acid in water. For the urine UPLC-HDMS (Waters Corp., Milford, USA) analysis, the optimal conditions for positive ion mode were as follows: capillary voltage of 2500 V, desolvation temperature of 350°C, sample cone voltage of 35 V, extraction cone voltage of 3 V, microchannel plate voltage of 1600 V, collision energy of 4 eV, source temperature of 110°C, cone gas flow of 50 L/h, and desolvation gas flow of 600 L/h. The mass spectrometry full-scan data were acquired in the positive ion mode from 100 Da −1000 Da with a 0.1s scan time. For the plasma UPLC-HDMS analysis, the desolvation gas flow was 600 L/h, whereas the other parameters were the same as for the urine. The MS data were generated and recorded using MassLynx (v4.1) (Waters Corp., Milford, USA), MarkerLynx Application Manager (Waters Corp., Milford, USA) was used for peak detection, and EZinfo 2.0 software (which is included in MarkerLynx Application Manager and can be applied directly) was used for the PCA, PLS-DA, and OPLS. Putative markers were extracted from S-plots that were constructed following the analysis using OPLS, and markers were chosen based on their contribution to the variation, and correlation within the data set. The potential biomarkers were matched with the structure message of metabolites acquired from available biochemical databases.

17.2.6 Proteomics Analysis Platform

17.2.6.1 Two-Dimensional Gel Electrophoresis and Image Analysis

The plasma obtained was disrupted in a homogenization buffer containing 30 mM Tris, 7 M urea, 2 M thiourea, 4% CHAPS, and 40 mM DTT, after which the lysates were centrifuged at 12,000 rpm for 60 min at 4°C. The supernatant was precipitated using the TCA sediment method to recover the total proteins, and the interfering components were removed using the 2-D Clean-Up Kit (GE Healthcare, Piscataway, NJ). The supernatant was removed, divided into several aliquots, and stored at −70°C until use. The protein concentration of each sample was measured using the Bradford method. A total of 350 mg of protein was loaded onto an 18 cm linear IPG strip (pH 3−10, Amersham Biosciences, Piscataway, NJ) for first-dimension isoelectric focusing (IEF). The strips were placed into a Protean IEF cell (Bio-Rad) and rehydrated at 50 V for 12 h, after which the proteins were separated based on their pI according to the following protocol: 250 V with linear climb for 60 min; 1,000 V with rapid climb for 60 min; 10,000 V with linear climb for 5 h; and 10,000 V with rapid climb until a level of 60,000 V-h was reached. After IEF, the IPG strips were equilibrated for 15 min in a buffer containing 50 mM Tris-HCl, pH 8.8, 30% glycerol, 7 M urea, 2% SDS, and 1% DTT; they were then further treated in a similar buffer (containing 4% iodoacetamide instead of DTT) for 15 min, and directly applied onto 12% homogeneous SDS-PAGE gels for 6 h electrophoresis using a Protean II xi cell system (Bio-Rad). Moreover, two kinds of electrophoresis conditions that were suitable for the separation of proteins with higher molecular weight (10 mA/gel for 30 min followed by 30 mA/gel for 5.5 h) and for the separation of proteins with lower molecular weight (10 mA/gel for 30 min followed by 20 mA/gel for 8 h) were used. The gels were then silver-stained using Bio-Rad Silver Stain Plus kit reagents (Bio-Rad) according to the manufacturer's instructions. The silver-stained gels were scanned using a GS-800 densitometer (Bio-Rad) and then analyzed using PDQuest software (Bio-Rad). Labscan Software (Applied Biosystems, Foster City, CA) was used for image analysis that included background abstraction, spot intensity calibration, spot detection, and matching. The intensity of each spot was quantified by calculating the spot volume after the gel image had been normalized. Each sample was processed in triplicate to ensure reproducibility, and the Student's t-test was used to evaluate the average change in protein abundance corresponding to each target spot across the gels. Protein spots with more than a threefold significant change (either increasing or decreasing) in density (paired t-test, $p \leq 0.05$) were considered to be differentially expressed, and were selected for further identification using MALDI-TOF MS/MS.

17.2.6.2 MALDI-TOF MS/MS and Database Analysis

Protein spots of interest with differential expression levels were excised from the 2D gels with an EXQuest spot cutter (Bio-Rad), washed three times in distilled water, destained with 15 mM potassium ferricyanide in 50 mM sodium thiosulfate for 2 min at room temperature, dehydrated in acetonitrile, and finally dried in a centrifugal vaporizer. After a 30-min incubation, the gels were digested for more than 12 h at 37°C. The peptides were then extracted twice using 0.1% TFA (trifluoroacetic acid) in 50% CAN dried under the protection of N_2. For MALDI-TOF MS/MS, the peptides were mixed with 0.5 μL MALDI matrix (5 mg/mL α-cyano-4-hydroxycinnamic acid diluted in 0.1% TFA and 50% ACN) and spotted onto the MALDI sample plates. The MS measurements were performed on a Proteomics Analyzer with delayed ion extraction (Applied Biosystems). The data were acquired in the delayed ion extraction mode using a 20 kV accelerating voltage, and a 100 ns extraction delay time. Using the individual PMF spectra, peptides with a signal-to-noise ratio exceeding 20 that passed through a mass exclusion filter were subjected to fragmentation analysis. The parameters for peak matching were, as follows: the minimum signal-to-noise ratio was 20, the mass tolerance was 0.2 Da, the minimum peak to match reference masses was 4, and the maximum outlier error was 50 ppm. The number of total shots for each PMF spectrum, and MS/MS were 2000 and 3000, respectively. Mass spectra were acquired using FlexControl 20=.0 software (Bruker Daltonics, Germany), and all the spectra were compiled and normalized. The FlexAnalysis 2.0 software was used for converting the raw data. Peptides were sequenced on a 4800 MALDI analyzer (Applied Biosystems, CA). Bioinformatics databases were searched for protein identification based on tryptic fragment sizes. Databases were searched for peptide identification using MASCOT Distiller 2.1 (Matrix Science, www.matrixscience.com). The protein score is $-10\text{Log}(p)$, where p is the probability that the observed match is a random event. The missed cleavage sites allowed were up to 1, and scores that were higher than 61 were considered significant.

17.2.6.3 Functional Analysis

We entered the names of the differentially expressed proteins into the BinGO platform to perform the GO functional enrichment analysis, with a special focus on the changes in the biological processes that are associated with HI after each treatment. We then classified the significantly enriched GO terms according to functional similarity. Finally, we visually represented the GO classification using a diagram.

17.2.7 Statistical Analyses

All statistical analyses were performed using the Student's t-test. Differences with a *p*-value of 0.05 or less were considered significant. Assays were performed in triplicate, and the results are expressed as mean ± SD.

17.3 RESULTS

17.3.1 Therapeutic Efficacy of DGR as Indicated by a Rat HI Model

The efficacy of combination therapy with D, G, and R was compared to the efficacy of monotherapy with each of the three components in a rat model of HI. There was significant variation between the biochemical indicators for the control, and model groups after carbon tetrachloride (CCl_4) treatment. It indicates that the HI model successfully replicates the disease. The model group had higher ALT, AST, ALP, r-GT, TG, TC, MDA, T-BIL, D-BIL, and TP values but had lower levels of SOD, and GSH-PX than the control animals. Each treatment group was treated back to baseline levels (i.e., those of the control group), which demonstrates that these drugs had a therapeutic effect in the rat HI model. Our data show that the DGR combination statistically intensified the therapeutic efficacy as compared with the control condition, and with monotherapy using either D, G, or R (data not shown). Interestingly, the DGR protocol decreased the levels of ALT, AST, ALP, T-BIL, D-BIL, GT, MDA, and TP but increased the levels of GSH-PX, TG, and TC. These results indicate that DGR combination therapy exerted a synergistic effect, and yielded better therapeutic effects than did the approaches that were based on the use of either D, G, or R as a single agent. These data provide evidence for the synergy that is created by coadministration. Among the various monotherapies, D showed the most potent therapeutic efficacy. Our data also indicates that D is the principal component of the formulae, whereas G and R serve as adjuvant ingredients.

17.3.2 DGR Reduces Histologic Changes and Hepatocyte Apoptosis

To determine the protective effects of DGR on liver tissue damage, histological, TUNEL, and immunohistochemical analyses were performed in liver tissue that was obtained from HI rats, and compared with tissue from control rats. Microscopic analyses of H&E- and TUNEL-stained liver sections showed that DGR significantly decreased hepatocyte necrosis, fibrotic area, and hepatocyte apoptosis levels, making them comparable to normal liver (Fig. 17.3A). The histopathological examination of liver sections that were stained with H&E revealed numerous apoptotic hepatocytes, and the accumulation of massive necrosis with intralobular hemorrhage in the livers of HI rats (Fig. 17.3A). Further analysis revealed multiple and extensive areas of portal inflammation, and hepatocellular necrosis in the HI group, and a moderate increase in inflammatory cell infiltration. In the portal areas, Kupffer cells were detected within the sinusoids. The degree of necrosis was clearly lower in the CCl_4^- treated rats that received YCHT, the DGR combination, or bitherapies using various combinations of D, G, and R (Fig. 17.3A). Minimal hepatocellular necrosis, and inflammatory cell infiltration, and mild portal inflammation were observed in rats that were treated with either the DGR combination or YCHT, as compared with animals that were treated with the control, or with monotherapies, or biotherapies composed of D, G, and R. It is important to note that only spotty necrotic hepatocytes were visible in the livers of the DGR-treated rats. Furthermore, the H&E-stained sections indicate that the hepatocyte size had changed in the

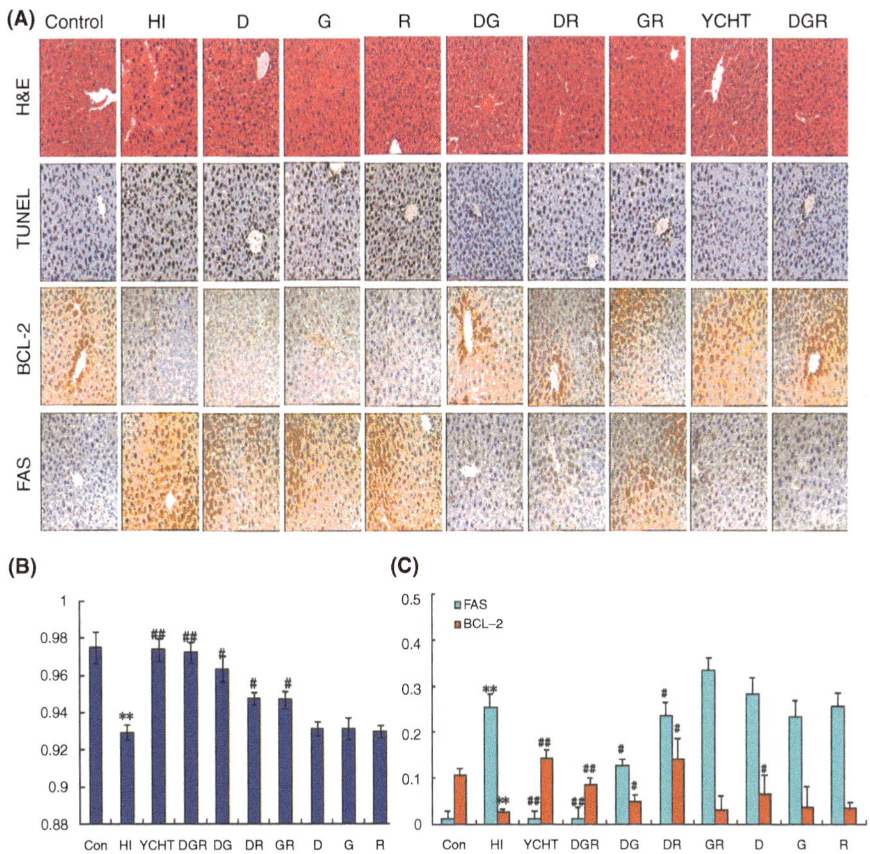

FIGURE 17.3 Histology, immunohistochemistry, and TUNEL assay.

(A) Representative pictures of liver histopathology (H&E staining), immunohistochemistry, and TUNEL assay of apoptotic hepatocytes after CCl_4 with DGR treatment in rats (original magnification: ×200). (B) Quantitative evaluation of the density of apoptotic hepatocytes using TUNEL staining. (C) Quantitative evaluation of the density of FAS- and BCL-2 positive cells by immunohistochemistry. The values are expressed as mean± SD, and were compared by ANOVA. $^*p < 0.05$ vs. the control group; $^{**}p < 0.01$ vs. the control group; $^{\#}p < 0.05$ vs. the HI group, $^{\#\#}p < 0.01$, vs. the HI group.

model rats, and that DGR restored normal cell size. These results suggest that the DGR combination prevents the destruction of liver tissue, and intensifies therapeutic efficacy.

Next, to investigate further the therapeutic and synergistic properties of the DGR combination, apoptotic hepatocytes were measured using TUNEL staining. Few TUNEL-positive hepatocytes were observed in the livers of CCl_4^- treated rats. However, numerous TUNEL-positive hepatocytes ($p < 0.01$) were found in the livers of animals that were pretreated with DGR, YCHT, or two agents (Fig. 17.3A). Interestingly, the rats that were treated with DGR had the highest density of apoptotic hepatocytes, and the density of apoptotic cells in the rats that were treated with D, G, and R bitherapies was higher than in the rats that were treated with monotherapies (Fig. 17.3B). These results demonstrate that the DG, DR, and RG combinations moderately elevated the density of apoptotic cells, whereas the maximum synergic effect was observed in rats that were treated with the DGR protocol. This is strong evidence that the synergistic effect of D, G, and R occurred at the level of hepatocyte apoptosis. To rigorously test this conclusion, we performed an extensive immunohistochemical analysis using several various antibodies in formalin-fixed and paraffin-embedded liver sections. This analysis was performed with antibodies against Fas and BCL-2 and revealed hepatocytes both in clusters of various sizes in the portal areas, and as single cells within the lobules (Fig. 17.3C). Interestingly, a statistically significant difference in the density of FAS- and BCL-2–positive cells of the model and controls was observed. A quantitative immunohistochemical analysis revealed that DG and DR, but not GR induced a certain degree of: (i) upregulation of BCL-2 (antiapoptotic) ($p < 0.05$); and (ii) decrease in the density of FAS-positive hepatocytes ($p < 0.05$). Surprisingly, however, the DGR combination induced a much stronger ($p < 0.01$) antiapoptotic effect (Fig. 17.1C). Although G, R, and the GR combination did not dramatically affect the staining pattern, these two compounds facilitated the effects of D (Fig. 17.3C) suggesting that the use of D to upregulate BCL-2 is more effective than when G and R are also used. The oral administration of DGR to HI rats may be more effective at inhibiting the extent of liver necrosis than may other treatment regimen. The immunohistochemical assay further confirmed the cooperativity of D and G with R in upregulating BCL-2, and downregulating FAS, with the strongest effect resulting from the DGR combination, thereby supporting the rationale of using YCHT to treat HI.

17.3.3 Multivariate Analysis

Using the optimal the Waters ACQUITY UPLC system together with a Waters Micromass quadrupole-time-of-flight Micro Synapt High Definition Mass Spectrometer (UPLC-HDMS) conditions, the representative base peak intensity (BPI) chromatograms of the urine or plasma samples that were collected from representative rats from each group are presented in Fig. 17.4A and B. The low molecular mass metabolites were well separated at 11 min because of the minor particles (sub-1.7 μm) of UPLC. PCA, PLS, and OPLS-DA were used to classify the metabolic phenotypes, and identify the differentiating metabolites. Analyzing the OPLS-DA score plots identified the control, and the HI rats based on differences in their metabolites, suggesting that their metabolic

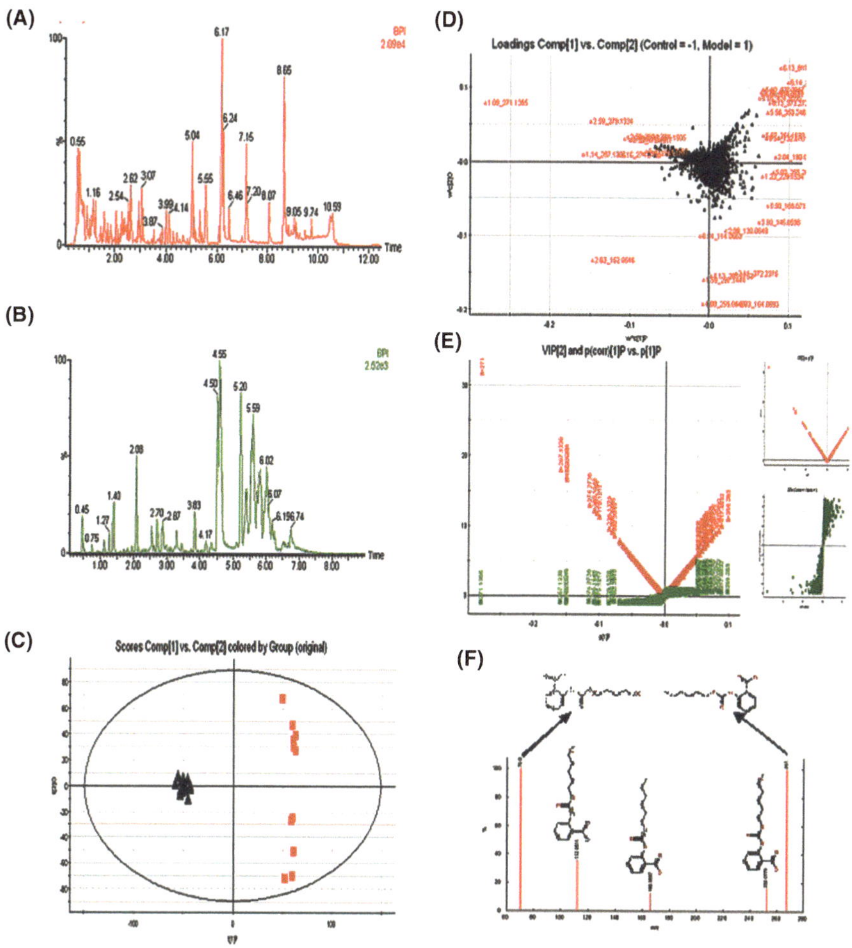

FIGURE 17.4 Metabolomic multivariate analysis.
(A) BPI chromatograms of urine samples; (B) BPI chromatograms of plasma samples; (C) Score plot of OPLS-DA for the control and model groups. (▲, control group; ■, model group); (D) The loading plots of PLS-DA for the control and model groups; (E) A combination plot of S- and VIP-values; (F) Chemical structure and mass fragment information of 2-((hexyloxy) carbonyloxy)benzoic acid.

profiles significantly changed as a result of CCl_4 administration (Fig. 17.4C). The PLS-DA loading plots displayed variables that were positively correlated with the score plots (Fig. 17.4D). S-plots and the VIP-value plots were combined for the structural identification of the biomarkers (Fig. 17.4). For example, this information for 2-((hexyloxy)carbonyloxy)benzoic acid is displayed in Fig. 17.4F. Finally, potential biomarkers of significant contribution were identified, with 20 and 12 in the urine and plasma, respectively (Fig. 17.5). As seen in Fig. 17.6, putative metabolic pathways were identified in the HI rats based on changes in the intermediates during substance metabolism.

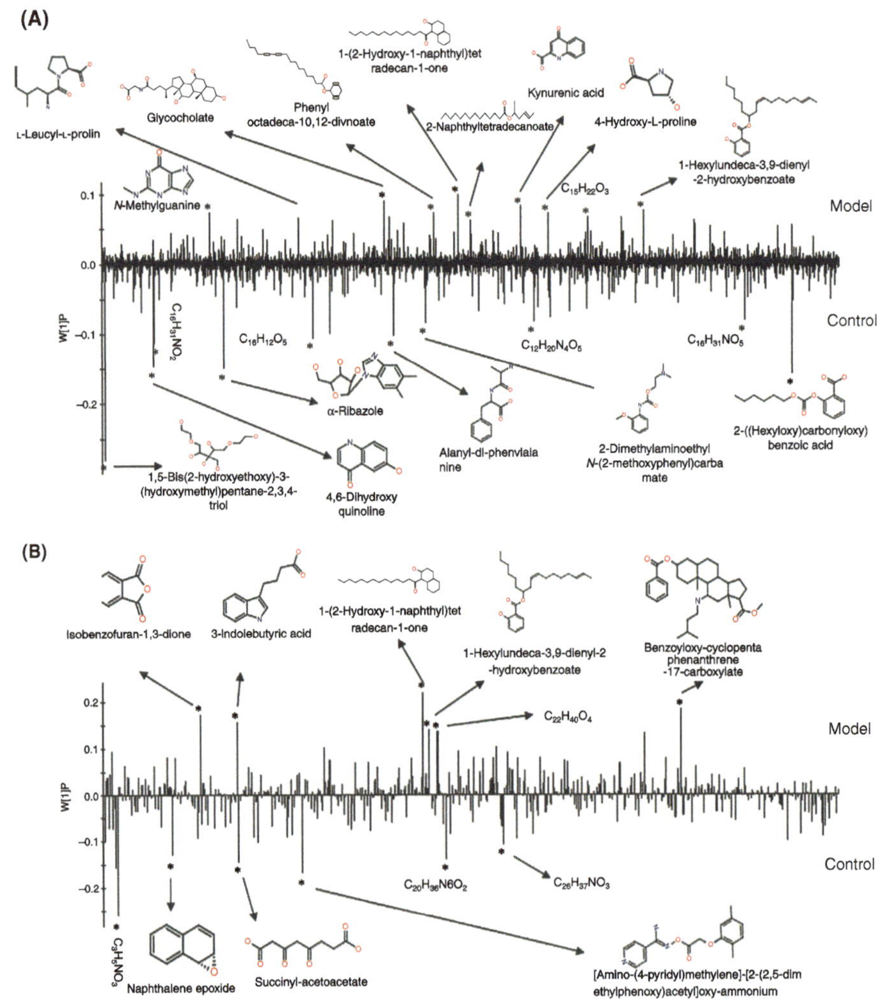

FIGURE 17.5 Loading plots of OPLS-DA model of potential biomarkers identified. (A) Urine; (B) plasma.

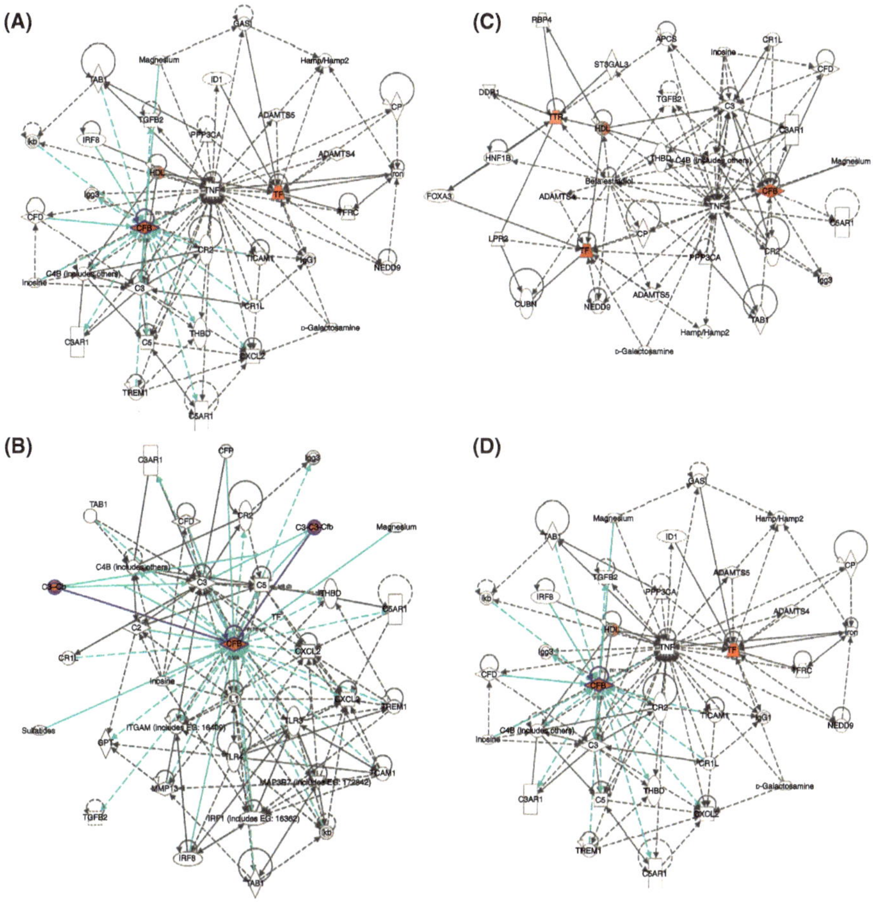

FIGURE 17.6 **Merged network combining major signaling networks associated with the metabolites differentially expressed by the ingenuity pathway analysis.**
(A) HI group; (B) G group; (C) DG group; (D) DGR group.

17.3.4 DGR Causes Synergistic Effects on Trajectory Analysis

In this study, a model of HI rats was constructed, and the dynamic metabolic profiles of treated rats with various outcomes were investigated using UPLC-HDMS, and a multivariate statistical analysis. With regard to PCA, as shown in Fig. 17.7A–C, the control and model groups were significantly different after CCl_4 treatment, thereby validating the HI model. Figure 17.7C shows that YCHT helped prevent HI, and maintained the animals in a normal state – there were no distinct clustering differences between the control group, and the YCHT treatment group. In contrast, distinct clustering differences were

(A)

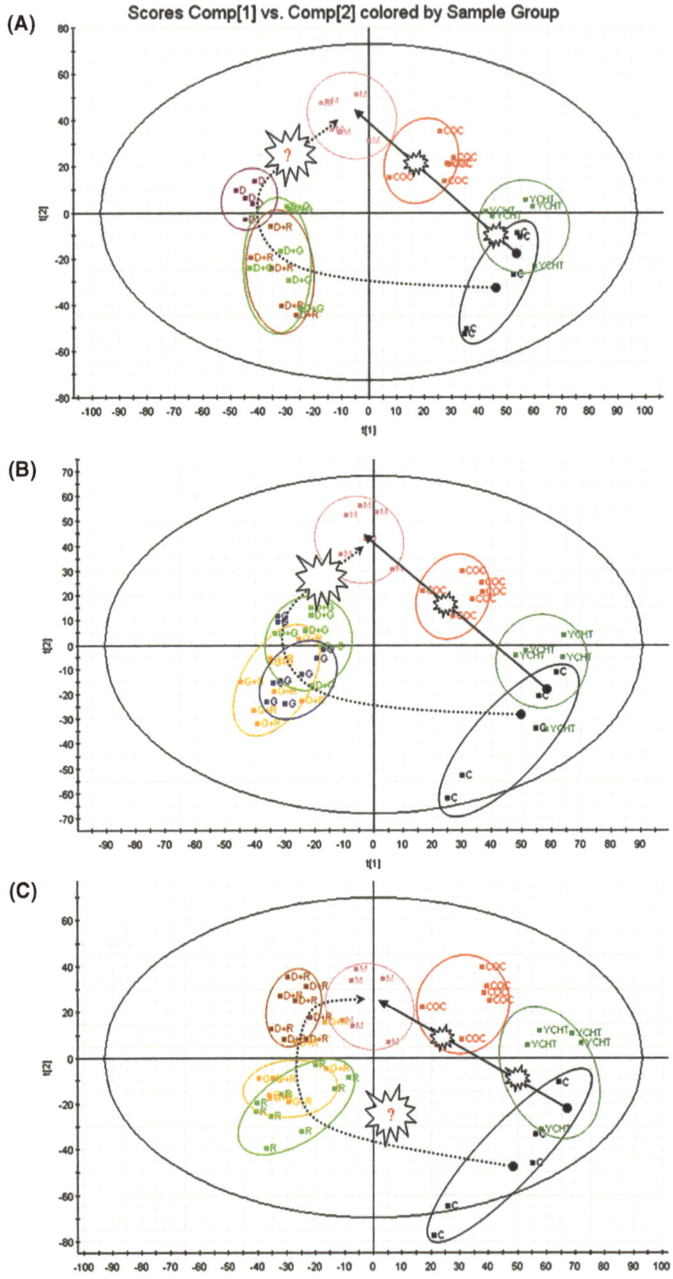

FIGURE 17.7 **Trajectory analysis of DGR effects on model rats.**

(A) PCA score plot for the model rats after D treatment; (B) PCA score plot for the model rats after G treatment; (C) PCA score plot for the model rats after R treatment. ■C, control; ■M, HI group; ■COC, DGR group; ■DG; ■DR; ■GR; ■G; ■D; ■R.

observed between the model group, and the groups that received either monotherapy or bitherapy using D, G, and/or R; each dosing regimen returned the values to their baseline levels (i.e., those of the control group) but the DGR combination further enhanced this effect. Therefore, the maximum synergistic effect in terms of HI prevention was observed in rats that were treated with the DGR combination as compared with animals that were treated with monotherapy or bitherapy using D, G, and/or R (Fig. 17.7A–C). Similarly, treatment with G or R alone inhibited the occurrence and development of HI, as shown by the clear differences between the control and the model groups. Among the monotherapy groups, the group that was treated with D showed the highest level of improvement. This suggests that the therapeutic effect of D on rats with liver injury can be improved by the addition of G and R, and provides compelling evidence that the synergistic effect of D, G, and R occurred at the metabolomic level.

17.3.5 DGR Triggers Dynamic Changes in Metabolic Biomarkers

A metabolomic trajectory analysis indicated that each dosing regimen returned the values to their baseline levels (i.e., those of the control group) but the maximum synergic effect in preventing liver injury was observed in rats that were treated with the DGR combination, not in the animals treated with the control, or with monotherapy, or bitherapy using D, G, and/or R (Fig. 17.7). These results clearly show that the components of DGR exert a mutual reinforcing effect. This raises the question of how the DGR combination exerts its synergic effects. To address this question, we further utilized the metabolomics approach to delineate metabolic changes in HI after each treatment. Potential biomarkers were identified, with 20 and 12 in the urine and plasma, respectively (data not shown). A dynamic change analysis of the levels of metabolic biomarkers was also used to determine whether D, G, and/or R exerted synergic effects. These metabolites, when identified together, are important to the host's response to HI by affecting energy metabolism, amino acid metabolism, nucleotide metabolism, and lipid metabolism. The metabolic pathways of the biomarkers for the HI model are presented in Fig. 17.6. Additionally, a dynamic change analysis of the biomarkers further validated the metabolomic findings. Clear differences between the treatment, and control groups were observed with each dose. Interestingly, the maximum synergic effect in preventing liver injury was observed in the rats that were treated with the DGR protocol, whereas the animals that were treated with the monotherapy or bitherapy using D, G, and/or R showed weaker results. Taken together, these results provide a comprehensive dynamic metabolic profile, and serve as the basis for the further study of the mechanisms that underlie synergism. Metabolic pathway analysis with IPA revealed that metabolites, which were identified together, are important for the host response to HI (Fig. 17.6). To reveal signal transduction pathways, and/or signaling networks associated with the metabolites differentially expressed, the identified metabolites were imported into the IPA software. According to the IPA knowledge base, major signaling networks, comprised of 38 nodes, were associated with this set of proteins. Integrated network included differentially expressed metabolites, and involved small molecular transport-associated major signaling pathways.

17.3.6 Proteomic Identification of the Target Proteins for HI

To better understand CCl_4-induced proteins in HI rats, we investigated dynamic proteomic differences between the model and control groups. After analyzing the 2-DE gels, the peptides were extracted from each differentially expressed protein spot using in-gel trypsin digestion, and the proteins were identified using MS/MS. Representative 2-DE images for the control, and model groups are shown in Fig. 17.8A and B. Among all of the protein spots that were separated using 2-DE, 42 spots appeared to be significantly changed in percent volume as

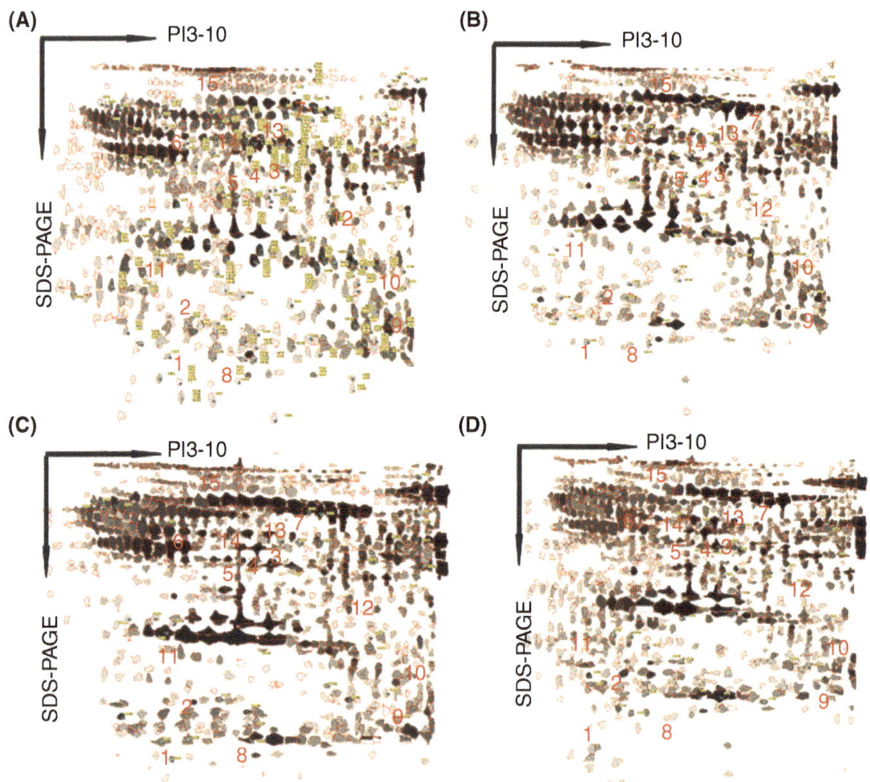

FIGURE 17.8 2-DE representative proteomic maps.
(A) Control group; (B) HI group; (C) YCHT-treated group; (D) DGR-treated group. The protein spots that are indicated with numbers represent the differentially expressed proteins.

identified by peptide mass fingerprints (PMFs) that were based on MALDI-TOF MS/MS and database searching. We found 15 plasma proteins (seven of which were upregulated and eight of which were downregulated) that were expressed in the animal model. The upregulated proteins in the HI group included: Ig kappa chain C, zinc finger protein 407, serotransferrin, haptoglobin, macroglobulin, alpha-1-antitrypsin, and complement factor B. The downregulated proteins included: fibrinogen alpha chain precursor, glyceraldehyde-3-phosphate dehydrogenase, albumin, 2,4-dienoyl-CoA reductase, transthyretin, α-1-inhibitor 3, vitamin D-binding protein, and prothrombin. The synergistically/additively-regulated proteins appear to be involved in metabolism, energy production, immunity, chaperoning, antioxidation, and signal transduction. These data help to clarify the molecular mechanisms that underlie the therapeutic, and synergistic properties of YCHT.

17.3.7 Functional Analysis of Novel Proteomic Biomarkers

To better understand at the proteomic level, the molecular mechanisms, and relevant pathways that underlie DGR's ability to treat HI, we employed a proteomics strategy that involved combining 2-DE, and MALDI-TOF MS/MS to perform functional enrichment analysis, with a special focus on the changes in the HI biological process after each treatment. We found seven to eight target proteins that were upregulated and downregulated, respectively, and that were highly specific to the HI model (data not shown). The detection of these proteins with distinct regulatory patterns provides evidence that novel biomarkers are actively involved in multifunctional pathways that are likely essential for HI. The biological functions of these critical proteins can be sorted into five groups: (A) generation and degradation of the extracellular matrix, including fibrinogen alpha chain precursors and macroglobulin; (B) the regulation of transcription and translation, as provided by zinc finger proteins; (C) acute phase reaction and immunity protection, as provided by Ig kappa chain C, alpha-1-antitrypsin, α-1-inhibitor 3, prothrombin, and vitamin D-binding protein; (D) oxygenation and cell apoptosis, to which 2,4-dienoyl-CoA reductase contributes; and (E) transport and metabolism, as provided by glyceraldehyde-3-phosphate dehydrogenase, haptoglobin, serotransferrin, and transthyretin. The characteristic functions of these differentially expressed proteins were enriched within clusters that were based on biological processes such as "immunity," "cellular apoptosis," "transport," "signal transduction," "cell growth and proliferation," and "metabolism." For example, the modulation of several key members of the immunity cluster was revealed by proteome analysis, as highlighted by the Ig kappa chain C, α-1-inhibitor 3, and prothrombin. As expected, currently available HI biomarkers rely on the measurement of substances that are key to the development, transport, and metabolism of HI, and thus have unlimited clinical application value in predicting HI. These proteins may, therefore, be considered candidates for the further investigation of synergistic mechanisms, and may be potential therapeutic targets for HI.

17.3.8 Synergistic Effects of DGR on the Relative Expression Level of Target Proteins

Because DGR exerts preferential and synergistic effects on the proteome, it was worthwhile to further investigate the protein expression profiles of the model, and of each dosing group using comparative proteomics analysis. We found that a total of 15 target proteins were synergistically/additively modified by DGR based on 2D gels. Figure 17.8B–D show the proteome maps (2-DE images) for HI, YCHT treated, and DGR treated rats, respectively. Image analysis revealed 15 differentially expressed proteins ($p < 0.01$ by Student's t-test with a fold change > 3.0). For the purpose of protein identification, differentially expressed proteins were isolated from the 2D gel, and analyzed using MS after in-gel digestion. Among these proteins, we found a striking number of uniquely HI-associated proteins. For example, the amount of zinc finger protein (spot 2) was 0.1811 in the control, and downregulated to 0.0693 in the model group. We found that both D and G upregulated the zinc finger protein at the proteomic level, whereas YCHT significantly enhanced zinc finger protein upregulation (0.1279). Interestingly, the DGR combination induced strong expression (0.1621), reflecting potential therapeutic, and synergistic properties of DGR in treating HI. Equally notable is the haptoglobin protein (spot 4), which showed strong upregulation in the model group, and downregulation in the YCHT, DGR, D, and G groups. In this study, the level of alpha-1-antitrypsin (spot 6) was upregulated in the model rats as compared with controls (indicating its importance in liver injury), downregulated in the YCHT, and DGR groups, and absent or not detectable in the DG, DR, GR, D, G, and R groups. We found that the expression of the vitamin D-binding protein (spot 14) was slightly increased in the groups that were treated with D (i.e., the D, DG, and DR groups) and was dramatically upregulated in the DGR treatment group. Although G, R, and the GR combination did not exert this significant effect, these two compounds (G and R) enhanced the effects of D, indicating that the expression of the vitamin D-binding protein by D can be further amplified by G and R. This is strong evidence of the synergistic effect of D, G, and R at the proteomic level by coordinating protein expression. These observations indicate that DGR had a synergistic/additive effect on transport, metabolism, and the modulation of immunological processes in HI rats, and that these effects were much more pronounced with cotreatment than with any single treatment. These data, together with the biochemical, immunohistochemical, and metabolomic analyses, suggest that D is the principal ingredient in this formulae, whereas G and R can serve as adjuvant components, and that by hitting multiple targets, the DGR combination exerts more profound therapeutic effects than either monotherapy, or bitherapy using D, G, and/or R.

17.4 DISCUSSION

Currently, rather than focusing on a single target, a paradigm shift is occurring that is sparking new interest in agents that modulate simultaneously multiple targets with fewer adverse effects and lower toxicity. Combination medicines with multieffect pathways, and multieffect targets tend to be more effective than single drugs, and may help to address many treatment-related challenges (Li et al., 2013). Modern medical therapy has long acknowledged the usefulness of combination therapies, which regulate simultaneously multiple nodes of the disease network, and have a synergistic effect on the treatment of multifactorial diseases (Shi et al., 2014; Li et al., 2012). However, continued progress will be essential if we are to develop effective, customized multidrug regimens. TCM, whose therapeutic efficacy is based on the combined action of a mixture of constituents, offers new treatment opportunities. In recent years, there has been a significant increase in the clinical use of combinatorial intervention in TCM to achieve synergistic interactions that are capable of producing a sufficient effect at low doses. Most Chinese therapeutic herbs that are traditionally used in cotreatment but not monotreatment series, exert significantly better pharmacological effects. However, the precise mechanism of synergistic action remains poorly understood. Based on syndromes and patient characteristics, and guided by the theories of TCM, formulae are designed to contain a combination of various kinds of crude drugs that, when combined, will achieve better clinical efficacy. One example is YCHT, whose efficacy in treating HI has recently been well established. Because the means of action of DGR on HI has not been sufficiently investigated, we used an "omics" platform, and identified the key mechanisms that underlie the observed effects. To gain insight into the complex biochemical mechanisms of this effective HI therapy, we conducted an investigation that incorporated modern biochemical analyses, immunohistochemistry, metabolomics, and proteomics, analyzing the effects of the D, G, R, DG, DR, RG, and DGR treatments on HI. It is well known that CCl_4 is a potent hepatotoxic agent that is rapidly metabolized to the CCl_3 free radical, which subsequently acts on liver cells to covalently conjugate membranous unsaturated lipids, leading to lipid peroxidation; thus, CCl_4 is widely used to model HI in rats. Based on our previous metabolomics findings, we know that YCHT can significantly prevent HI by interfering with the trajectory changes in the biomarkers, exerting an overtly hepatoprotective effect. We have now tested the therapeutic, and synergistic properties of YCHT as an effective treatment for HI using a rat model. We employed UPLC-HDMS, MALDI-TOF MS/MS, and 2-DE to systematically analyze the synergism of DGR at the levels of the proteome and metabolome; thus exploring some of the key molecular mechanisms that underlie these synergistic effects. Our approach extends the well-established concept of combinatorial therapeutics, and identifies new potential strands of investigation that involve multicomponent combinations that target multiple pathways.

In this study, we show that the combined *in vivo* use of the active components of YCHT (namely, D, G, and R) exerts more profound therapeutic effects than any component used individually in a rat model of HI. DGR significantly intensified the therapeutic efficacy as indicated by our modern biochemical analysis. The immunohistochemical assay further supports the cooperation of D and G with R in upregulating BCL-2, and downregulating FAS, with the strongest effect occurring with the DGR combination, thereby supporting the rationale of using YCHT for treating HI. These observations provide a mechanistic insight into the synergistic effect. Our metabolomic trajectory analysis indicated that each dosing group could be regulated back to baseline levels (i.e., to the control group) but the maximum synergic effect on liver injury was observed in rats treated with the DGR combination, as opposed to mono- or bitherapies. These observations support the rationale of the formulae wherein the compounds mutually reinforce each other. We also found that DGR activated an array of factors that are involved in energy, amino acid, nucleotide, fatty acid, cofactor, and vitamin metabolism, and we suggest that these effects may form the basis of DGR synergy. Using a metabolomics approach, this study has thereby indicated that metabolomics can be used to elucidate the synergistic mechanisms that govern TCM. Our study has identified robust biomarkers that are related to HI, with a special focus on the changes in the relative expression levels of target proteins after each treatment. It is important to note that based on our results, DGR targets not only immunity, and metabolism but also targets key regulatory pathways that are used in transport, signal transduction, cell growth, and proliferation, thereby helping to restore the normal function of the liver. Our data will help to indicate the molecular mechanisms of synergism at the proteomic level.

According to the rules of TCM theory, the monarch, minister, assistant, and servant herbs that comprise the formulae can work together harmoniously to achieve an ideally synergistic and therapeutic outcome. Understanding the synergistic effects of YCHT represents an even greater challenge than usual. This is because in YCHT, the multilayer regulation structure involves three compounds with overlapping but distinct target properties. However, determining their targets can shed new light on synergism, and efficient therapeutic strategies. Although the YCHT formulae, was designed by TCM doctors in the premolecular era, its mode of function can be revealed using modern biochemical analyses. In summary, we show that TCM formulae can be analyzed using biomedical and chemical research approaches at the biochemistry, proteome, and metabolome levels. Here, we report that the DGR-based treatment regimen yields encouraging outcomes, thus reinforcing its potential use as a frontline therapy for HI. This study can be considered a useful pilot trial in the effort to evaluate traditional formulae on a larger scale, and to bridge Western and Eastern medicines in this era of systems biology. Furthermore, these observations indicate that traditional Chinese formulae usually require multiple components to exert their effects, possibly laying the foundation for promising new schemes, and patterns of drugs that are derived from TCM.

References

Drews, J., 2000. Drug discovery: a historical perspective. Science 287 (5460), 1960–1964.

Gao, M., Nettles, R.E., Belema, M., Snyder, L.B., Nguyen, V.N., Fridell, R.A., Serrano-Wu, M.H., Langley, D.R., Sun, J.H., O'Boyle, 2nd, D.R., Lemm, J.A., Wang, C., Knipe, J.O., Chien, C., Colonno, R.J., Grasela, D.M., Meanwell, N.A., Hamann, L.G., 2010. Chemical genetics strategy identifies an HCV NS5A inhibitor with a potent clinical effect. Nature 465, 96–100.

Keith, C.T., Borisy, A.A., Stockwell, B.R., 2005. Multicomponent therapeutics for networked systems. Nat. Rev. Drug Discov. 4, 71–78.

Lee, J.Y., Kim, J.Y., Park, G.W., Cheon, M.H., Kwon, K.H., Ahn, Y.H., Moon, M.H., Lee, H.J., Paik, Y.K., Yoo, J.S., 2011. Targeted mass spectrometric approach for biomarker discovery and validation with nonglycosylated tryptic peptides from N-linked glycoproteins in human plasma. Mol. Cell. Proteomics 10, M111.009290.

Li, J., Wu, R.G., Meng, F.Y., Wang, Z., Wang, C.M., Wang, Y.Y., Zhang, Z.J., 2012. Synergism and rules from combination of baicalin, jasminoidin and desoxycholic acid in refined Qing Kai Ling for treat ischemic stroke mice model. PLoS ONE 7 (9), e45811.

Li, S., Wu, C., Chen, J., Lu, P., Chen, C., Fu, M., Fang, J., Gao, J., Zhu, L., Liang, R., Shen, X., Yang, H., 2013. An effective solution to discover synergistic drugs for anti-cerebral ischemia from traditional Chinese medicinal formulae. PLoS ONE 8 (11), e78902.

Liao, Z.G., Liang, X.L., Zhu, J.Y., Zhao, G.W., Yang, M., Wang, G.F., Jiang, Q.Y., Chen, X.L., 2010. Correlation between synergistic action of Radix Angelica dahurica extracts on analgesic effects of Corydalis alkaloid and plasma concentration of dl-THP. J. Ethnopharmacol. 129 (1), 115–120.

Perelson, A.S., Essunger, P., Cao, Y., Vesanen, M., Hurley, A., Saksela, K., Markowitz, M., Ho, D.D., 1997. Decay characteristics of HIV-1-infected compartments during combination therapy. Nature 387, 188–191.

Pibiri, F., Kozikowski, A.P., Pinna, G., Auta, J., Kadriu, B., Costa, E., Guidotti, A., 2008. The combination of huperzine A and imidazenil is an effective strategy to prevent diisopropyl fluorophosphate toxicity in mice. Proc. Natl. Acad. Sci. USA 105, 14169–14174.

Shi, S.H., Cai, Y.P., Cai, X.J., Zheng, X.Y., Cao, D.S., Ye, F.Q., Xiang, Z., 2014. A network pharmacology approach to understanding the mechanisms of action of traditional medicine: Bushenhuoxue formula for treatment of chronic kidney disease. PLoS ONE 9 (3), e89123.

Tang, Q.F., Ji, Q., Qiu, Y.Y., Cao, A.L., Chi, Y.F., Liang, B., Peng, W., Yin, P.H., 2014. Synergistic effect of Zuo Jin Wan on DDP-induced apoptosis in human gastric cancer SGC-7901/DDP Cells. Evid. Based Complement. Alternat. Med. 2014, 724764.

van der Greef, J., Martin, S., Juhasz, P., Adourian, A., Plasterer, T., Verheij, E.R., McBurney, R.N., 2007. The art and practice of systems biology in medicine: mapping patterns of relationships. J. Proteome. Res. 6, 1540–1559.

Wang, Z.Y., Chen, Z., 2008. Acute promyelocytic leukemia: from highly fatal to highly curable. Blood 111, 2505–2515.

Wang, L., Zhou, G.B., Liu, P., Song, J.H., Liang, Y., Yan, X.J., Xu, F., Wang, B.S., Mao, J.H., Shen, Z.X., Chen, S.J., Chen, Z., 2008a. Dissection of mechanisms of Chinese medicinal formula

Realgar-Indigo naturalis as an effective treatment for promyelocytic leukemia. Proc. Natl. Acad. Sci. USA 105, 4826–4831.

Wang, X., Sun, W., Sun, H., Lv, H., Wu, Z., Wang, P., Liu, L., Cao, H., 2008b. Analysis of the constituents in the rat plasma after oral administration of Yin Chen Hao Tang by UPLC/Q-TOF-MS/MS. J. Pharm. Biomed. Anal. 46, 477–490.

Wang, X., Sun, H., Zhang, A., Jiao, G., Sun, W., Yuan, Y., 2011. Pharmacokinetics screening for multi-components absorbed in the rat plasma after oral administration traditional Chinese medicine formula Yin-Chen-Hao-Tang by ultra performance liquid chromatography-electrospray ionization/quadrupole-time-of-flight mass spectrometry combined with pattern recognition methods. Analyst 136, 5068–5076.

Wang, X., Zhang, A., Wang, P., Sun, H., Wu, G., Sun, W., Lv, H., Jiao, G., Xu, H., Yuan, Y., Liu, L., Zou, D., Wu, Z., Han, Y., Yan, G., Dong, W., Wu, F., Dong, T., Yu, Y., Zhang, S., Wu, X., Tong, X., Meng, X., 2013. Metabolomics coupled with proteomics advancing drug discovery toward more agile development of targeted combination therapies. Mol. Cell. Proteomics 12 (5), 1226–1238.

Xiao, T., Ying, W., Li, L., Hu, Z., Ma, Y., Jiao, L., Ma, J., Cai, Y., Lin, D., Guo, S., Han, N., Di, X., Li, M., Zhang, D., Su, K., Yuan, J., Zheng, H., Gao, M., He, J., Shi, S., Li, W., Xu, N., Zhang, H., Liu, Y., Zhang, K., Gao, Y., Qian, X., Cheng, S., 2005. An approach to studying lung cancer-related proteins in human blood. Mol. Cell. Proteomics 4, 1480–1486.

Yohannes, E., Chang, J., Tar, M.T., Davies, K.P., Chance, M.R., 2010. Molecular targets for diabetes mellitus-associated erectile dysfunction. Mol. Cell. Proteomics 9 (3), 565–578.

Zhang, Z., Zhang, Y., Ning, G., Deb, D.K., Kong, J., Li, Y.C., 2008. Combination therapy with AT1 blocker and vitamin D analog markedly ameliorates diabetic nephropathy: blockade of compensatory renin increase. Proc. Natl. Acad. Sci. USA 105, 15896–15901.

Zhang, Q.Y., Mao, J.H., Liu, P., Huang, Q.H., Lu, J., Xie, Y.Y., Weng, L., Zhang, Y., Chen, Q., Chen, S.J., Chen, Z., 2009. A systems biology understanding of the synergistic effects of arsenic sulfide and Imatinib in BCR/ABL-associated leukemia. Proc. Natl. Acad. Sci. USA 106, 3378–3383.

Zhang, L., Ren, X., Alt, E., Bai, X., Huang, S., Xu, Z., Lynch, P.M., Moyer, M.P., Wen, X.F., Wu, X., 2010a. Chemoprevention of colorectal cancer by targeting APC-deficient cells for apoptosis. Nature 464, 1058–1061.

Zhang, X.W., Yan, X.J., Zhou, Z.R., Yang, F.F., Wu, Z.Y., Sun, H.B., Liang, W.X., Song, A.X., Lallemand-Breitenbach, V., Jeanne, M., Zhang, Q.Y., Yang, H.Y., Huang, Q.H., Zhou, G.B., Tong, J.H., Zhang, Y., Wu, J.H., Hu, H.Y., de Thé, H., Chen, S.J., Chen, Z., 2010b. Arsenic trioxide controls the fate of the PML-RARalpha oncoprotein by directly binding PML. Science 328, 240–243.

Zhang, A., Sun, H., Yuan, Y., Sun, W., Jiao, G., Wang, X., 2011. An *in vivo* analysis of the therapeutic and synergistic properties of Chinese medicinal formula Yin-Chen-Hao-Tang based on its active constituents. Fitoterapia 82, 1160–1168.

Zheng, P.Z., Wang, K.K., Zhang, Q.Y., Huang, Q.H., Du, Y.Z., Zhang, Q.H., Xiao, D.K., Shen, S.H., Imbeaud, S., Eveno, E., Zhao, C.J., Chen, Y.L., Fan, H.Y., Waxman, S., Auffray, C., Jin, G., Chen, S.J., Chen, Z., Zhang, J., 2005. Systems analysis of transcriptome and proteome in retinoic acid/arsenic trioxide-induced cell differentiation/apoptosis of promyelocytic leukemia. Proc. Natl. Acad. Sci. USA 102, 7653–7658.

CHAPTER 18

Metabolic Profiling and Biomarkers of Type 2 Diabetes and the Effective Evaluation of the Tianqi Jiangtang Capsule

Xijun Wang, Shuxiang Zhang, Hui Sun, Aihua Zhang

CONTENTS

18.1 INTRODUCTION

The prevalence of diet-induced obesity is increasing globally, and posing significant health problems for millions of people in the world. Diet-induced obesity is a major contributor to the global pandemic of Type 2 diabetes (T2D), which is a manifestation of Xiaoke Syndrome in TCM (Rubin, 2013; DeFuria et al., 2013). Typically, a civilization disease, particularly T2D, represents one of the most significant global health problems because it is associated with a large economic burden on the health systems of many countries (Seaquist, 2014). However, little is known about this inherited metabolic relation. In order to give a new insight into the diabetic process itself, as well as conduct clinical instruction, it is necessary to clarify the global metabolic alteration that characterizes its progression. The burden of T2D is growing worldwide, and with it an urgent requirement for better tools to detect, diagnose, and monitor the disease. The new platform of metabolomics, focuses on a holistic investigation of living systems to external stimuli, based on the global metabolite profiles in biological samples, provides variation of whole metabolic networks for characterizing pathological states, as well as giving mechanistic insights into the biochemical effects of TCM (Jiang et al., 2011; Tian et al., 2014; Gao et al., 2014).

The Tianqi Jiangtang Capsule (TJC) has been efficiently and widely used to treat T2D or Xiaoke Syndrome in TCM. It is composed of 10 herbs, including: *Astragalus membranaceus (Fisch.) Bunge., Dried root of Trichosanthes kirilowii Maxim., Panax ginseng C. A. Mey., Dendrobium nobile Lindl, Fructus Ligustri Lucidi, Coptis chinensis Franch., Cortex Lycii, Cornus officinalis Sieb. et Zucc., Eclipta prostrate*, and *Rhus chinensis Mill* (Zhang et al., 2010). The detailed therapeutic mechanism, based on the metabolomics of TJC against T2D, remains unknown. By utilizing this comprehensive biochemical profiling approach, we seek to identify metabolites with different concentrations in T2D, and thereby allowing new insights into the pathophysiological progression of this important metabolic disease. Various analytical techniques,

Chinmedomics. http://dx.doi.org/10.1016/B978-0-12-803117-9.00018-4

with multivariate data analysis, such as partial least squares-discriminant analysis (PLS-DA) have been applied in metabolomics-based metabolism studies. UPLC coupled with MS has become one of the widely applied techniques in metabolomics owing to its high sensitivity and reproducibility. Herein, urinary metabolomics based on UPLC-MS was applied to investigate the metabolic profiles, and the potential biomarkers in a rat model of T2D, which may facilitate understanding the pathological changes of T2D when treated by TJC.

18.2 MATERIALS AND METHODS

18.2.1 Animal Handling

After acclimatization, animals were randomly divided into three groups. Wistar rats were treated with a high fat emulsion (10 mL/kg) by ig for 13 consecutive days to induce T2D. The high fat emulsion was prepared by our laboratory, briefly: lard 20 g, methyl thiouracil 1 g, cholesterol 5 g, sodium glutamate 1 g, sugar 5 g, fructose 5 g, propylene glycol 30 mL were mixed, and an added water volume to 100 mL. The rats in the control group were given distilled water for 20 consecutive days, and anesthetized via an intraperitoneal injection of 1% pentobarbital sodium (0.15 mL/100 g body weight) on day 21. The rats in the T2D group were orally administrated high fat emulsion for 20 consecutive days (at 8:00 p.m.), and on day 21, the rats were anesthetized as in the control group. Simultaneously, the TJC group was administrated high fat emulsion, and TJC treatments for 20 consecutive days, and was then anesthetized as in the control group.

Blood was collected from the abdominal aorta, and plasma, and serum were separated via centrifuging at 6000 rpm for 20 min at 4°C. The serum was used for biochemical assay according to the manufacturer's instructions of commercial kits. The activities and levels of fasting serum glucose (FSG), triglycerides (TG), total cholesterol (TC), malondialdehyde (MDA), superoxide dismutase (SOD), free fatty acid (FFA), and the rate constant for plasma glucose disappearance (KITT) were determined by using commercially available kits. All procedures completely complied with the manufacturer's instructions.

All the rats of each group were housed in metabolic cages (1 per cage). Urine was collected daily (at 6:00 a.m.) from metabolism cages at ambient temperature throughout the whole procedure, and centrifuged at 13,000 rpm at 4°C for 5 min, and the supernatants were stored frozen at −80°C until analysis. All these samples were thawed at room temperature before analysis, and centrifuged at 13,000 rpm for 5 min. An aliquot of 5 μL was injected for UPLC/MS analysis after being filtered through a 0.22 μm membrane filter.

18.2.2 Metabolic Profiling Platform

18.2.2.1 Chromatography

UPLC/ESI-Q-TOF/MS was used for the global analysis of urine samples. Chromatographic analysis was performed in a Waters ACQUITY UHPLC system controlled with Masslynx (v4.1) (Waters Corp., Milford, USA). An aliquot of 6 μL of the sample solution was injected onto an ACQUITY UPLC BEH C_{18} column (50 × 2.1 mm, 1.7 μm, Waters Corp., Milford, USA) at 35°C, the flow rate was 0.5 mL/min, and the injection volume was 2 μL. The optimal mobile phase consisted of a linear gradient system of: (A) 0.1% formic acid in water; and (B) 0.1% formic acid in acetonitrile, 0–1 min, 99–90% A; 1–4 min, 90–80% A; 4–6 min, 80–65% A; 6–9 min, 65–1% A; 9–11 min, 1% A; 11–11.5 min, 1–99% A, 11.5–13 min, 99% A.

18.2.2.2 Mass Spectrometry

The mass spectrometry was operated by electrospray ionization in the negative ionization mode. The optimal conditions of analysis were as follows: the source temperature was set at 110°C, desolvation gas temperature was 350°C, cone gas flow was 50 h, desolvation gas flow was 600 L/h; the capillary voltage was 2.3 kV, the sampling cone voltage was 35 V, microchannel plate voltage was 2450 V, and the extraction cone voltage was 3.0 V. The data acquisition rate was set to 0.14 s/scan, with a 0.1 s inter scan delay. Data were collected in centroid mode from 100 Da–1000 Da. For accurate mass acquisition, a lock-mass of leucine enkephalin at a concentration of 0.2 ng/mL was used via a lock spray interface at a flow rate of 100 μL/min monitoring for negative ion mode ($[M + H]^+ = 556.2615$) to ensure accuracy during the MS analysis.

18.2.2.3 Multivariate Data Analysis and Data Processing

LC-MS raw data were imported into the MassLynx™ software for peak detection and alignment. The intensity of each ion was normalized with respect to the total ion count, to generate a data matrix that consisted of the retention time, *m/z* value, and the normalized peak area. The multivariate data matrix was analyzed by EZinfo software 2.0 (Waters Corp., Milford, USA). Metabolite candidates of interest were extracted from the PLS-DA loading plots based on their contribution to the variation, and their correlation within the data set. With regard to the identification of potential biomarkers, the ion spectrum was matched with the structure message of metabolites acquired from biochemical databases. The reconstruction and pathway analysis of potential biomarkers was performed with MetPA software based on the above database source. Student's t-test was performed to identify features with differential abundances across groups.

Table 18.1 Biochemistry Results of Type 2 Diabetes Rat

Group	FSG (mmol/L)	TG (mmol/L)	T-CHO (mmol/L)	MDA (nmol/mL)	SOD (U/mL)	FFA (μmol/L)	KITT	Urine (mL/12 h)
Control	4.36 ± 0.61	0.33 ± 0.09	1.55 ± 0.45	11.56 ± 5.77	84.83 ± 23.58	987.35 ± 216.04	0.69 ± 0.30	12.93 ± 4.08
Model	23.63 ± 6.77***	0.69 ± 0.49*	2.87 ± 1.05**	17.33 ± 6.33*	50.16 ± 15.68**	1299.65 ± 348.90*	0.37 ± 0.25*	48.88 ± 11.68***

Notes: FSG, fasting serum glucose; TG, triglycerides; TC, total cholesterol; MDA, malondialdehyde; SOD, superoxide dismutase; FFA, free fatty acid; KITT, rate constant for plasma glucose disappearance.
*Significant difference from control at $p < 0.05$;
**Significant difference from control at $p < 0.01$;
***Significant difference from control at $p < 0.001$.

18.3 RESULTS

18.3.1 Multivariate Statistical Analysis of Rat Urine

The biochemistry parameters of the control group, and the T2D group were summarized in Table 18.1. The biochemical results observed in the T2D group did show significant difference, compared to the control group. In this study, a PLS-DA method was established, and employed to identify biomarkers that were related to T2D development. As Fig. 18.1 illustrates, there is a distinguished classification between the clustering of the control and the T2D groups. According to the results of S-plot, a total of 6912 variable ions (Fig. 18.2) were significantly different between the control, and the model groups. Finally, six of them in negative mode were identified by searching MS and MS/MS

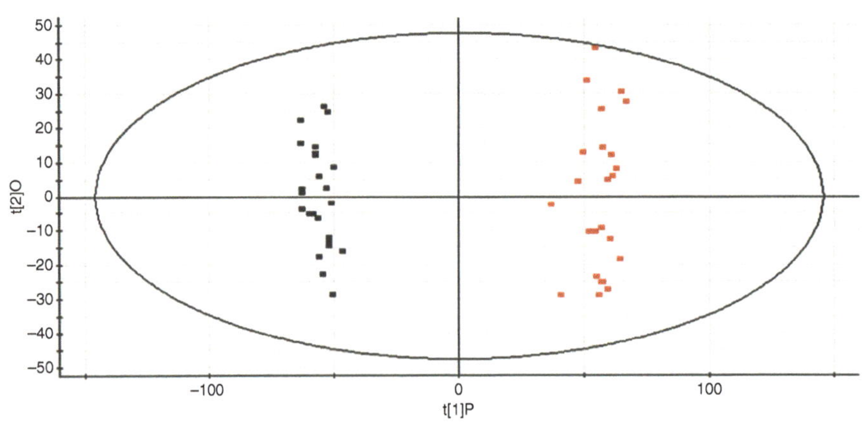

FIGURE 18.1 PLS-DA plot derived from the UPLC/MS profiles of rat urine samples demonstrating separation of the control group (black) and the T2D group (red) rats.

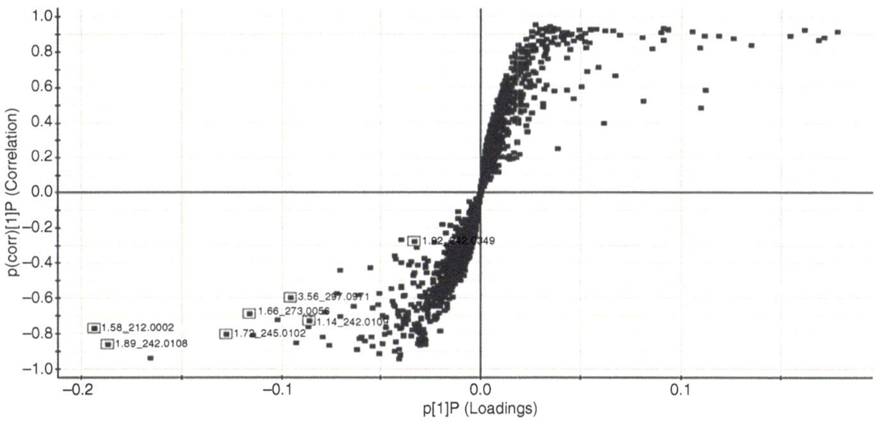

FIGURE 18.2 S-plot of PLS in rat urine samples represents the impact of the metabolites on the clustering results.

Urine samples from the control group, and the T2D group rats were subjected to UPLC/MS. The PLS model was then used to generate loading S-plot showing ions important to the clustering of samples. Box data points indicate the ions that are most responsible for the variance in the score plot.

fragments in metabolites database, and finally confirmed by commercial standards (Table 18.2).

18.3.2 Identification of Metabolite Candidates

The information including the retention time, the exact mass, and the MS/MS data were supplied by the robust UPLC-MS platform. The precise molecular mass was determined within a reasonable degree of measurement error using

Table 18.2 List of Potential Urinary Biomarkers of Urine Samples from Type 2 Diabetes

No.	Retention Time	Measured Mass	Elemental Composition	Identified Metabolites	*p*-value
1	1.89	242.0108	$C_9H_9NO_5S$	*N*-acetyl-2-carboxy benzenesulfonamide	0.00
2	1.58	212.0002	$C_4H_7NO_7P$	4-Phospho-L-aspartate	0.01
3	1.66	273.0056	$C_6H_8O_{10}P$	3-Dehydro-L-gulonate 6-phosphate	0.01
4	1.72	245.0102	$C_9H_{10}O_6S$	2-(4-Hydroxy-3-methoxy-phenyl) acetaldehyde sulfate	0.01
5	1.92	338.0849	$C_{15}H_{17}NO_8$	6-Hydroxy-5-methoxyin-dole glucuronide	0.02
6	3.56	297.0971	$C_{14}H_{18}O_7$	2-Phenylethanol glucuronide	0.03

Q-TOF, and the potential element composition, degree of unsaturation, and fractional isotope abundance of the compounds were also obtained. Metabolite identification was conducted with high resolution MS, and MS/MS fragments, as well as database analyses. We searched for the presumed molecular formula in the ChemSpider, Human Metabolome Database, KEGG, and Small Molecule Pathway Database to confirm possible chemical compositions. According to the protocol described above, six endogenous metabolites were identified and summarized in Table 18.2.

18.3.3 Biomarker Network and Metabolic Pathway Reconstruction

The related pathways of the biomarkers were investigated by searching the KEGG and HMDB, and a network of some biomarkers was established. Metabolic pathway analysis with MetPA revealed that potential biomarkers are mainly involved in the pathway of pentose, and glucuronate interconversions, starch, and sucrose metabolism, and tyrosine metabolism that changed specifically in the setting of T2D (Fig. 18.3). Two metabolites (2-phenylethanol

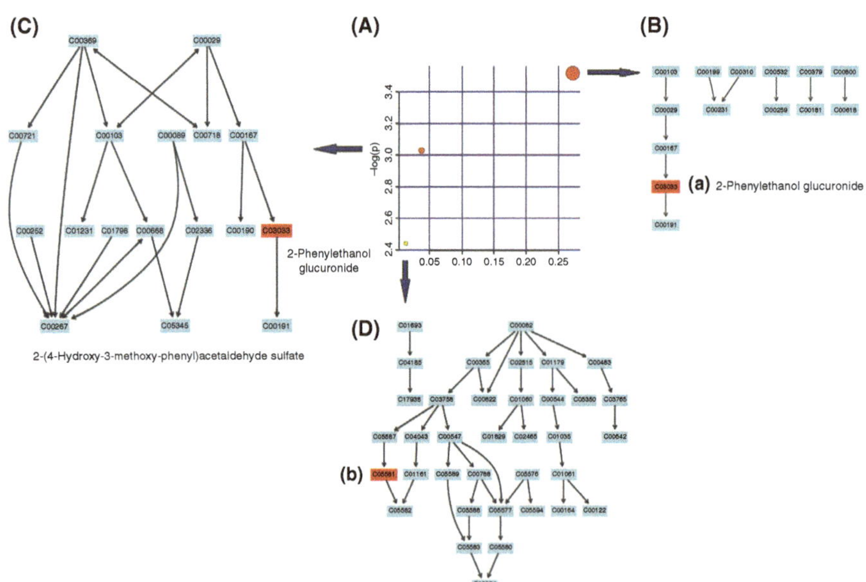

FIGURE 18.3 A systemic view of metabolic pathways that are associated with T2D in this study. (A) Identifying network pathway by MetPA software; (B) putative metabolic pathways of pentose and glucuronate interconversions; (C) starch and sucrose metabolism; (D) pyrimidine metabolism –were inferred from the rat urine of intermediates during substance metabolism. The map was generated using the reference map by KEGG. Red denotes the affected metabolites related to the pathway. (a) 2-Phenylethanol glucuronide; and, (b) 2-(4-Hydroxy-3-methoxy-phenyl) acetaldehyde sulfate.

glucuronide, 2-(4-hydroxy-3-methoxy-phenyl) acetaldehyde sulfate) identified from these pathways were involved in T2D development, which indicated that a dysfunction of multipathways was involved in the pathological process of T2D. The detailed construction of the metabolism pathways with higher scores was shown in Fig. 18.3. Results suggested that these target pathways showed the marked perturbations over the time-course of T2D, and could contribute to the development of T2D.

18.3.4 Trajectory Analysis of Metabolic Profiling

Typically, the trajectory analysis of PCA score plots for the T2D induced by high-fat treatment in the positive mode can really reflect the differences between the model and the control groups, and showed the metabolic profiles in the control, and model groups were separated clearly (Fig. 18.4). PCA can effectively demonstrate the differences between the treated and the control groups: demonstrates a clear separation. The PCA plot of the T2D group deviated from the control groups on day 13 and reached the maximum trend, and then on day 15 to day 20 gradually returned back to the normal state. The tracks of the metabolic profiles at different time points also clearly demonstrate the time-dependent changes in the urine metabolites.

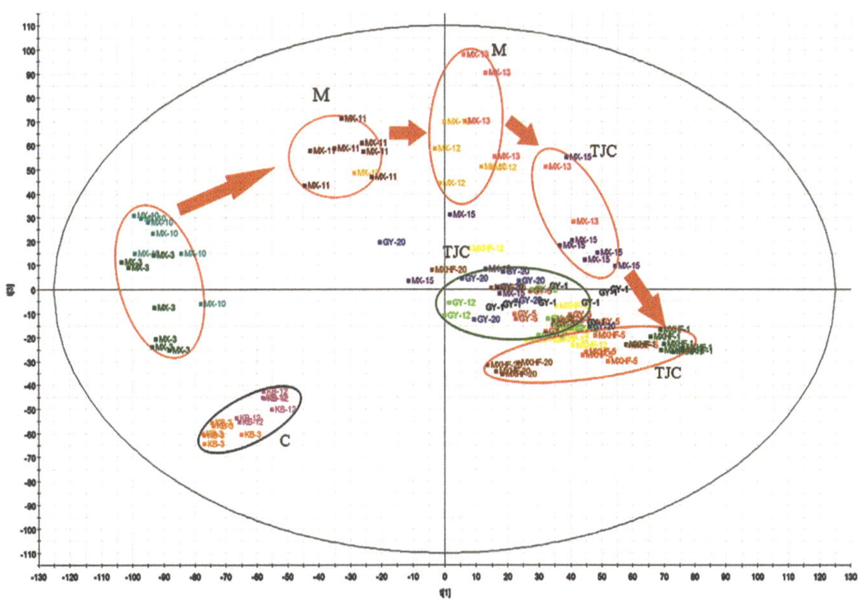

FIGURE 18.4 **PCA score plot of trajectory analysis after Tianqi Jiangtang Capsule treatment.** (C, control group; M, model group; TJC, Tianqi Jiangtang Capsule group).

18.4 DISCUSSION

Metabolomics has been studied extensively in human diseases, and has resulted in significant advances in the understanding of the pathophysiology of diseases. Diabetes represents one of the most important global health problems, and approximately 90% of patients with diabetes have T2D, and its incidence remains the highest in the world (Zhang et al., 2014a). Recently, a variety of biomarkers reflecting T2D pathologies have been developed to gain new insights into metabolic pathways, and pathophysiological mechanisms (Sun et al., 2014; Zhang et al., 2014b). Because some of the processes identified above may result in metabolic "signatures" in the urine that would be useful for T2D diagnosis as well as therapeutic responsiveness. Although, great effort has been put forth to uncover the complex molecular mechanisms exploited in the pathogenesis of the T2D disease, a satisfactory explanation remains to be discovered.

In our study, an UPLC-MS-based urine metabolomics approach coupled with multivariate statistical methods provides a powerful approach to clearly differentiate patients with T2D from matched controls, and to identify the potential biomarkers. The mass to charge ratio (m/z), retention time, and the abundance data generated were subjected to PLS multivariate data analysis. The loading S-plot generated from PLS is a convenient way of visualizing those ions with the highest contribution to the separation between the control, and the T2D groups, in relation to their correlation to the model. Results indicate that PLS-DA revealed a very good visual separation between the T2D, and control samples. Interestingly, six distinct metabolites were identified from these pathways, many are in various stages of progress at the T2D. Further study of these metabolites may facilitate the development of noninvasive biomarkers, and the development of more efficient therapeutic strategies for T2D. Further investigations are also underway to clarify the precise pathogenesis of why T2D induced these results. Our results also indicated that the disturbed metabolic track in model rats was gradually restored to normal after administration of TJC. Importantly, TJC reversed the elevated biomarkers to the control levels from the T2D, which provided a systematic view, and a theoretical basis for its prevention or treatment. Furthermore, it is noteworthy that two metabolites together are important for the host response to T2D through metabolism pathways. T2D-related metabolites were tightly correlated with pentose and glucuronate interconversions, starch and sucrose metabolism, pyrimidine metabolism, and tyrosine metabolism network; that are strongly associated with T2D development. These biochemical changes are helpful to understand the key features of T2D. In addition, these metabolic features provided useful clues for future mechanism exploration, and identification of the therapeutic targets of T2D.

18.5 CONCLUSIONS

Emerging high-throughput metabolomics technologies have been widely applied, aiming at the discovery of candidate biomarkers for disease, and may help to promote understanding of the mechanism of T2D occurrence on the metabolic level. Herein, we illustrate how metabolomics can be utilized to explore the mechanisms of T2D that affect different "key pathway". Currently, T2D is one of the most common diseases in the world, but at the present time it is difficult to determine the precise pathophysiology. Here, we applied the metabolomics approach, based on UPLC/MS to systematically investigate T2D. Interestingly, three distinct pathways, namely, pentose and glucuronate interconversions, starch and sucrose metabolism, and tyrosine metabolism, etc. were found to be associated with T2D – according to ingenuity pathway analysis – which will require further validation. The study indicates that TJC treatment can ameliorate T2D through partially regulating the perturbed metabolism. Our work demonstrates that LC-MS-based metabolomic strategy is a promising new tool to evaluate the therapeutic effects, and the mechanism of complex TCM prescriptions.

References

DeFuria, J., Belkina, A.C., Jagannathan-Bogdan, M., Snyder-Cappione, J., Carr, J.D., Nersesova, Y.R., Markham, D., Strissel, K.J., Watkins, A.A., Zhu, M., Allen, J., Bouchard, J., Toraldo, G., Jasuja, R., Obin, M.S., McDonnell, M.E., Apovian, C., Denis, G.V., Nikolajczyk, B.S., 2013. B cells promote inflammation in obesity and type 2 diabetes through regulation of T-cell function and an inflammatory cytokine profile. Proc. Natl. Acad. Sci. USA 110 (13), 5133–5138.

Gao, X., Guo, M., Li, Q., Peng, L., Liu, H., Zhang, L., Bai, X., Wang, Y., Li, J., Cai, C., 2014. Plasma metabolomic profiling to reveal antipyretic mechanism of Shuang-huang-lian injection on yeast-induced pyrexia rats. PLoS One 9 (6), e100017.

Jiang, P., Dai, W., Yan, S., Chen, Z., Xu, R., Ding, J., Xiang, L., Wang, S., Liu, R., Zhang, W., 2011. Potential biomarkers in the urine of myocardial infarction rats: a metabolomic method and its application. Mol. Biosyst. 7 (3), 824–831.

Rubin, H., 2013. Promotion and selection by serum growth factors drive field cancerization, which is anticipated *in vivo* by type 2 diabetes and obesity. Proc. Natl. Acad. Sci. USA 110 (34), 13927–13931.

Seaquist, E.R., 2014. Addressing the burden of diabetes. JAMA 311 (22), 2267–2268.

Sun, H., Zhang, S., Zhang, A., Yan, G., Wu, X., Han, Y., Wang, X., 2014. Metabolomic analysis of diet-induced type 2 diabetes using UPLC/MS integrated with pattern recognition approach. PLoS One 9 (3), e93384.

Tian, J.S., Peng, G.J., Gao, X.X., Zhou, Y.Z., Xing, J., Qin, X.M., Guan-Hua, D., 2014. Dynamic analysis of the endogenous metabolites in depressed patients treated with TCM formula Xiaoyaosan using urinary 1 H NMR-based metabolomics. J. Ethnopharmacol., 1–10, 158PA.

Zhang, S.X., Sun, H., Sun, W.J., Jiao, G.Z., Wang, X.J., 2010. Proteomic study of serum proteins in a type 2 diabetes mellitus rat model by Chinese traditional medicine Tianqi Jiangtang Capsule administration. J. Pharm. Biomed. Anal. 53 (4), 1011–1014.

Zhang, A.H., Qiu, S., Xu, H.Y., Sun, H., Wang, X.J., 2014a. Metabolomics in diabetes. Clin. Chim. Acta 429, 106–110.

Zhang, A.H., Sun, H., Yan, G.L., Yuan, Y., Han, Y., Wang, X.J., 2014b. Metabolomics study of type 2 diabetes using ultra-performance LC-ESI/quadrupole-TOF high-definition MS coupled with pattern recognition methods. J. Physiol. Biochem. 70 (1), 117–128.

CHAPTER 19

Metabolic Biomarkers of Alcohol Liver Damage and the Intervention Effect of Yinchenhao Tang

Hui Sun, Haitao Lv, Aihua Zhang, Xijun Wang

CONTENTS

19.1 INTRODUCTION

Alcohol liver disease (ALD) remains one of the most common causes of chronic liver disease in the world, and is the leading cause of cirrhosis, and hepatocellular carcinoma worldwide. It is reported that several mediator systems are correlated with the development of ALD, from fatty liver to advanced liver injury, such as inflammation and necrosis. Alcohol fatty liver is more susceptible to many inflammatory stimuli, the bacterial endotoxins mainly involved in the inflammatory process. Moreover, oxidative stress or ischemic damage also seems to aggravate alcohol-induced hepatic inflammation. Recently, herbal drugs have played a specific role in the therapy of liver disorders. Yinchenhao Tang (YCHT) a classical TCM formula is an aqueous extract from the mixture of three herbs, which are clinically used for hepatitis C, primary biliary cirrhosis, and cholestatic liver diseases. The purpose of the study is the use of a metabolomic method to investigate and characterize metabolic profiling of alcohol liver damage, and the positive intervention of YCHT in rats. The overall study flowchart is shown in Fig. 19.1.

19.2 EXPERIMENTAL

19.2.1 Chemicals and Reagents

YCHT as recorded in *Shanghanlun* accurately consists of: 18.0 g of the flos of *Artemisia capillaries* Thunb (Compositae) (YCH), 9.0 g of the fruit of *Gardenia jasminoides* Ellis (Rubiaceae) (ZZ), and 6.0 g of the root of *Rheum officinale* Baill (Polygonaceae) (DH). The flos of *Artemisia capillaries* Thunb (Compositae) was purchased from the China Company of the Xinneitian Pharmaceutical Company of Japan (Tianjin, China); the fruit of *Gardenia jasminoides* Ellis (Rubiaceae), and the root of *Rheum officinale* Baill (Polygonaceae) were purchased from the Harbin Tongrentang Drug Store (Harbin, China). The experimental samples in the form of freeze-dried powders were produced according to the method recorded in *Shanghanlun*'. First, 18 g of YCH was boiled in 1400 mL

Chinmedomics. http://dx.doi.org/10.1016/B978-0-12-803117-9.00019-6

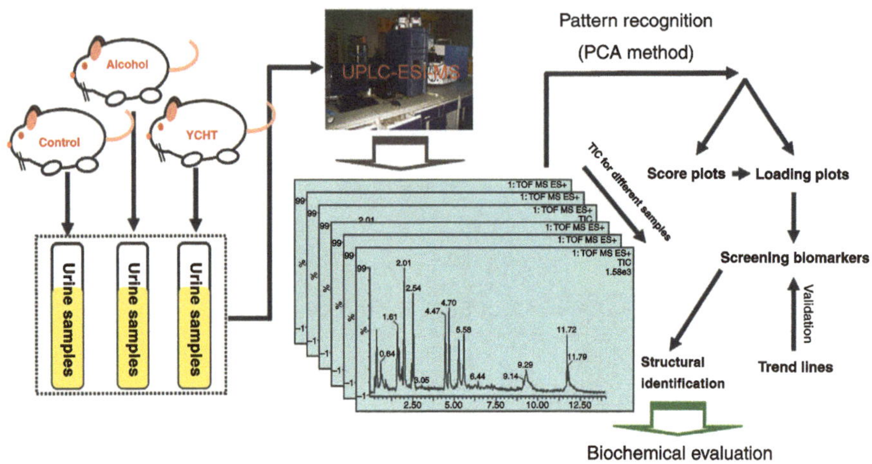

FIGURE 19.1 Flowchart of metabolomic analysis in this study.

of H_2O until the solution volume was reduced to 700 mL, then 9.0 g of ZZ, and 6.0 g of DH were added, and the solution was kept boiling for 10 min. The mixture was filtered through a five-layer bandage; finally, the filtrate was prepared as freeze-dried powders for experimental usage.

Acetonitrile (HPLC grade) was purchased from Sigma–Aldrich (MO, UK); formic acid (analytical grade) was purchased from Beijing Reagent Company (Beijing, China); the deionized water was purified by Milli-Q system (Millipore, Bedford, MA, USA); leucine enkephalin was purchased from Sigma–Aldrich (MO, UK); alcohol was purchased from the Beijing Reagent Company (Beijing, China); the assay kits of alanine aminotransferase (ALT), aspartate aminotransferase (AST), γ-glutamyltranspeptidase (γ-GT), triglyceride (TG), and alcohol dehydrogenase (ADH) were purchased from the Zhongsheng Beikong Biological Technology Ltd Corp. (Beijing, China); the assay kits of malondialdehyde (MDA), and glutathion (GSH) were purchased from the Nanjing Jiancheng Bioengineering Institute (Nanjing, China).

19.2.2 Animal Handling and Sampling

Thirty male Wistar rats (age: 6 weeks, body weight: 200 ± 20 g) were purchased from the Shanghai Slac Laboratory Animal Ltd Corp. (Shanghai, China). All animals were kept at an animal room temperature of: 25 ± 2°C, humidity: 60 ± 5%, and 12 h dark to light cycle for 1 week before the experiment. All rats were randomly divided into: control group (CG), ethanol group (EG), high dosage group of YCHT (HG), middle dosage group of YCHT (MG), and low dosage group of YCHT (LG). All animals were allowed to acclimatize in metabolism cages for 3 days prior to treatment. The CG rats were orally administrated by water between 11:00 a.m. and 12:00 a.m. from day 1 to day 7. The EG rats were orally administrated by water between 11:00 a.m. and 12:00 a.m. from day 1 to day 2, and were orally administrated by 50% alcohol at a dose of 5 g/kg/day between 11:00 a.m. and 12:00 a.m. from day 3 to day 7. The HG, MG and LG rats were orally administrated by freeze-dried powders of YCHT at a dose of 4.8, 2.8, and 0.6 g/kg/day, respectively, between 8:00 and 9:00 from day 1 to day 7, simultaneously, all the rats of HG, MG, and LG were orally administrated by 50% alcohol at a dose of 5 g/kg/day between 11:00 and 12:00 a.m. from day 3 to day 7. The control morning urine was collected from CG rats from 18:00 p.m. on day 7 to 6:00 a.m. on day 8. The individually dosed morning urine was collected respectively from the EG, HG, MG, and LG rats from 18:00 p.m. on day 7, and 6:00 a.m. on day 8. Food (standard diet) and water (uniform labeled water) were available between 12:00 and 6:00 throughout the study. Samples were stored at −20°C prior to analysis; for UPLC/MS samples were prepared by centrifugation at 13,000 rpm (5°C).

19.2.3 UPLC-MS Conditions

The UPLC/MS analysis was carried out using a Waters ACQUITY UPLC system (Waters Corp., Milford, USA) coupled with a Waters Micromass Q-tof micro™ Mass Spectrometer (Manchester, UK). The data were obtained using MassLynx 4.1 (Waters Corp., Milford, USA) and the MarkerLynx Application Manager (Waters Corp., Milford, USA) was used for the peak detection, and for principal component analysis. For the reversed-phase UPLC analysis, the ACQUITY UPLC™ BEH C_{18} column (50 × 2.1mm i.d., 1.7 μm, Waters Corp., Milford, USA) was used. The column temperature was maintained at 40°C. The flow rate of the mobile phase was 400 μL/min. The injection volume was fixed at 5 μL. Mobile phase A consisted of 0.1% formic acid in water, while mobile phase B consisted of 0.1% formic acid in acetonitrile. The column was eluted with a linear gradient (Table 19.1).

An ESI interface was used, and the profile data for positive ions from *m*/*z* 50–900 were recorded. The source temperature was set at 110°C with a cone gas

Table 19.1 UPLC Gradient Elution Program

Time (min)	A/% (0.1% Formic H_2O)	B/% (0.1% Formic ACN)	Curve
Initial	98	2	Initial
2	98	2	6
3	80	20	6
7	5	95	6
9	5	95	6
11	98	2	6
15	98	2	6

flow of 100 L/h, a desolvation gas temperature of 300°C, and a desolvation gas flow of 600 L/h was employed. The capillary voltage was set at 3.2 kV in positive ion mode, and 2.6 kV in negative ion mode, and the cone voltage was set 35 V. A scan time of 0.4 s with an interscan delay of 0.1 s was used throughout, with collision energy of 4 eV. A lock-mass of leucine enkephalin at a concentration of 0.2 ng/μL, in acetonitrile (0.1% formic acid): H_2O (0.1% formic acid) [50:50; v/v] for positive ion mode ($[M + H]^+ = 556.2771$) were employed via a lock spray interface. Data was collected in centroid mode, the lock spray frequency was set at 5 s and the lock mass data was averaged over 10 scans for correction. A "purge–wash–purge" cycle was employed on the auto-sampler, with 90% aqueous formic acid used for the wash solvent, and 0.1% aqueous formic acid used as the purge solvent, this ensured that the carry-over between injections was minimized. The mass spectrometric data was collected in full scan mode from *m/z* 50–900, from 0–10 min in positive ion mode.

19.2.4 Clinical Chemistry and Histopathology

On day 8, all rats were scarified; the collected samples of rat serum, and the liver tissue of every group were characterized on an Automated Synchron LX20 Beckman–Coulter analyzer by following the clinical chemical parameters according to the corresponding method: ALT, AST, γ-GT, TG, ADH, MDA, and GSH. The liver histological sections of control, alcohol treated, and YCHT treated rats after being stained with hematoxylin and eosin (H and E) were examined by light microscopy.

19.3 RESULTS

19.3.1 Clinical Chemistry and Histopathology

The activities of the plasma enzymes: ALT, AST, γ-GT, and ADH all significantly increased after being exposed to alcohol for 5 days compared to CG, and all these increases were obviously intervened by YCHT in different groups, the intervene in HG is most obvious in (Fig. 19.2). The contents of TG and MDA were significantly increased, the activities of hepatocytes enzymes of GSH were obviously decreased after being exposed to alcohol for 5 days, but changes were significantly regulated by only YCHT in HG (Fig. 19.3). Histopathological changes of rats' liver are shown in Fig. 19.4, alcohol treated rats obviously showed feathery degeneration, micro, and macro cellular fatty changes, periportal fibrosis, and vascular congestion compared to the CG group. In the YCHT treated group, the pathological abnormalities were largely improved, and the effect of HG is the most preferred, which almost showed normal histology with mild congestion of the central vein. It is concluded that the animal model of alcohol liver injury in this study was successfully copied, and that YCHT demonstrated a positive intervention effect, and that the action of HG is

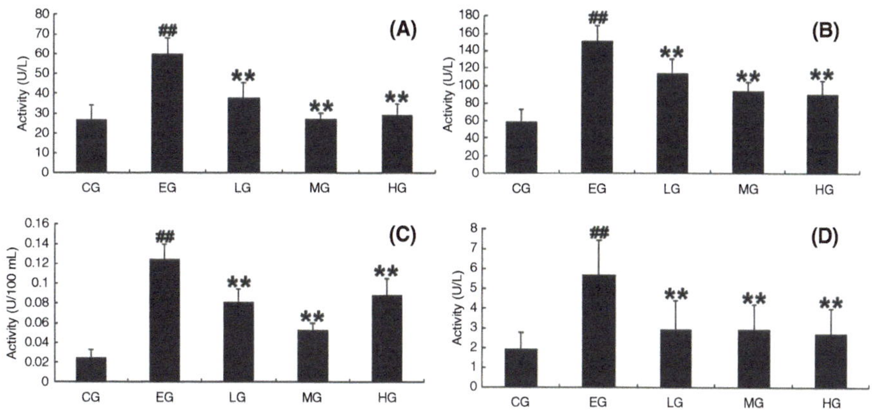

FIGURE 19.2 Effect of alcohol on the activities of AST, ALT, γ-GT, and ADH in rat serum.
(The data were expressed as mean ± SD. # # Significant difference compared with CG groups ($p < 0.01$); ** significant difference compared with EG groups ($p < 0.01$). (A) ALT; (B) AST; (C) γ-GT; (D) ADH.)

the most strong, so subsequently, metabolic profiling should only address the high dose group rats' urine samples.

19.3.2 Validation of UPLC Conditions

The recommended flow rate into the electrospray ion source is limited within 0.4 mL/min in order to reach the best ionization efficiency, and to avoid the ion

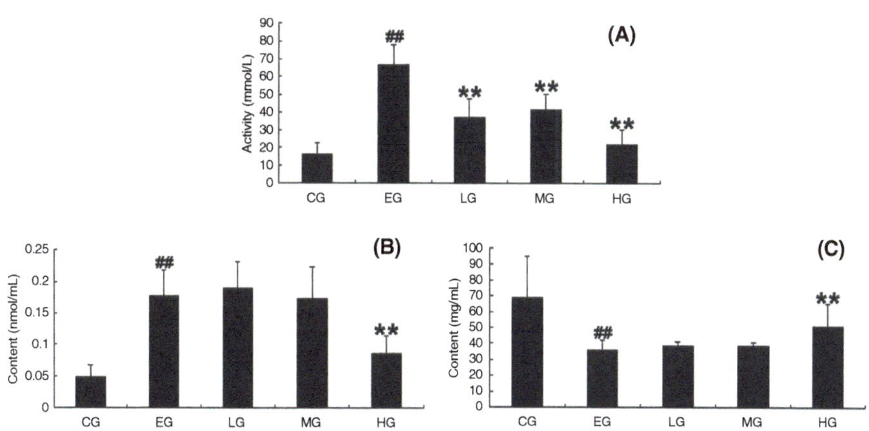

FIGURE 19.3 Effect of alcohol on content of TG and MDA, and activity of GSH in rat hepatocytes.
(The data were expressed as mean ± SD. # # Significant difference compared with CG groups ($p < 0.01$); ** significant difference compared with EG groups ($p < 0.01$). A, TG; B, MDA; C, GSH.)

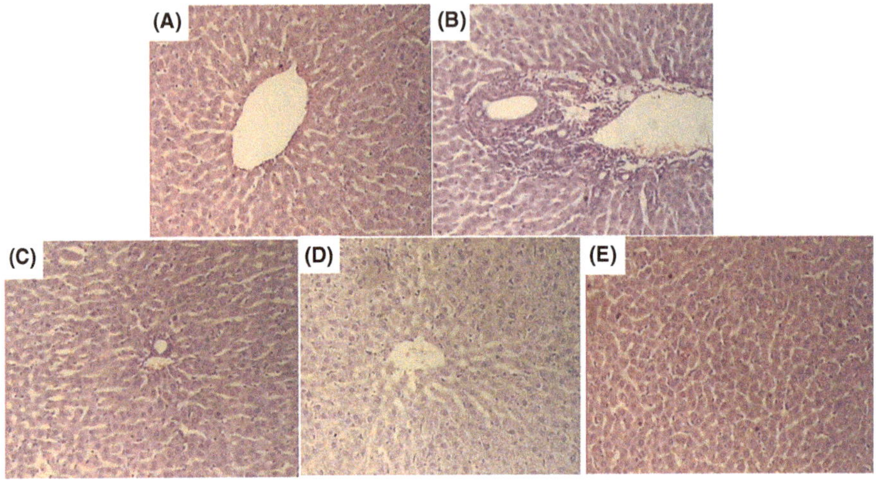

FIGURE 19.4 Histopathological observation.
(A) Control rat liver; (B) alcohol treated rat liver; (C) alcohol + LG treated rat liver; (D) alcohol + MG treated rat liver; (E) alcohol + HG treated rat liver.

suppression, which will seriously influence the analytical sensitivity. Considering sensitivity and resolution meanwhile, the ultimate flow rate was optimized at 0.4 mL/min in this study. Furthermore, the effects of an acid modifier (e.g., formic acid) with different concentrations (1, 0.1, and 0.01%) were characterized, the optimal peak resolution, and shape was obtained by 0.1% formic acid. The LC column temperature was set at 25°C, also this is to reduce the column pressure generating from a higher flow rate. The gradient elution was adopted to obtain better separation. Six batches of urinary samples from the same rat were injected on the UPLC column for evaluating the reproducibility of the chromatography. The retention time of visible peaks varied with the RSD values within the range of 0.05–0.8% during the analysis, suggesting that the retention time reproducibility of the system was good for large-scale sample analysis. For testing the ability of the UPLC system, the control, alcohol treated, and YCHT treated rats' urine were analyzed. Typical chromatograms of these samples were displayed in Fig. 19.5. Visual examination of these chromatograms was viewed distinctly among the different samples.

19.3.3 Optimization of MS Conditions

The desolvation gas flow rate was set at 600 L/h, and desolvation temperature at 300°C, respectively, in order to remove a redundant solvent resulting from a higher flow rate of 0.4 mL/min for mass spectrometer. Cone voltage was optimized to the corresponding values ranging from 15 V–90 V owing to the broad range of the molecular weight of saponins (50–900 Da). The results

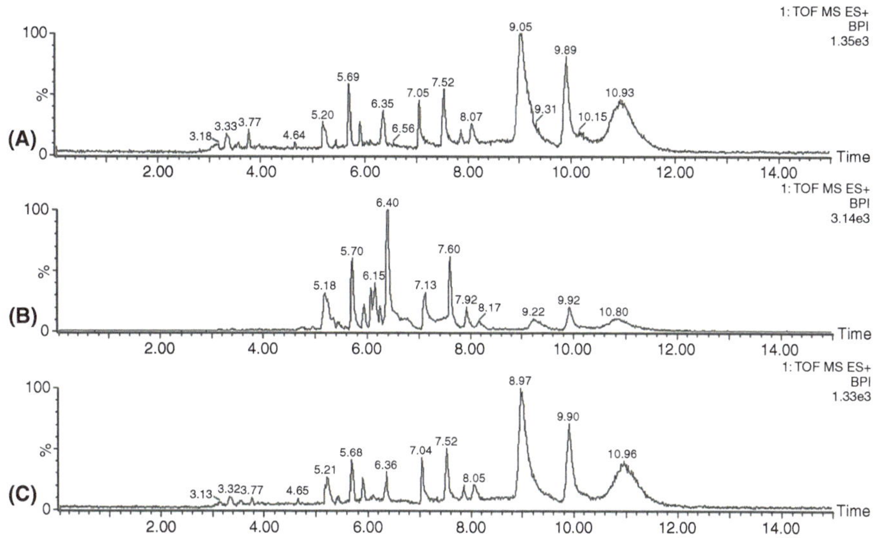

FIGURE 19.5

BPI chromatograms of: (A) control rat; (B) alcohol treated rat; (C) YCHT treated rat urine samples at day 7 using reversed-phase UPLC-ESI-TOF-MS.

indicated that a cone voltage of 35 V was sufficient to produce product ions in the ES^+ mode for metabolites. As the LC peaks become narrower for the UPLC system, the mass spectrometer is required to acquire the data faster. The faster duty cycle of the TOF instrumentation should allow rapid data collection, and good quality MS/MS analysis. In this study, the LC eluent was analyzed in positive ion electrospray without splitting eluent. The data were collected with alternating collision energy, at 5 eV for the parent ion information, and a range of 15–25 eV for fragment ion information. The data acquisition rate of 100 ms per channel with a 20 ms interchannel delay was used. The resolution was greater than 5000 in both high and low collision energy modes. The Q-TOF instrument was switching with a frequency of 100 ms, and we can see that the frequency was sufficiently rapid to keep the chromatographic information whilst still obtaining good MS/MS data.

19.3.4 Analysis of Metabolic Profiling

Figure 19.5 shows the representative BPI chromatograms for the reversed-phase of the UPLC-ESI-TOF-MS analysis of the control rats, alcohol treated rats, and YCHT treated rats urine samples collected at day 9. Although some differences can be visually noted among the three sets of detail illustrated in Fig. 19.6, more subtle changes can be visualized using the PCA approach at day 9. From the PCA scores plot, it is clear that a distinct clustering of control rat, alcohol treated rat, and the YCHT treated rat urine samples has

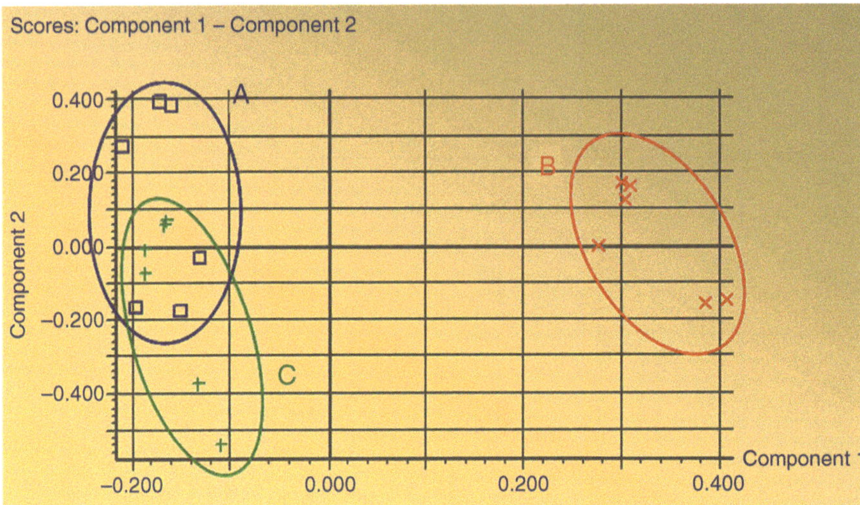

FIGURE 19.6 **Score plot from PCA of reversed-phase UPLC-ESI-TOF-MS data obtained from: (A) control rat; (B) alcohol treated rat; (C) YCHT treated rat urine samples collected from day 7.** (Symbols: control rat (□); alcohol treated rat (×); and YCHT treated rat (+).)

been achieved. The corresponding loadings plot indicated the ions, and their chromatographic retention times that were attributable to the clustering observed in the scores plot. A large number of peaks could be detected using the positive ion mode, as compared to using the negative ion mode, therefore, the positive ion mode was selected in this study. Among them, three dominant ions *m*/*z* 664.3126 (9.00 min), 155.3547 (10.96 min), and 708.2932 (9.01 min), respectively (Fig. 19.7), and the amount of three ions from the alcohol treated group was statistically different from that of the control, and the YCHT treated group. The trend lines for three ions are shown in Fig. 19.8. These three compounds were considered to be intrinsic compounds, and were detected from the control, and the YCHT treated groups. The existence ratio of *m*/*z* 155.3547 in the control, and YCHT treated rats was approximately threefold higher when compared to that in the alcohol treated rats at day 9. The existence ratio of *m*/*z* 664.3126 in the alcohol treated rats was approximately fivefold higher, when compared to that in the control group rats, and YCHT treated group at day 9, and the existence ratio of *m*/*z* 708.2932 in the alcohol treated rats was approximately zerofold higher when compared to that in the control, and YCHT treated group rats at day 9. These results suggest that three compounds may be used as biomarkers for the diagnosis of alcohol hepatotoxicity. The structure of *m*/*z* 664.3126 was preliminarily identified as ceramide (d18:1/25:0). The structures of *m*/*z* 155.3547 and *m*/*z* 708.2932 are now in progress.

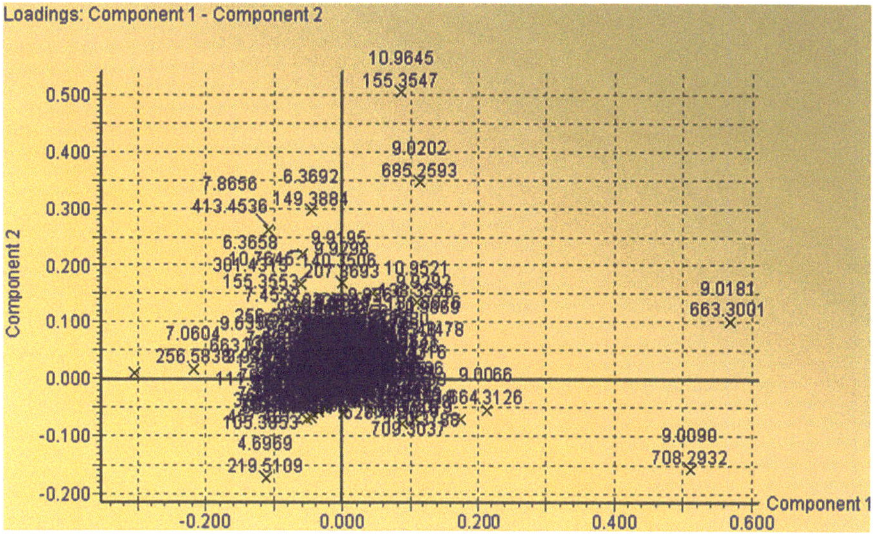

FIGURE 19.7 Loading plot from PLS-DA of UPLC-MS data obtained from urine samples collected at day 7.

19.4 DISCUSSIONS

Based on the evaluation of clinical chemistry and histopathology, a rapid and validated UPLC/MS method has been developed to investigate the metabolic profiling of ALD, and the intervention effect of YCHT in rats. It is clear that a distinct clustering of alcohol treated group and control group rats urine samples has been demonstrated in the scores plot; however, a weak clustering of control group rats, and the YCHT treated group rats urine samples was found, and these results preliminarily suggest that YCHT may have interfered partially with concurrence of ALD. The corresponding loadings plot indicated that the ions, and their chromatographic retention times were attributable to the clustering observed in the scores plot. Among these ions, three dominant ions *m/z* 664.3126 (9.00 min), 155.3547 (10.96 min), and 708.2932 (9.01 min) were observed. These results obviously demonstrate that the three compounds are marker compounds, which are related to developing of alcohol liver injury. The structure of *m/z* 664.3126 was preliminarily identified as ceramide (d18:1/25:0) (d18:1/25:0) (Fig. 19.9). The structures of *m/z* 155.3547 and *m/z* 708.2932 are now in progress.

Ceramide (d18:1/25:0) is generated by *de novo* synthesis, sphingomyelin hydrolysis, and the recycling of sphingolipids. It is known to be involved in the regulation of several cellular processes, including differentiation, growth suppression, cell senescence, and apoptosis. Of these, the role of ceramide

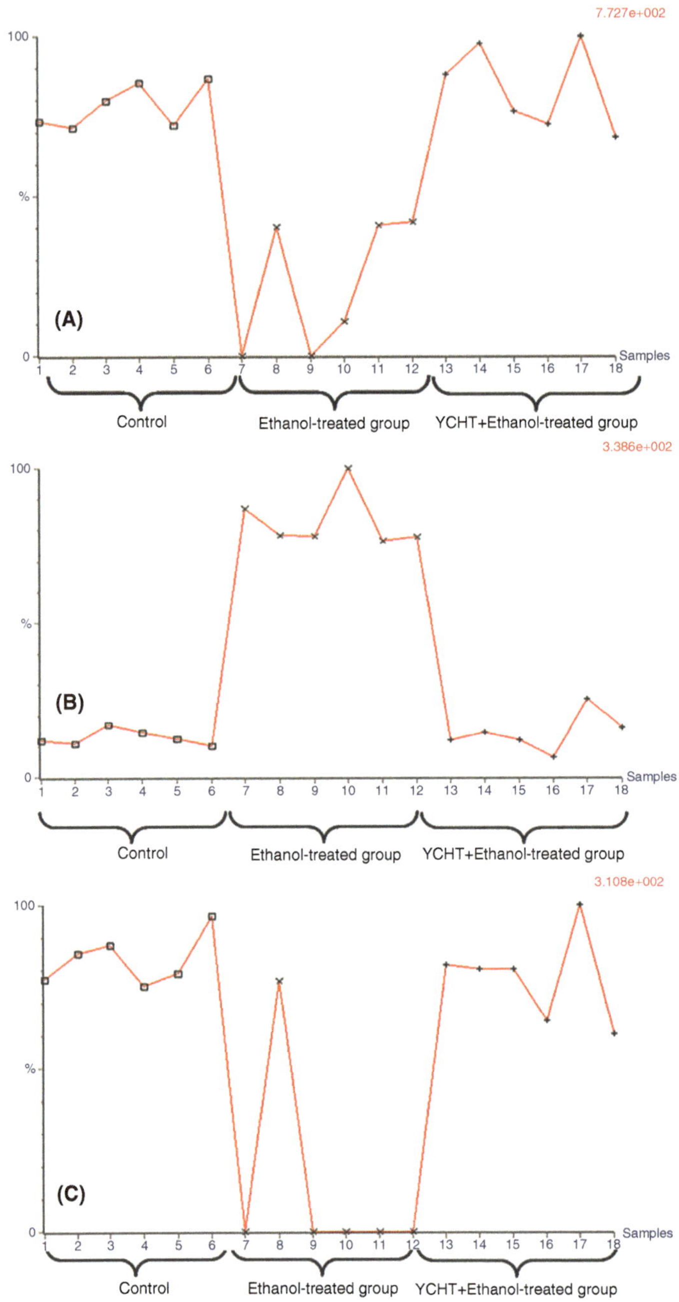

FIGURE 19.8 **Trend lines for the *m*/*z* 155.3547, 664.3126, and 708.2932 ions in urinary samples among control; alcohol-treated rat; and YCHT-treated rats.**
(Symbols: control rat (□); alcohol-treated rat (×); and YCHT-treated rat (+).)

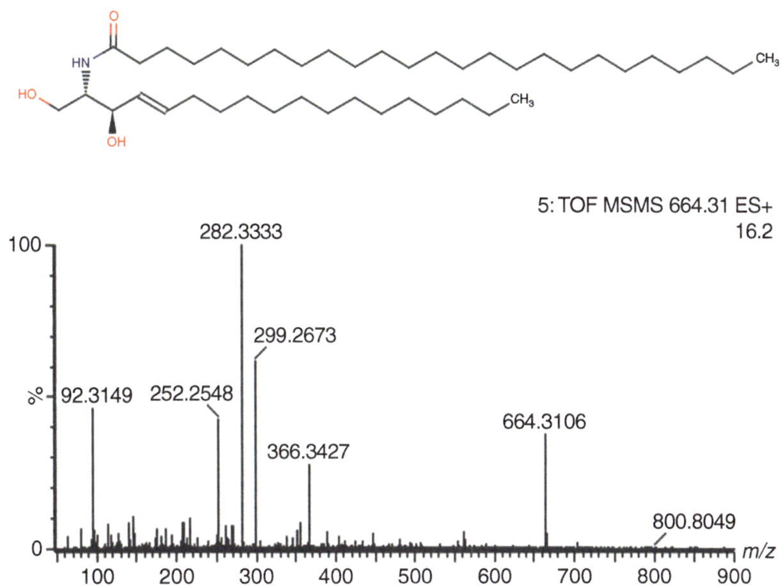

FIGURE 19.9 Chemical structure and mass fragments information of ceramide (d18:1/25:0).

(d18:1/25:0) induced cell apoptosis has attracted much attention in recent years. Ceramide (d18:1/25:0) has been also extensively studied for its role in cellular response to stress and inflammation. Key inducers and mediators of host acute phase response (APR), such as bacterial lipopolysaccharide (LPS), Interleukin 1b (IL-1b), and tumor necrosis factor α (TNF-α) induce early generation of ceramide (d18:1/25:0) in the liver. The observation found that astrocytes prepared from rats prenatally exposed to alcohol are more susceptible to the cytotoxic effect of TNF-α, suggesting that alcohol exposure could increase the sensitivity of astrocytes to the cytotoxic effect of TNF-α and may be attributable to the altered metabolism of sphingolipids, resulting in the increased production of ceramide (d18:1/25:0), and was further excreted into urine, and elevated the level of urine ceramide. It suggested that the alcohol treated, strongly affected the sphingomyelin signaling pathway in the liver by activation of TNF-α (Fig. 19.10). There was TNF-αactivation in the content of ceramide (d18:1/25:0) in the alcohol treated group, as compared to the control group, and the YCHT treated group. These factors can ultimately lead to hepatocyte necrosis. Consequently, the urine ceramide (d18:1/25:0) may be used as a biomarker for the diagnosis of alcohol liver injury.

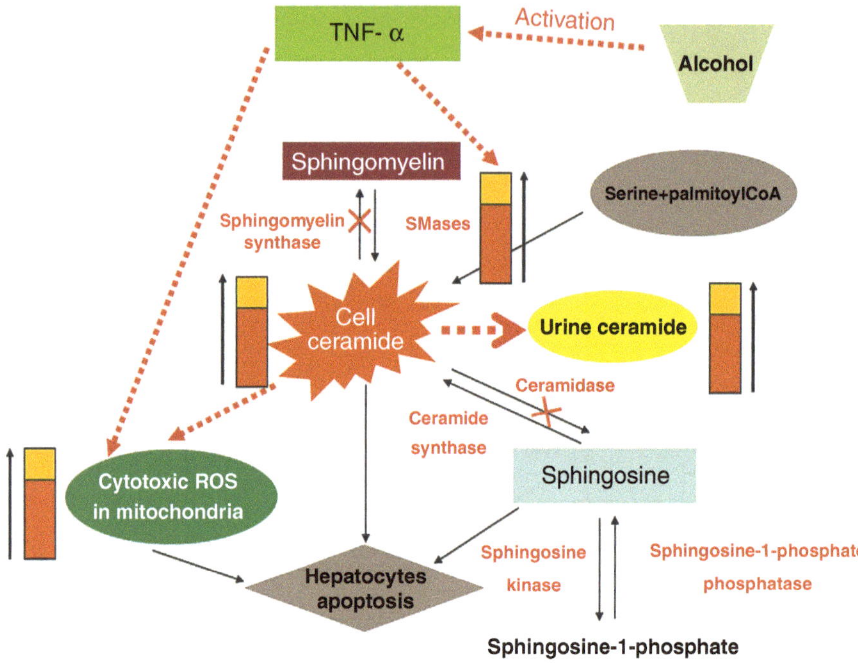

FIGURE 19.10 The proposed metabolic pathway for an explanation of the relationship between ALD, and the increased content of ceramide (d18:1/25:0) in urine.

19.5 CONCLUSIONS

In this study, the UPLC/MS combined with the PCA technology has been used for urine metabolic profiling, and the data interpretation is based on the validated assessments of clinical chemistry and histopathology. Unsupervised PCA have been applied to sample classification. This method was applied to urine samples collected from control, the alcohol treated, and the YCHT treated rat groups. From the PCA scores plot, a distinct clustering of the three groups based on their urine samples has been clearly demonstrated. Consequently, we have succeeded in detecting three potential biomarkers as being associated to ALD, the trends lines revealed the contents of those biomarkers in the alcohol treated rats, when compared to the control and the YCHT treated rat groups, and there is an obvious difference. The results also demonstrated that the alcohol treated rats had perturbed normal metabolic processes, and that YCHT could ameliorate the perturbed metabolic process.

CHAPTER 20

Metabolic Profiling and Biomarkers Analysis of Insomnia, and the Intervention Effects of Suanzaoren Decoction, and its Related Active Ingredients

Xijun Wang, Bo Yang, Aihua Zhang, Hui Sun

CONTENTS

20.1 INTRODUCTION

The sedative–hypnotic medications, including benzodiazepines and nonbenzodiazepines, are the most common treatments for insomnia. However, concerns regarding patterns of inappropriate use, dependence, and adverse effects have led to caution in prescribing these sedative–hypnotic medications. Traditional Chinese Medicine (TCM) has been commonly used for treating insomnia in Asian countries for centuries. A traditional Chinese herb remedy, Suanzaoren decoction (SZRD) has been efficiently and widely used as a tranquilizing agent to treat insomnia for thousands of years in Asia (Xie et al., 2013). SZRD which was reported originally in the *Synopsis of the Golden Chamber* written by Zhang Zhongjing (Han Dynasty) and is the most commonly prescribed Chinese herbal formulae with the effects of hypnosis, sedation, and anticonvulsion, and has been used to treat insomnia and anxiety in clinical practice (Chen et al., 2008). It consists of five herbal medicines: *Semen Ziziphi Spinosae* (18 g), *Poria* (10 g), *Rhizoma Chuanxiong* (5 g), *Rhizoma Anemarrhenae* (10 g) and *Radix Glycyrrhizae* (3 g) (Li and Bi, 2006). Although SZRD has been used to improve sleep disruption in insomniac patients, its precise mechanism is still unclear. The major component of SZRD is suanzaoren (*Semen ziziphi spinosae*), which is the dried seed of *Ziziphus jujuba Mill* var. spinosa (Bunge) Hu ex. H.F. Chou (*Rhamnaceae*). The pharmacological studies have showed that suanzaoren possesses multiple activities, such as hypnotic–sedative, hypotensive, antihypoxia, antihyperlipidemia, and hypothermic effects. Triterpene saponin Jujuboside A (JuA, Fig. 20.1A) and B (JuB, Fig. 20.1B) extracted from suanzaoren are considered to be the major pharmacologically active compounds responsible for insomnia treatment (Xie et al., 2011; Wang et al., 2012a,b).

Sleep is essential for basic survival, and insufficient sleep leads to a variety of dysfunctions. Problems with sleep are one of the commonest reasons for seeking medical attention. Knowledge gained from basic research into sleep

Chinmedomics. http://dx.doi.org/10.1016/B978-0-12-803117-9.00020-2

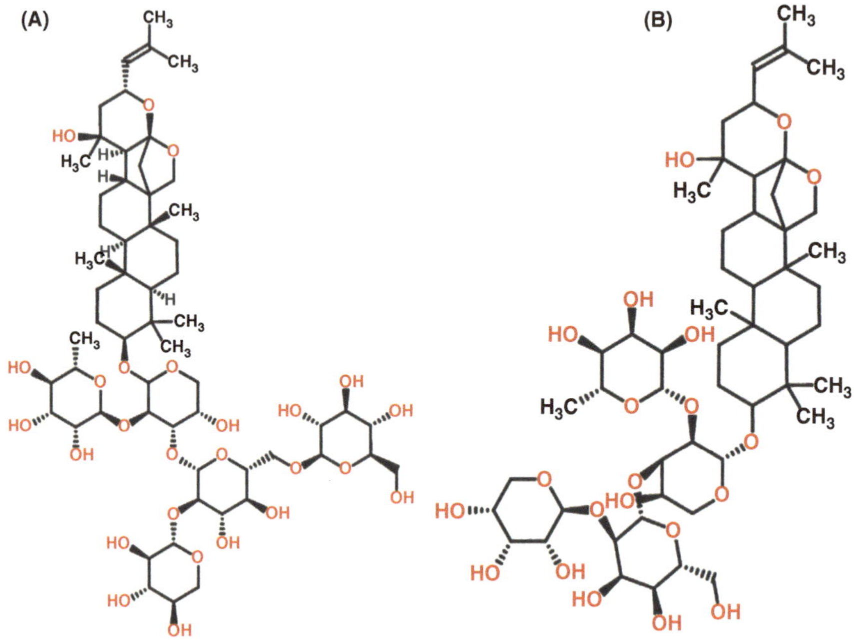

FIGURE 20.1 Chemical structure of Jujuboside A and Jujuboside B.

in animals has led to marked advances in the understanding of human sleep, with important diagnostic, and therapeutic implications. Insomnia is an extremely common symptom in the context of psychiatric disorders. Sleep disturbance is a common health problem among adults, and enhancing sleep quality is an issue of significant importance to healthcare providers. Insomnia can also impair attention and memory, and its growing prevalence, especially among "knowledge workers," motivates a detailed exploration of its impact on human behavior. Several behavioral and imaging studies have examined the effects of sleep deprivation on attention, memory, and emotion processing. As sleep quality worsens into insomnia, individuals may seek assistance from medication. However, sedative and hypnotic drugs possess serious adverse side effects. Many Chinese medicinal products have been traditionally used for the treatment of insomnia; however, few literature has reported about the detailed therapeutic mechanisms of how these Chinese medicines trigger the promotion of sleep. In trying to approach the rationale of formulae design in TCM, here we use the treatment of insomnia with SZRD as a working model. The aim of the current study was to obtain a systematic view of the dissection of the mechanisms of SZRD, and its related active ingredients as effective treatments for sleeping disorder. We wished to identify multiple metabolites that could facilitate the understanding of the actions of SZRD, and aid their incorporation into the future improvement of TCM therapy.

20.2 MATERIALS AND METHODS

20.2.1 Chemicals and Reagents

Acetonitrile (HPLC grade) was purchased from Merck (Darmstadt, Germany); methanol (HPLC grade) was purchased from Fisher Corp. (Michigan, USA); distilled water was purchased from Watson's Food & Beverage Co. Ltd (Guangzhou, China); formic acid was of HPLC grade, and was produced from Honeywell Company (Morristown, New Jersey, USA); leucine enkephalin was purchased from Sigma–Aldrich (MO, USA). All other reagents were of analytical grade. JuA and JuB (purity ≥ 98%) was obtained from Chengdu Mansite Pharmaceutical Co. Ltd (China), *Semen Ziziphi Spinosae, Poria, Rhizoma Chuanxiong, Rhizoma Anemarrhenae,* and *Radix Glycytthizae* were purchased from Harbin Tongrentang Drug Store (Harbin, China). SZRD was prepared in our laboratory according to the method recorded in *Synopsis of the Golden Chamber,* and the decoction was transformed into freeze-dried powder. According to the original composition and preparation method of SZRD recorded in *Synopsis of the Golden Chamber,* SZRD was prepared by the following procedure: *Semen Ziziphi Spinosae* (18 g) was crushed to powder (40 mesh size) and immersed in 1000 mL of deionized water for 1 h, and then decocted to boil, keeping for 30 min until the volume of water reduced to about 400 mL, then 6 g *Poria,* 6 g *Rhizoma Chuanxiong,* 6 g *Rhizoma Anemarrhenae,* and 3 g *Radix Glycytthizae* were added, and kept boiling for 10 more min. The extracted solution was filtered through 6-layer gauzes, and made to a concentration of 1 g crude drug per milliliter, and finally the solution was freeze-dried. Nutrient medium preparation: Maizena (14 g), saccharobiose (10 g), agar (1 g), and distilled water (150 mL) are stirred, and boiled for 2–3 min, then add in yeast powder (1 g), and propionic acid (0.8 mL) miscing. SZRD was added in nutrient medium.

20.2.2 Animals

The wild-type *Drosophila melanogaster* (Canton S) was provided by Beijing University. *Drosophila melanogaster* adult of the wild-type strain Canton-S were collected within 24 h after eclosion, and kept at 25 °C, humidity (55 ± 5%), under 12:12 h light-dark cycle conditions either in population vials that housed 50 flies, and contained 5 mL of 5% sucrose, 1% agar, or in individual monitor tubes that housed a single fly and contained 150 μL of 5% sucrose, 1% agar, for the duration of the experiment. Individual flies were acclimatized for 5 days. The locomotor activity was measured using the Drosophila Activity Monitoring System (DAMS, Trikinetics Inc., USA).

20.2.3 Animals Sampling and Handling

Seven-day drosophila were randomly divided into 5 groups with 10 drosophila in each (male, 5; female, 5): the control, model, SZRD, JuA, and JuB groups. The drosophila in the model, SZRD, JuA, and JuB groups were performed by light sleep deprivation method. In the experiments with sleep deprivation, at night 7, the drosophila were sleep deprived by handling starting at 19:00 p.m. light onset, and restored the 10 min light period for each whole time point from 20:00 p.m. to 7:00 a.m. on day 8. The whole procedure of light sleep deprivation was held for 3 consecutive days. The drosophila in the control and the model groups were administrated with normal nutrient medium in the whole procedure for 10 consecutive days, and anesthetized by CO_2 on day 10. Flies were anaesthetized by CO_2 and killed on day 8 (8:00 a.m.). Drosophila was shattered, and extracted from methanol, and by centrifugation at 3000 rpm for 10 min at 4°C and stored at −20°C for metabolomic analysis.

20.2.4 UPLC-MS Analysis Platform

The UPLC–MS analysis was performed on a Waters ACQUITY UPLC system (Waters Corp., Milford, USA) coupled with a Waters Micromass Q-tof micro™ Synapt High Definition Mass Spectrometer (Manchester, UK) equipped with electrospray ionization. For the reversed-phase UPLC analysis, the ACQUITY UPLC™ BEH C_{18} column (100 × 2.1 mm i.d., 1.7 μm, Waters Corp., Milford, USA) was used. The column temperature was maintained at 45°C; the flow rate of the mobile phase was 0.40 mL/min; the injection volume was fixed at 5.0 μL. Mobile phase A consisted of 0.1% formic acid in acetonitrile, whilst mobile phase B consisted of 0.1% formic acid in water. The column was eluted with a linear gradient of 1−50% B over initial−2.0 min, 50−70% B over 2.0−7.0 min, 70−99% B over 7.0−11.0 min, and held at 99% B for 2.0 min, returned to 1% B for 0.5 min, and then held for 1.5 min at an eluent flow rate of 0.40 mL/min. Twenty milligrams of the freeze-dried powder were extracted with 20 mL methanol for 30 min under ultrasonics. The methanol extraction was centrifuged at 10,000 rpm for 5 min at 4°C, and the supernatant was filtered through a 0.22 μm-filter, the filtrate was used for UPLC analysis.

For the UPLC–HDMS analysis, the optimal conditions were as follows: capillary voltage of 2500 V, desolvation temperature of 250°C, sample cone voltage of 35 V, extraction cone voltage of 4.0 V, microchannel plate voltage of 1600 V, collision energy of 4 eV, source temperature of 120°C, cone gas flow of 50 L/h, and desolvation gas flow of 700 L/h for positive ion mode; capillary voltage of 2300 V, desolvation temperature of 300°C, sample cone voltage of 45 V, extraction cone voltage of 4.0 V, microchannel plate voltage of 1600 V, collision energy of 4 eV, source temperature of 100°C, cone gas flow of 50 L/h, and a desolvation gas flow of 700 L/h for negative ion mode. The mass spectrometric full-scan data were acquired in the positive ion mode from 100 Da−1000 Da with a 0.1 s scan time. Data were centroided, and mass was corrected during acquisition using an external reference (Lock-Spray™) consisting of a 0.2 ng/mL solution of leucine enkephalin infused at a flow rate of 100 μL/min via a lockspray interface, generating a reference ion for positive ion mode ($[M + H]^+ = 556.2771$) and negative ion mode ($[M - H]^- = 554.2615$) to ensure accuracy during the MS analysis.

20.2.5 Data Processing and Analysis

All data were normalized to the summed total ion intensity per chromatogram, and the resultant data matrices were introduced to EZinfo 2.0 software for PCA, PLS-DA, and OPLS analyses. Metabolite peaks were assigned by MS/MS analysis, or interpreted with available biochemical databases. Other statistical analyses used include: one-way analysis of variance (ANOVA), least significant difference (LSD) test, and independent sample t-test. They were performed with SPSS 17.0. Statistical differences are considered significant when the test p-value is less than 0.05.

20.3 RESULTS

20.3.1 Metabolic Profiling and Biomarkers Analysis of Insomnia

Using the optimal reversed-phase UPLC-HDMS conditions described above, the representative basic peak intensity (BPI) chromatograms of drosophila samples collected are presented in Fig. 20.2. Low molecular mass metabolites could be separated well in the short time of 14 min due to the minor particles (sub-1.7μm) of UPLC. The OPLS-DA approach separated samples into two blocks and was applied to obtain better discrimination as a result of sleep deprivation between the control and the model groups, indicating that the insomnia model was successfully reproduced (Fig. 20.3A and B). The corresponding PLS-DA loadings plot indicated that differentiating metabolites were attributable to the clustering observed in the scores plot (Fig. 20.4).

To exhibit the responsibility of each ion for these variations more intuitively, S-plots, and VIP-value plots were combined (Fig. 20.5). The green point graph

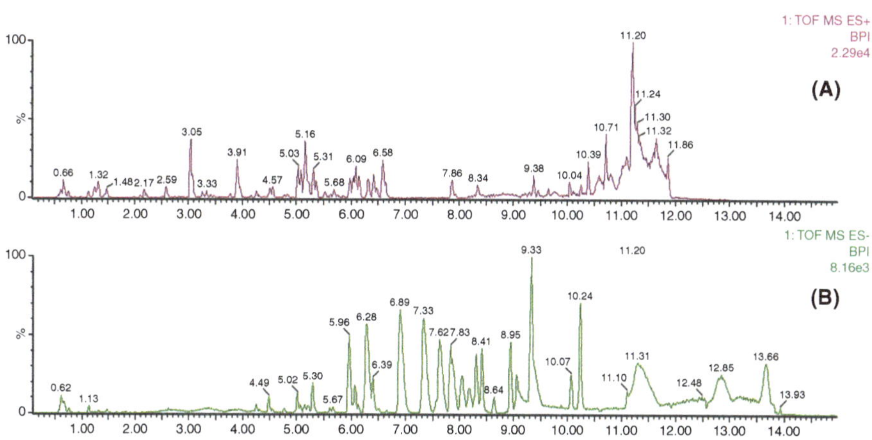

FIGURE 20.2 BPI chromatograms derived from UPLC-HDMS in: (A) positive ion mode; and (B) negative ion mode.

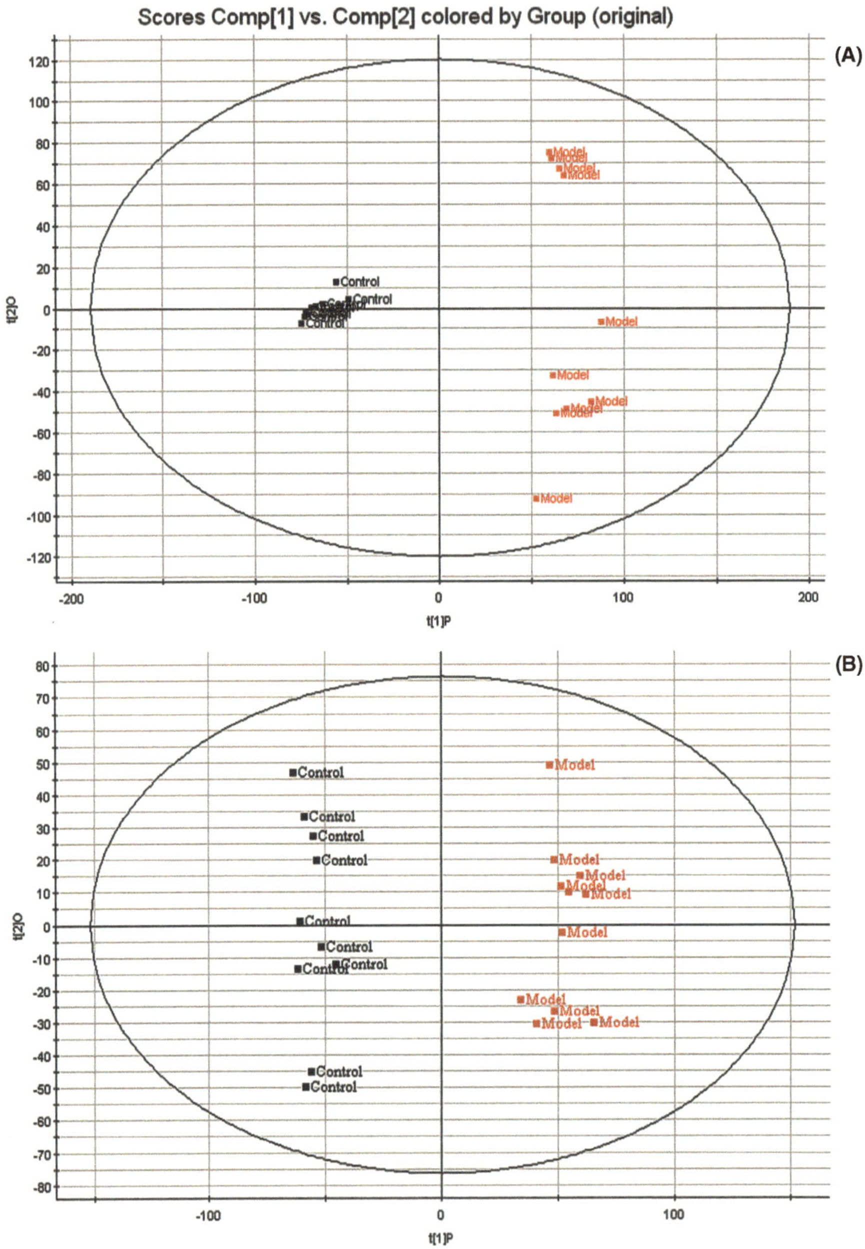

FIGURE 20.3 The OPLS-DA model results for the control and the model groups.
One data point stands for one subject: (A) positive mode; (B) negative mode; ■, control group; ■, model group.

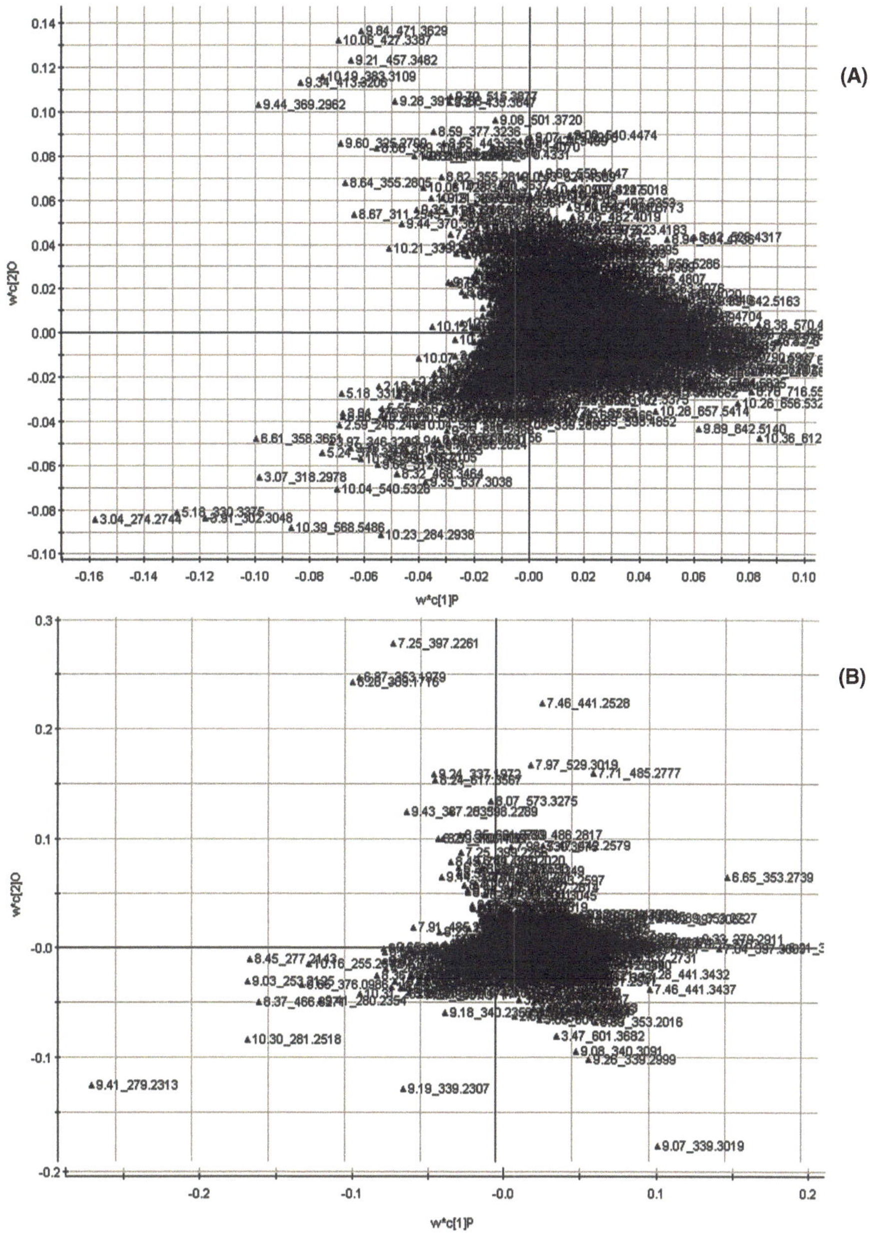

FIGURE 20.4 **Loading plot of PLS-DA of insomnia drosophila.**

The loading plot represents the impact of the metabolites on the clustering results correlated with score plots. ((A) positive ion mode; (B) negative ion mode.)

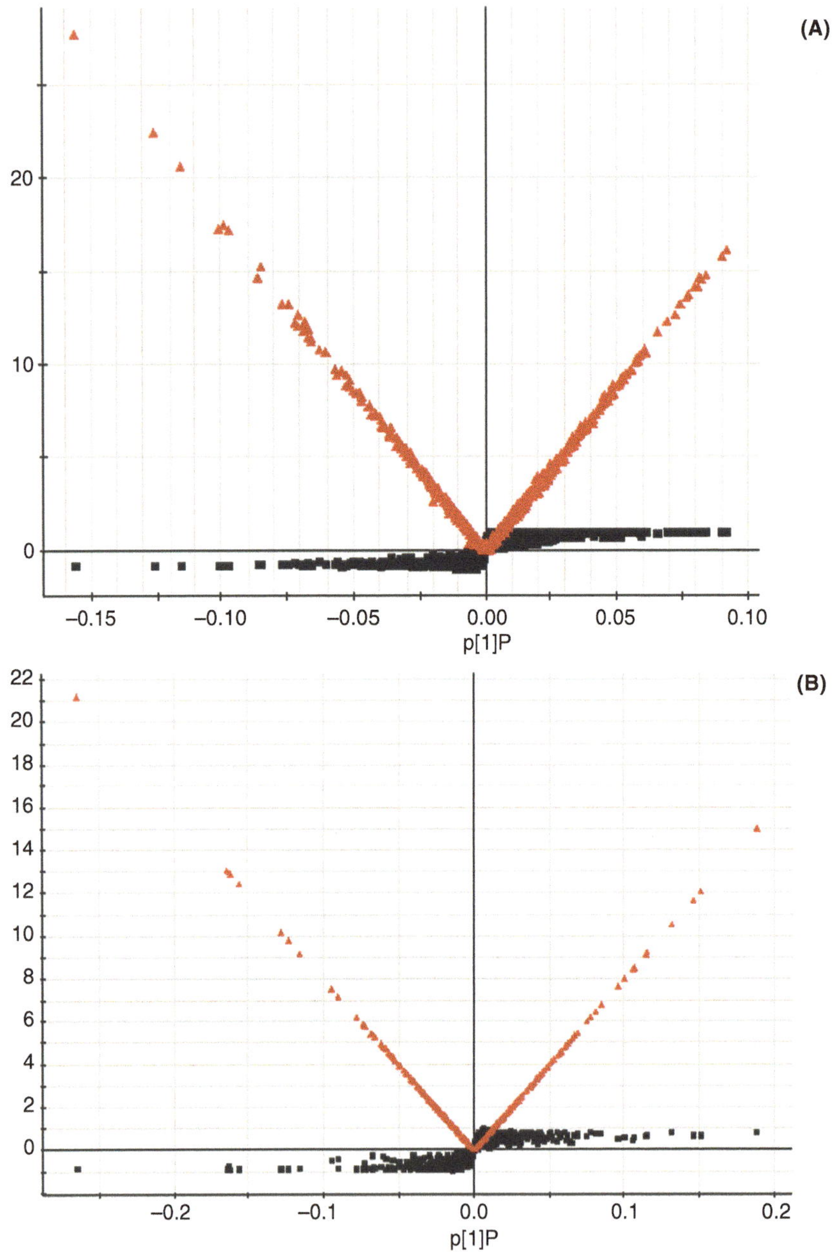

FIGURE 20.5 Combination plot of S-plot and VIP values.
The red point graph is the VIP-value plot, which represents the value of each of the metabolites. The farther away from the origin, the higher the VIP value of the metabolites was. The black points and red points were in one-to-one correspondence in the combination. ((A) positive mode; (B) negative mode.)

is the S-plot, and most of the ions were clustered around the origin point; only a few of them scattered in to the margin region, and just these few ions contributed to the clustering observed in the score plot, and were also the differentiating metabolites. The red point graph is the VIP-value plot, which represents the value of each ion. The farther away from the origin, the higher the VIP value of the ions was. The green points and red points were in one-to-one correspondence in the combination. Finally, potential biomarkers of significant contribution were characterizated: 11 in positive mode and 9 in negative mode, respectively (Table 20.1). The precise molecular mass was determined within measurement errors (<5 ppm) by Q-TOF, and meanwhile, the potential elemental composition, degree of unsaturation, and fractional isotope abundance of compounds were obtained. The presumed molecular formula was searched in Chemspider, HMDB, METLIN, and other databases to identify the possible chemical constitutions, and to determine the potential structures of the ions. All of the potential biomarkers indentified are shown in Fig. 20.6.

Table 20.1 Potential Biomarkers Identified of Insomnia in Positive and Negative Ion Mode

No.	VIP	Formula	Observed $[M+H]^+$	Observed $[M-H]^-$	R_t	Actual-M	Proposed Compound
1	27.72	$C_{16}H_{35}NO_2$	274.2744	–	3.04	273.2668	*N*-Lauryl diethanolamine
2	22.45	$C_{18}H_{39}NO_3$	318.2978	–	3.07	317.2930	2-Aminooctadecane-1,3,4-triol
3	20.64	$C_{18}H_{39}NO_2$	302.3048	–	3.91	301.2981	(2S,3R)-2-aminooctadecane-1,3-diol
4	17.48	$C_{20}H_{43}NO_2$	330.3375	–	5.18	329.3294	2,2'-(Hexadecylimino)
5	17.24	$C_{22}H_{47}NO_2$	358.3651	–	6.61	357.3607	2,2'-(Octadecylimino)diethanol
6	14.56	$C_{21}H_{38}O_4$	355.2805	–	8.64	354.2770	Ricinoleic acid, methyl ester, acetate
7	14.19	$C_{36}H_{69}NO_5S$	628.4999	–	8.83	627.4896	Ditetradecyl *N*-(4-sulfany lbutanoyl)-L-aspartate
8	13.75	$C_{10}H_{12}N_2O$	176.0850	–	9.21	175.0949	Serotonin
9	12.24	$C_{28}H_{44}N_2O_3$	457.3482	–	9.44	456.3352	15-Hydroxy-*N*-methyl-9-oxo-*N*-(2-pyridin-2-ylethyl)prost-13-en-1-amide
10	11.91	$C_{29}H_{46}N_2O_3$	471.3629	–	9.84	470.3508	*N*-octadecanoy L-tryptophan
11	11.78	$C_{13}H_{16}N_2O_2$	232.1212	–	10.19	232.1212	Melatonin
12	21.17	$C_{18}H_{32}O_2$	–	279.2911	9.33	280.2402	Linoleic acid
13	13.01	$C_{15}H_{26}N_4OS$	–	309.2186	6.21	310.1827	1'-(Cyclopropylcarb-amothioyl)-1,4'-bi piperidine-4'-carboxamide
14	12.41	$C_{17}H_{30}N_4O_2S$	–	353.2073	6.65	354.2089	2-{[4-(3-Methoxypropyl)-5-propyl-4H-1,2,4-triazol-3-yl]sulfanyl}-1-(3-methylpiperidin-1-yl)ethanone
15	12.05	$C_{23}H_{46}N_2OS$	–	397.3085	7.04	398.3331	*N*-(hexadecyl-carbamothioyl) hexanamide
16	11.65	$C_{28}H_{46}N_2S$	–	441.3432	7.46	442.3382	5-Heptadecyl-*N*-phenyl-2,3-dihydro-1H-pyrrole-1-carbothioamide
17	9.22	$C_{31}H_{53}N_3$	–	466.3271	8.37	467.3239	4'-(Hexadecahydropyren-1-yl)-2,2':6',2"-terpiperidine
18	8.46	$C_{11}H_{12}N_2O_3$	–	219.0848	8.45	220.0848	5-Hydroxy-L-tryptophan
19	7.99	$C_{19}H_{31}ClN_2O$	–	337.2019	9.04	338.2125	1-(3-Chlorophenyl)-3-dodecylurea
20	7.19	$C_{20}H_{32}O_5$	–	351.2250	4.65	352.2249	Prostaglandin D2

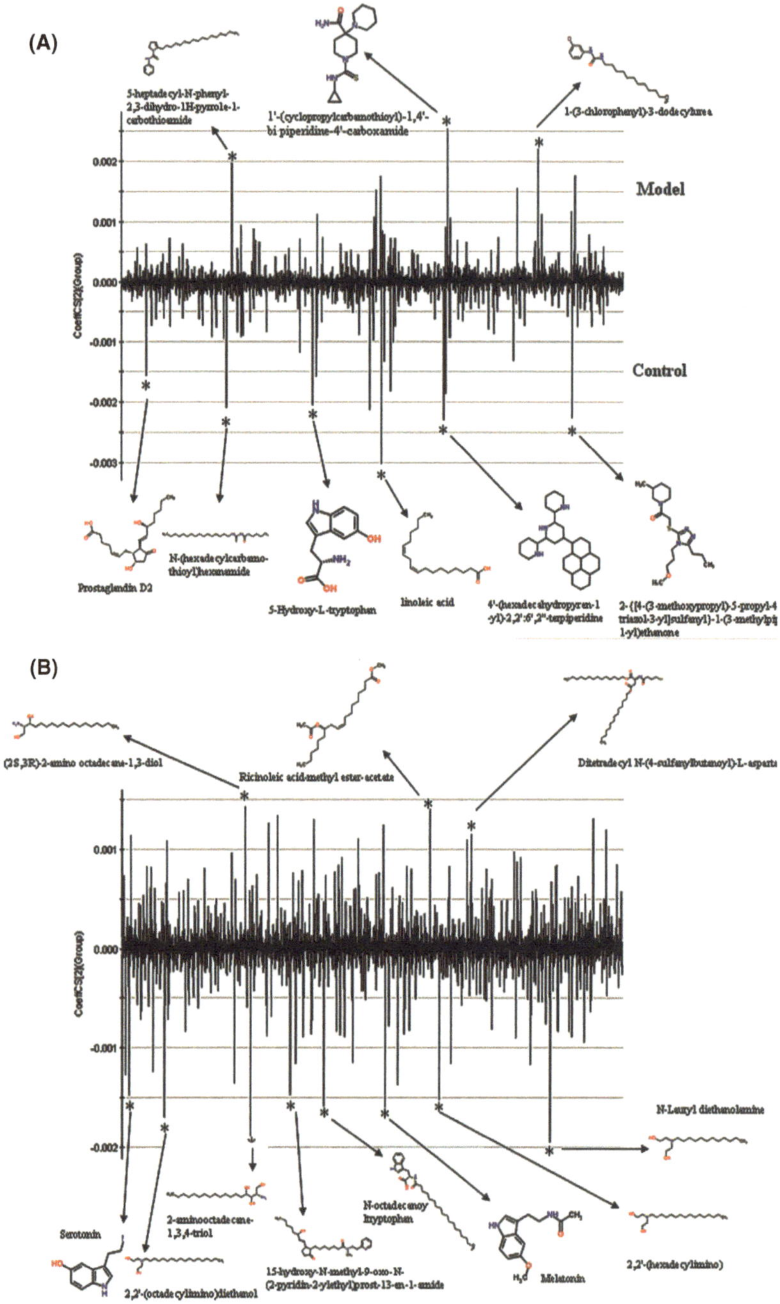

FIGURE 20.6 Loading plots of the OPLS-DA model of potential biomarkers identified in the negative model (A) and positive model (B).

20.3.2 Intervention Effects of Suanzaoren Decoction

We utilized the metabolomic approach further to delineate metabolic changes of insomnia postdosing SZRD treatment. Using metabolomic trajectory analysis it has been found that the SZRD dosing regiment can be regulated back towards their baseline levels, to that of control group. With regard to the information analyst of PCA, shown in Fig. 20.7, the control and the insomnia drosophila groups were significantly divided into two classes after sleep deprivation

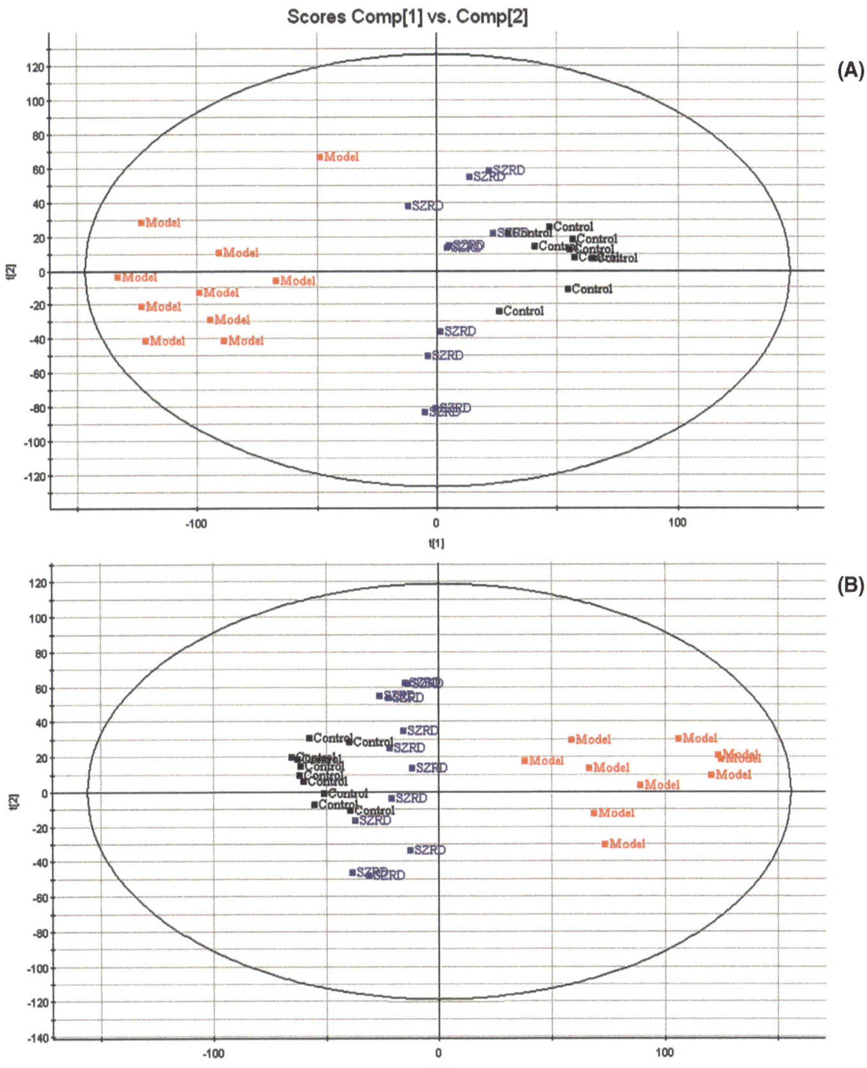

FIGURE 20.7 PCA score plots of trajectory analysis for the insomnia drosophila after SZRD treatment. (A) positive mode; and (B) negative mode.

treatment, indicating that the model of insomnia was successfully reproduced. The SZRD regiment developed the distant tendency to the diverse extent of the model group, demonstrating that SZRD could have some therapeutic effect on the insomnia. Interestingly, SZRD exhibited a good prevention role of insomnia, and kept the groups in the normal situation because there was no distinct clustering differences between the control group, and the SZRD treatment. Of note, distinct clustering differences were observed after SZRD treatment in comparison with the model group; it has been found that the dosing regiment can be regulated back towards their baseline levels to that of control group.

In order to more clearly investigative the treatment effects of SZRD, the relative intensity level of 20 bio markers three groups was analyzed (Fig. 20.8) and, we have found that the content of the key markers closer to the normal group. Thus, SZRD may regulate the metabolism of these markers to efficient use for sleeping disorder. Additionally, a correlation coefficient analysis was applied to investigate the connections between biomarkers and corresponding groups (Fig. 20.9). The plot shows how each biomarker behaves in relation to the group, and represents the feature of the group. Variables situated upper are positively correlated to the group, and those situated opposite are negatively correlated to the group. The markers 4, 10, 14, 16, such as, 2,2′-(hexadecylimino), 2-{[4-(3-methoxypropyl)-5-propyl-4H-1,2,4-triazol-3-yl]sulfanyl}-1-(3-methylpiperidin-1-yl)ethanone, 5-heptadecyl-*N*-phenyl-2,3-dihydro-1H-pyrrole-1-carbothioamide, 5-heptadecyl-*N*-phenyl-2,3-dihydro-1H-pyrrole-1-carbothioamide, have negative correlation with the control group; others have positive correlation with the control group, indicating the normal sleep characteristic.

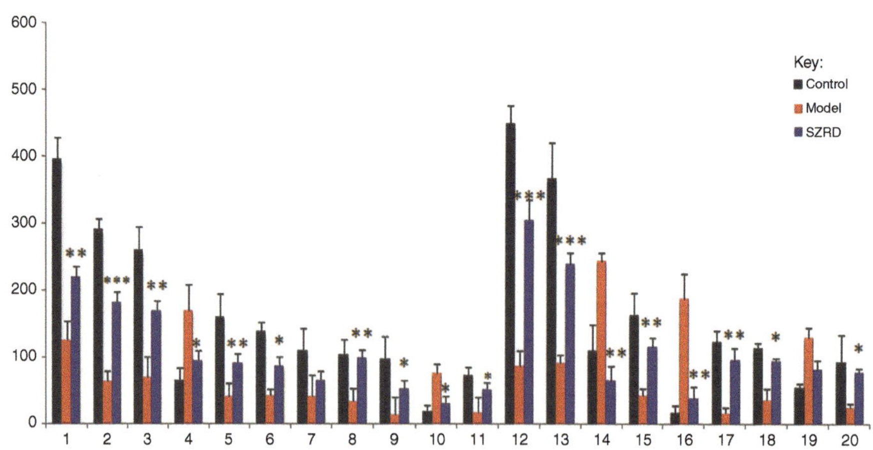

FIGURE 20.8 Differential expression levels (mean) of biomarkers in different groups. The "asterisk" indicated the statistical significance of the biomarkers changes by Student's t-test: *, $p < 0.05$; **, $p < 0.01$; ***, $p < 0.001$.

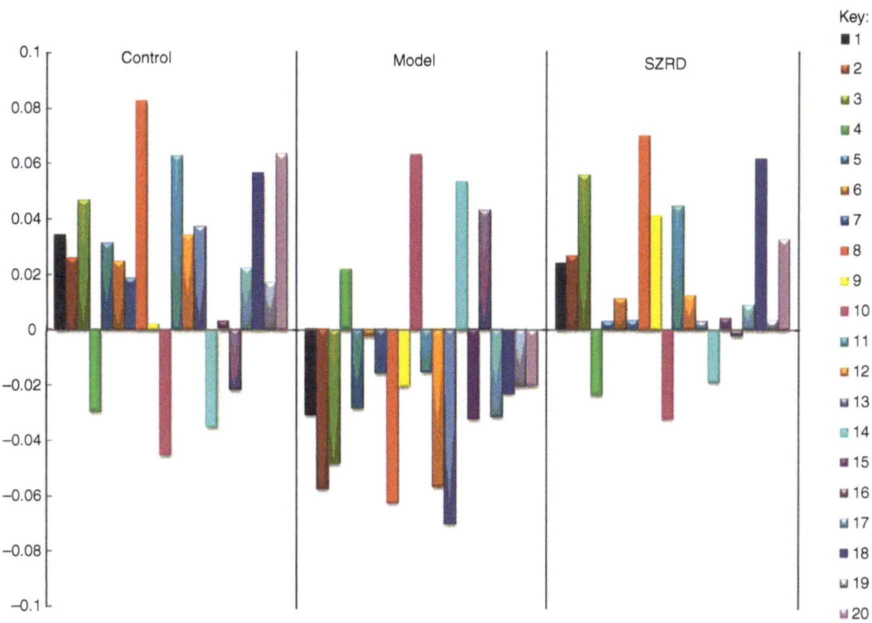

FIGURE 20.9 Correlation coefficient analysis between groups with corresponding markers in different groups.

Variables are presented in three groups: control, model, and SZRD. Values of correlations are shown in the vertical axis (the upper for positive correlations, and the lower for negative correlations) and the corresponding markers are represented to the right of the bars. Numbers are consistent with Tables 18.1, 18.2.

There are correlations between markers 8, 11, 18, 20 with the control group, whilst being relatively high when compared to the other variables, therefore, the change in serotonin, melatonin, 5-Hydroxy-L-tryptophan, and prostaglandin D2 has the strongest association with normal characteristics, and are enough to suggest that these markers can principally represent the insomnia model in this study. The markers 4, 10, 14, 16 have the positive correlation with the model group; others have the negative correlation with the model group, and shows that the overall metabolic profile of Drosophila caused a significant disturbance via sleep deprivation. The markers 4, 10, 14, 16 have negative correlation with the SZRD group; others have positive correlation with the SZRD group, consistent with the control group. Interestingly, these results provide evidence for the dissection of the mode of action of clinically well established SZRD for insomnia.

Insomnia is a common problem that for many sufferers persists chronically, and may result from a wide range of causes. Specific treatments address particular underlying medical disorders. General therapeutic approaches, including pharmacologic and behavioral strategies, may have broad applicability to insomnia patients. Many different medications and substances have been used in an attempt to improve sleep. Insomnia is a common problem requiring appropriate recognition and management. Despite recent advances in the development of newer hypnotics in Western medicine, a significant proportion of patients with insomnia, both locally and internationally, consume TCM. Sedative–hypnotic medications, including benzodiazepines and nonbenzodiazepines, are usually prescribed for insomniac patients; however, the addiction, dependence, and adverse side effects of those medications have drawn much negative attention. In contrast, SZRD has been efficiently used for insomnia relief in China, although its mechanism remains unclear. This study was designed to further elucidate the underlying mechanism of SZRD on sleep regulation. The use of metabolomics for the screening of biomarker patterns, and the elucidation of biochemical processes during the postgenomic era has increased contemporaneously with progress in global systems biology. In the present study, an insomnia model was constructed, and dynamic metabolic profiles of treated drosophila with SZRD were investigated with UPLC–HDMS combined with multivariate statistical analysis. We utilized this approach to delineate the hypnotic property of SZRD. Interestingly, metabolomic trajectory analysis has found that the dosing SZRD regiment can be regulated back towards their baseline levels, to that of the control group; showing potent therapeutic efficacy. Importantly, we found that SZRD activated an array of factors involved in amino acid metabolism and lipid metabolism, et al.

Metabolic analysis of insomnia was inferred from changes in the intermediates during substance metabolism in the present study. Further analysis revealed that metabolites identified together are important for the host response to insomnia through the immune system and the nervous system, et al. Melatonin is a biogenic amine that is found in animals, plants, and microbes. In mammals, melatonin is produced by the pineal gland, which is located in the center of the brain but outside the blood–brain barrier. The secretion of melatonin increases in darkness, and decreases during exposure to light, thereby regulating the circadian rhythms of several biological functions (Singh and Jadhav, 2014). In particular, melatonin regulates the sleep–wake cycle by chemically causing drowsiness, and lowering the body temperature. Melatonin is also implicated in the regulation of mood, learning and memory, immune activity, dreaming, fertility, and reproduction. Melatonin is also an effective antioxidant. Most of the actions of melatonin are mediated through the binding, and activation of melatonin receptors. Individuals with insomnia disorders may have lower than normal levels of melatonin, and that the deficits were associated with low activity of the melatonin. Reduced melatonin production has also been proposed as a likely factor in the significantly higher insomnia rates in night workers. Serotonin (5-hydroxytryptamine, 5-HT) is a biochemical messenger and regulator, synthesized from the essential amino acid L-tryptophan. Serotonin in the nervous system acts as a local transmitter at synapses, and as a paracrine or hormonal modulator of circuits upon diffusion, allowing a wide variety of "state-dependent" behavioral responses to different stimuli (Nishida et al., 2013). Serotonin is widely distributed in the nervous system of vertebrates and invertebrates. In vertebrates, which display a wider and much more sophisticated behavioral repertoire, serotonin also modulates sleep, the arousal state, sexual behavior, and others, and deficiencies of the serotonergic system causes disorders, such as depression, obsessive–compulsive disorder (OCD), phobias, posttraumatic stress disorder (PTSD), epilepsy, and generalized anxiety disorder. Serotonin has three different modes of action in the nervous system: as transmitter, acting locally at synaptic boutons; upon diffusion at a distance from its release sites, producing paracrine effects; and by circulating in the blood stream, and producing hormonal effects. Serotonin has been claimed to help alleviate insomnia, depression, and headaches (Szabadi, 2014). So, serotonin might be a useful biomarker for insomnia. Prostaglandin D2 (or PGD2) is a prostaglandin that is actively produced in various organs, such as the brain, spleen, thymus, bone marrow, uterus, ovary, oviduct, testis, prostate, and epididymis, and is involved in many physiological events. It is a potent endogenous somnogens, and plays a critical role in the regulation of physiological sleep. PGD2 binds to the prostaglandin D2 receptor, which is a G-protein-coupled receptor. Its activity is mainly to promote sleep, regulate body temperature, olfactory function, hormone release, and nociception in the central nervous system; (preventing platelet aggregation, and inducing vasodilation, and bronchoconstriction). PGD2 is also released from mast cells as an allergic and inflammatory mediator, through receptor-mediated G-protein linked signaling pathways. In mammalian systems, it is efficiently converted into more stable arachidonate metabolites, such as PGD2, PGE2, and PGF2a by the action of three groups of enzymes, PGD synthases, PGE synthases, and PGF synthases, respectively. PGD2 has been proposed to be essential for the initiation, and maintenance of sleep, and plays a pivotal role in the regulation of the physiological sleep (Qu et al., 2006). The presence of PGD2 suggests that sleep was disturbed. So, PGD2 might appear to contribute to insomnia, and has a close relationship with insomnia. 5-hydroxy-L-tryptophan is synthesized by the pineal gland. Daily rhythms in pineal methoxyindole metabolism have been described in rodents and humans (5-hydroxy-L-tryptophan level are coincident with serotonin levels in rodents pineal) and 5-Hydroxy-L-tryptophan at its highest during the daylight hours, and it falls markedly soon after the onset of darkness, coincident with increases in the levels of pineal melatonin, and the activities of pineal serotonin-*N*-acetyltransferase, and hydroxyindole-*O*-methyltransferase. The fact that the levels of 5-hydroxy-L-tryptophan and melatonin, vary in parallel suggests that the major factor generating the methoxyindole rhythms is not serotonin-*N*-acetyltransferase activity, but perhaps a change in the availability (for metabolism) of "stored" serotonin. 5-hydroxy-L-tryptophan is frequently seen in sleep disruption, nightly restlessness, sundowning, and other circadian disturbances diseases. After sleep deprivation, the presence of of 5-hydroxy-L-tryptophan, suggests a potential monitoring role for insomnia progression. Thus, 5-hydroxy-L-tryptophan might be a sensitive indicator biomarker of insomnia. However, other markers also warrant the further mechanistic exploration of the SZRD formulae.

20.3.3 Intervention Effects of Jujuboside A on Insomnia

Herein, PCA analysis (Fig. 20.10A and B) and Heatmap visualization (Fig. 20.10C) of metabolomic data showed distinct segregation between

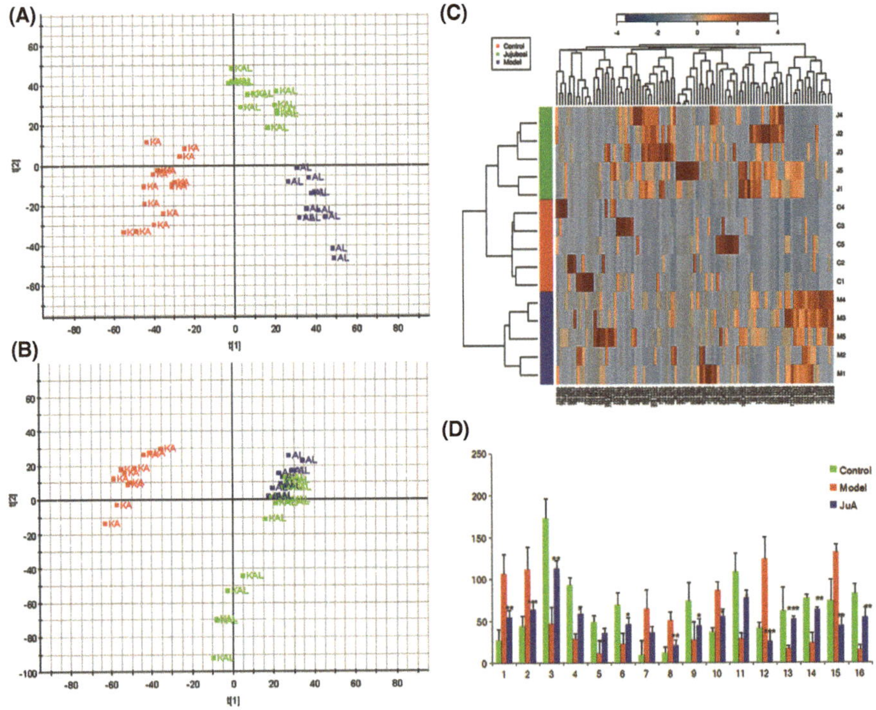

FIGURE 20.10

Trajectory analysis of PCA score plots for the insomnia drosophila post-Jujuboside A treatment in positive model (A) and negative model (B). AL, control group; KA, model group; KAL, JuA group. Heatmap visualization of the intervention effects of Jujuboside A through metabolic pathway analysis (C). Rows, samples; columns, metabolites. Color key indicates metabolite expression value: blue, lowest; red, highest. Changes in the relative concentrations of the target metabolites identified post-Jujuboside A treatment. A two-tailed, parametric t-test was used to determine the significance of the change in relative concentrations for each metabolite. Bars represent the mean relative metabolite concentration, and standard deviations.*, $p < 0.05$; **, $p < 0.01$; ***, $p < 0.001$.

the control, and sleep deprivation-treated animals. However, the parallel observation that PCA and Heatmap analysis, including samples among the control and the JuA groups resulted in similar clustering indicating a consistent metabolic pattern. The lists of discriminating ions were further screened to identify the ions that consistently contributed to the separation of the control, and the sleep deprivation-treated animals. The significantly up-regulated 5-methyltetrahydrofolate, androstenol, uridine triphosphate, prostaglandin D2, fructose 6-phosphate, dopamine, pantothenic acid, and the down-regulated adenosine triphosphate, serotonin, uridine diphosphate glucose, antepan, enkephalin, 5-hydroxy-L-tryptophan, melatonin, NADPH, adenosine were observed in the model group following sleep deprivation, compared with the control group (Fig. 20.10D). This difference of metabolites may denote their potential as targeted biomarkers for differentiating insomnia and normal states. Additionally, in order to more clearly characterize sleep effects of JuA, changes in the relative concentrations of target metabolites identified in the different groups were analyzed, we have found that the content of these key markers is closer to the normal group. PCA and Heatmap showed that JuA exhibited preventive efficacy against insomnia by adjusting multiple metabolic pathways to their normal state. Compared with the alterations of insomnia-related metabolites, most of them were reset to a healthier level after JuA administration. In this study, the relative concentration of 16 metabolites was significantly affected by JuA treatment. According to the parametric t-test, we found that the relative concentration of these metabolites could be reversed after taking the JuA.

20.3.4 Jujuboside B Against Insomnia

The trajectory analysis of PCA score plots (3-D) for the insomnia drosophila after JuB treatment showed clear segregation (Fig. 20.11A). Acquired data were subjected to computational systems analysis to further investigate the effects of JuB on the insomnia drosophila metabolite profiles. The parallel hierarchical clustering analysis (Fig. 20.11B), and heatmap visualization (Fig. 20.11C) for the intervention effects of JuB showed distinct segregation. These models were capable of distinguishing models from healthy subjects, and showed that JuB exhibited preventive efficacy against insomnia by adjusting multiple metabolic pathways to their normal state. The significantly up-regulated serotonin, dopamine, arachidonic acid, glutathionylspermidine, and the down-regulated S-adenosyl methionine, phenylalanine, adenosine monophosphate, 5-hydroxy-L-tryptophan, melatonin, adenosine, tryptophan, and trypanothione were observed in the model group following sleep deprivation compared with the control group (Fig. 20.11D). In this study, the relative concentration of 12 endogenous metabolites was significantly affected by JuB treatment. Interestingly, according to the parametric t-test, we found that the relative concentration of these metabolites could be reversed after taking the JuB. Compared

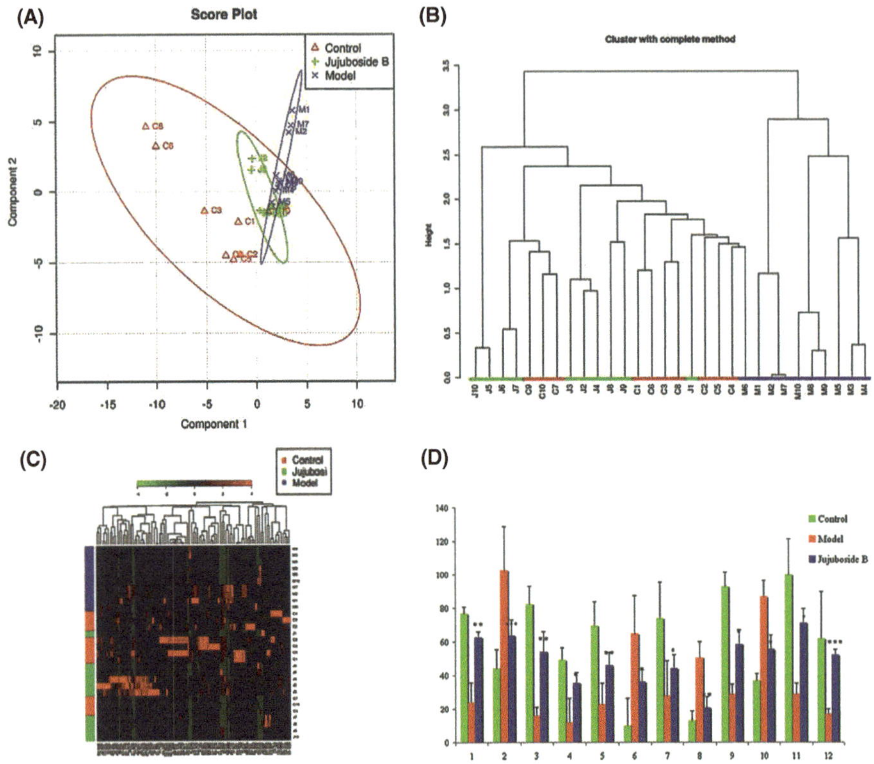

FIGURE 20.11 Systems analysis with MetaboAnalyst's data annotation tools.
PCA score trajectory plots: (A) hierarchical clustering analysis; (B) heatmap visualization; (C) the intervention effects of Jujuboside B for insomnia drosophila; (D) changes in the relative concentrations of target metabolites identified by PCA of different groups. A two-tailed, parametric t-test was used to determine the significance of the change in relative concentrations for each metabolite. Bars represent the mean relative metabolite concentration, and standard deviations.*, $p < 0.05$; **, $p < 0.01$; ***, $p < 0.001$.

with the alterations of insomnia-related metabolites, most of them were reset to a healthier level post-JuB administration. Thus, JuB may regulate metabolism of these markers efficiently for sleep disorder.

Of the 12 distinct metabolites identified from these pathways, many are in the various stages of progress of the insomnia. Some significantly changed metabolites like serotonin, 5-hydroxy-L-tryptophan, melatonin, and tryptophan have been found and used to explain the tryptophan mechanism. The detailed construction of the tryptophan metabolism pathways with higher scores are shown in Fig. 20.12. These results suggest that these target pathways show the marked perturbations over the entire time-course of insomnia, and could contribute to the development of insomnia. Serotonin (5-hydroxytryptamine,

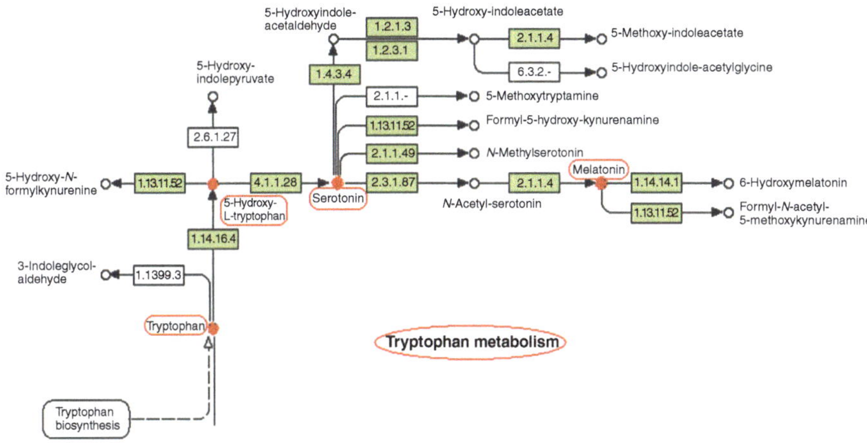

FIGURE 20.12 Construction of the tryptophan metabolism pathways in drosophila.
The map was generated using the reference map by KEGG (www.genome.jp/kegg/). Green boxes indicate enzymatic activities with putative cases of analogy in drosophila.

5-HT) widely distributed in the nervous system of vertebrates and invertebrates, is a biochemical messenger and regulator, and acts as a local transmitter at synapses, and is synthesized from the essential amino acid L-tryptophan. 5-HT has three different modes of action in the nervous system: as transmitter, acting locally at synaptic boutons; upon diffusion at a distance from its release sites, producing hormonal effects. 5-HT has been claimed to help alleviate insomnia, depression, and headaches. 5-Hydroxy-L-tryptophan is the immediate precursor of the neurotransmitter serotonin, and is frequently seen in sleep disruption, nightly restlessness, sundowning, and other circadian disturbances diseases. Tryptophan is an essential amino acid that is the precursor of serotonin. Serotonin is a brain neurotransmitter, platelet clotting factor, and neurohormone found in organs throughout the body. Tryptophan and serotonin requires nutrients, such as vitamin B6, niacin, and glutathione. Some disorders of the excess metabolism of tryptophan in the body may contribute to mental retardation. Increased tryptophan fragments correlate with increased tryptophan degradation, which occurs with depression, mental retardation, hypertension, and anxiety states. Melatonin is produced by the pineal gland, which is located in the center of the brain but outside the blood–brain barrier, and regulates the circadian rhythms of several biological functions, including the sleep–wake cycle. Melatonin is also implicated in the regulation of mood, learning, memory, immune activity, dreaming, fertility, and reproduction. In particular, most of the actions of melatonin are mediated through the binding, and the activation of melatonin receptors. Indeed, future metabolomic studies in human populations with insomnia will be needed to validate the biomarkers found in the animal model. These results implicate the JuB effects may be mediated through receptor, neurotransmitter, enzymes, signal transduction, and electron carrier. It provides strong evidence that the hypnotic effect of JuB occurred at the level of global metabolomics.

20.4 CONCLUSIONS

The metabolomic approach was used to delineate the metabolic changes in insomnia postdosing with SZRD regiment treatment. In this study, a model of insomnia drosophila was constructed, and dynamic metabolic profiles were investigated with UPLC-MS, combined with multivariate statistical analysis. The potential biomarkers of significant contribution was characterized as follows: 9 in negative ion mode, and 11 in positive ion mode. Of note, the SZRD regiment, and its related active ingredients developed the distant tendency to the diverse extent of the model group, and could be regulated back toward their baseline levels, to that of the control group; demonstrating that SZRD could have a good prevention role with regard to insomnia, and keep animals in the normal situation. Analysis of the correlations revealed that: serotonin, melatonin, 5-hydroxy-L-tryptophan, and prostaglandin D2 have the strongest association with insomnia characteristics. Compared with the alterations of the insomnia-related metabolites, most of them were reset to a healthier level after SZRD, and JuA and JuB administration. We found that SZRD, and JuA and JuB increases sleep activity, and exhibits a binding affinity for serotonin receptors. These results implicate the hypnotic effect of SZRD, JuA, and JuB, and its effects may be mediated through serotonergic activation. It provided strong evidence that the hypnotic effect of SZRD occurred at the level of global metabolomics. By using a metabolomic approach, this study also exemplifies that metabolomics could provide a very promising methodology to elucidate the therapeutic mechanisms of TCM.

References

Chen, C.Y., Chen, Y.F., Tsai, H.Y., 2008. What is the effective component in suanzaoren decoction for curing insomnia? Discovery by virtual screening and molecular dynamic simulation. J. Biomol. Struct. Dyn. 26 (1), 57–64.

Li, Y.J., Bi, K.S., 2006. Study on the therapeutic material basis of traditional Chinese medicinal preparation suanzaoren decoction. Chem. Pharm. Bull. (Tokyo) 54 (6), 847–851.

Nishida, K., Takechi, K., Akiyama, T., Carstens, M.I., Carstens, E., 2013. Scratching inhibits serotonin-evoked responses of rat dorsal horn neurons in a site- and state-dependent manner. Neuroscience 250, 275–281.

Qu, W.M., Huang, Z.L., Xu, X.H., Aritake, K., Eguchi, N., Nambu, F., Narumiya, S., Urade, Y., Hayaishi, O., 2006. Lipocalin-type prostaglandin D synthase produces prostaglandin D2 involved in regulation of physiological sleep. Proc. Natl. Acad. Sci. USA 103 (47), 17949–17954.

Singh, M., Jadhav, H.R., 2014. Melatonin: functions and ligands. Drug Discov. Today, 2014;19(9):1410–1418.

Szabadi, E., 2014. Selective targets for arousal-modifying drugs: implications for the treatment of sleep disorders. Drug Discov. Today 19 (5), 701–708.

Wang, X., Yang, B., Zhang, A., Sun, H., Yan, G., 2012a. Potential drug targets on insomnia and intervention effects of Jujuboside A through metabolic pathway analysis as revealed by UPLC/ESI-SYNAPT-HDMS coupled with pattern recognition approach. J. Proteomics 75 (4), 1411–1427.

Wang, X., Yang, B., Sun, H., Zhang, A., 2012b. Pattern recognition approaches and computational systems tools for ultra performance liquid chromatography–mass spectrometry-based comprehensive metabolomic profiling and pathways analysis of biological data sets. Anal. Chem. 84 (1), 428–439.

Xie, Y.Y., Xu, Z.L., Wang, H., Kano, Y., Yuan, D., 2011. A novel spinosin derivative from Semen Ziziphi Spinosae. J. Asian Nat. Prod. Res. 13 (12), 1151–1157.

Xie, C.L., Gu, Y., Wang, W.W., Lu, L., Fu, D.L., Liu, A.J., Li, H.Q., Li, J.H., Lin, Y., Tang, W.J., Zheng, G.Q., 2013. Efficacy and safety of Suanzaoren decoction for primary insomnia: a systematic review of randomized controlled trials. BMC Complement. Altern. Med. 13, 18.

CHAPTER 21

Metabolic Biomarkers of Nonbacterial Prostatitis, and the Treatment Evaluation of *Phellodendri Amurensis* Cortex and its Main Components

Hui Sun, Huiyu Wang, Aihua Zhang, Xijun Wang

21.1 INTRODUCTION

Nonbacterial prostatitis (NBP) is the most common urological diagnosis in men under 50 years of age, and is the third most common urologic diagnosis in men over 50 years of age. An estimated 50% of all men experience prostatitis-like symptoms at some point during their lifetime. A suitable preclinical animal model of prostatitis-induced pain or prostatodynia can help us to understand this clinical issue sufficiently. Capsaicin is a polysaccharide that is very commonly used to induce inflammation, and subsequent pain in various inflammatory pain models (Chuang et al., 2008). The rat model of capsaicin-associated nonbacterial prostatitis can be useful for elucidating the mechanisms of the pathogenesis of NBP (Chuang et al., 2007). Wistar rats spontaneously develop NBP with advancing age, which makes them a good animal model for the laboratory investigation of NBP (Hu et al., 2014). NBP has become an important clinical issue. However, the effective therapies for treating NBP have yet to be found, and this has contributed to an increased use by sufferers, of natural products. Plant-derived compounds possess a wide range of pharmacological properties, and their action has been the subject of considerable interest in recent years.

Phellodendri Amurensis Cortex (Guan huangbai (GHB); Fig. 21.1A) was commonly used as an herbal medicine in China for the treatment of NBP, dysentery, urinary infection, rheumatoid arthritis, tumors, and other diseases (Wang et al., 2013a,b; Xian et al., 2011). Recently, alkaloids, and limonoids, such as berberine (Fig. 21.1B) and obaculactone (Fig. 21.1C) were found to be the major components of GHB (Wang et al., 2014). However, the effects and the exact mechanisms by which GHB, and its main components affect NBP have not yet been fully described. In light of this, we performed a metabolomic approach to evaluate, and investigate the beneficial effects of herbal GHB, and its main components against NBP, via a more holistic approach that uses metabolomics.

CONTENTS

(Continued)

Chinmedomics. http://dx.doi.org/10.1016/B978-0-12-803117-9.00021-4

CONTENTS

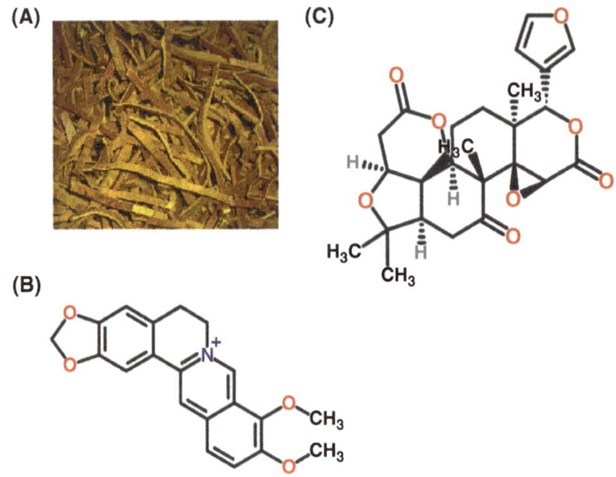

FIGURE 21.1
(A) *Phellodendri Amurensis* Cortex; (B) chemical structure of berberine; (C) obaculactone.

An animal model for NBP in rats was developed with the use of intraprostatic injection of capsaicin, and the reliability, and validity of this model was tested by changes in the extent of the inflammatory changes of the prostate.

21.2 EXPERIMENTAL METHODS

21.2.1 GHB Extract

The raw herbs were purchased from Shiyitang Drug Store (Harbin, China). Next, 500 g GHB was cut into small pieces, and extracted three times under reflux in the 10 times amount of 75% methanol, every time for 1 h, recovering methanol under reduced pressure, until no alcohol flavor, freeze-dried, and set aside. Flour rate was about 19.8%. Weighing 1.215 g powder, with 15 mL 0.5% CMC at ultrasonic treatment for 30 min, made a crude drug concentration 0.081 g/mL solution; dose 1.0 mL/100 g, orally administered to rats. Capsaicin (97%) was obtained from (Aladdin, USA); Tween 80 was obtained from Bodi Chemical Co., Ltd (Tianjin, China). Standards of berberine and obaculactone (98%) were purchased from the Tianjin Kermel Chemical Reagent Co., (Tianjin, China). Nexrutine extra was obtained from Solanova, USA. All other reagents were of analytical grade.

21.2.2 Animal Handling

Male Wistar rats (weighing 280–300 g) were provided by the GLP Centre of Heilongjiang University of Chinese Medicine. All rats were randomly divided into seven groups as follows: sham group (KB, $n = 12$); model group (surgery group, $n = 14$); nexrutine group (positive control group, M1, $n = 14$); GHB group (M2, $n = 14$); berberine group (M3, $n = 14$); obaculactone group (M4, $n = 14$); berberine + obaculactone group (M5, $n = 14$). All animals were allowed to acclimatize in metabolism cages for 1 week prior to treatment. After orally administered GHB at a dose of 3.402 g/kg once daily for the following 7 days; rats in the normal control group were given saline alone, under the same conditions.

For the injection of capsaicin, the rats were anesthetized with 0.25 mL/100 g sodium pentobarbital (0.2%) and were fixed in a supine position. Then, the lower abdomen above the penis of rats was shaved, and the skin in this area sterilized using three applications of 10% povidone–iodine solution. A small midline incision was made in the sterile area, then the bladder, and the prostate carefully exposed. With a 30-gauge needle, 50 μL sterile suspension of 3% capsaicin (Aladdin, USA) was injected into both right and left ventral lobes of the prostate gland. For the sham group, commensurable sterile normal saline was injected. After the injection, a 2% lidocaine solution was applied to the wound, and then the wound was closed in layers. Preparation of the animal model requires histopathology, PCR testing, and a metabolomics evaluation. For histological analysis, one part of the prostate was fixed in buffered 10% formaldehyde for 24 h, embedded in paraffin, cut with a microtome, and stained with hematoxylin–eosin. After 1 week of drug treatment, the prostatic proinflammatory cytokine TNF-α level, and histological findings were noted. Urine was collected from metabolism cages at an ambient temperature throughout the whole procedure, and centrifuged at 13,000 rpm at 4°C for 10 min, and the supernatants were stored frozen at −80°C until metabolomic analysis.

21.2.3 Metabolic Profiling Platform

21.2.3.1 Chromatography

Chromatography was performed using a Waters ACQUITY™ ultra performance liquid chromatography system (Waters Corp., Milford, USA) controlled with Masslynx (v4.1), (Waters Corp., Milford, USA). An aliquot of 5 μL of sample solution was injected onto an ACQUITY UPLC HSS T3 column (50 × 2.1 mm i.d., 1.8 μm, Waters Corp., Milford, USA) at 45°C and the flow rate was 0.4 mL/min. The mobile phases were composed of 0.1% formic acid in acetonitrile (solvent A), and 0.1% formic acid in water (solvent B), the gradient was used as follows: a linear gradient of 1−10% A over initial–2.0 min, 10−20% A over 2–4 min, 20−60% A over 4−7.5 min, 60−99% A over 7.5−8.5 min, 99% A over 8.5−10.5 min, 99−1% A over 10.5−11.0 min, 1% A over 11−13 min.

21.2.3.2 Mass Spectrometry

The eluent was introduced into the high-definition mass spectrometer (Waters Corp., Milford, USA) analysis, and the optimal conditions of analysis were as follow: In positive ion mode, the capillary voltage was 3.0 kV, the sampling cone voltage was 30 V, and extraction cone voltage was 4.0 V, the source temperature was set at 110 °C, desolvation gas temperature was 350 °C, desolvation gas flow was 600 L/h; in negative ion mode, the capillary voltage was 2.6 kV, the sampling cone voltage was 30 V, and extraction cone voltage was 3.5 V, the source temperature was set at 120 °C , desolvation gas temperature was 350 °C, desolvation gas flow was 650 L/h. The data acquisition rate was set to 0.4 s/scan, with a 0.1 s inter scan delay. Data were collected in centroid mode, and the mass range was set at *m*/*z* 50–1000 using an extended dynamic range. For accurate mass acquisition, a lock-mass of leucine enkephalin at a concentration of 0.2 ng/mL was used via a lock spray interface at a flowrate of 100 μL/min monitoring for positive ion mode ($[M + H]^+$ = 556.2771) and negative ion mode ($[M-H]^-$ = 554.2615) to ensure accuracy during the MS analysis.

21.2.3.3 Data Processing

Raw data of all the determined samples were further processed by Markerlynx (v4.1 software) (Waters Corp., Milford, USA) for peak detection and peak alignment. The method parameters for data processing were set as follows: retention time range 0–8.5 min; mass range 50–1000 Da; retention time tolerance 0.2 min; mass tolerance 0.1 Da; noise elimination level 6; peak intensity threshold 50 in positive ion mode, and 150 in negative ion mode. The peak width at 5% height, and the peak-to-peak baseline noise were automatically determined. All data matrices were introduced to EZinfo 2.0 software for pattern recognition analysis.

21.2.3.4 Multivariate Data Analysis

Each sample at ninth was represented by a total ion current (TIC) chromatogram. The three-dimensional matrix of LC-MS raw data, including retention time and *m*/*z* pairs (variable indices), sample names (observations), and normalized ion intensities (variables), were exported to EZinfo software (v2.0) (Waters Corp., Milford, USA) package for multivariate data analysis to cope with the highly multivariate, noisy, collinear, and possibly incomplete data. From the PCA and OPLS-DA, various metabolites could be identified as being responsible for the separation, and were therefore viewed as potential biomarkers.

21.2.3.5 Biomarkers Identification

The identification of potential biomarkers was determined by Q-TOF. The MS collision energy is 35 eV, and the data was obtained in the positive, and negative ion mode, and MarkerLynx (v4 software) was used for data analysis. The identities of the specific metabolites were confirmed by an elements information comparison of their mass spectra using the elemental composition information provided by the software. Furthermore, MassFragment conducts rapid putative ion identification, and a putative biologically relevant analysis via the incorporation of the small molecule databases. The pathway analysis of the potential biomarkers was performed with MetaboAnalyst based on the database source.

21.2.3.6 Network and Pathway Analysis

The MetPA software was employed to all significant up-regulated and down-regulated metabolites, and related biological pathways. The potential markers identified were compared with the accurate mass charge ratio in some databases to discover the related pathways. T-test and fold-alter were used to determine the statistical significance in the pathways. p-value <0.05 was considered to be the criteria for statistical significance, and would be selected.

21.2.4 Statistical Analysis

The PCA was used to uncover unknown trends in the treated groups. Statistically significant differences in mean values were tested by using two-tailed, two-sample Student's t-test, and $p < 0.05$ was considered statistically significant. Prior to multivariate analysis, the resultant data matrices from the two analytical techniques were mean-centered and pareto-scaled.

21.3 RESULTS

21.3.1 Histopathology

Extensive infiltration of inflammatory cells in the lumina, mononuclear cells in the stroma of the gland, and epithelial degeneration were observed in the NBP group, suggesting the presence of nonbacterial prostatitis. In the microscopic findings from the GHB-treatment group, there were fewer inflammatory cells in the lumina, and the epithelial cells of the gland, and stroma showed more improvement compared to the control group. In the nexrutine, berberine, and obaculactone, berberine + obaculactone treated groups, the inflammatory cells in the lumina, and the epithelial cells of the gland and stroma, showed more improvement than the control group, but less improvement than the GHB group (Fig. 21.2). Treatment with the herbal medicine GHB led to a decreased inflammation level from our NBP rat model. These effects are presumed to be the result of the anti-inflammatory effects of GHB.

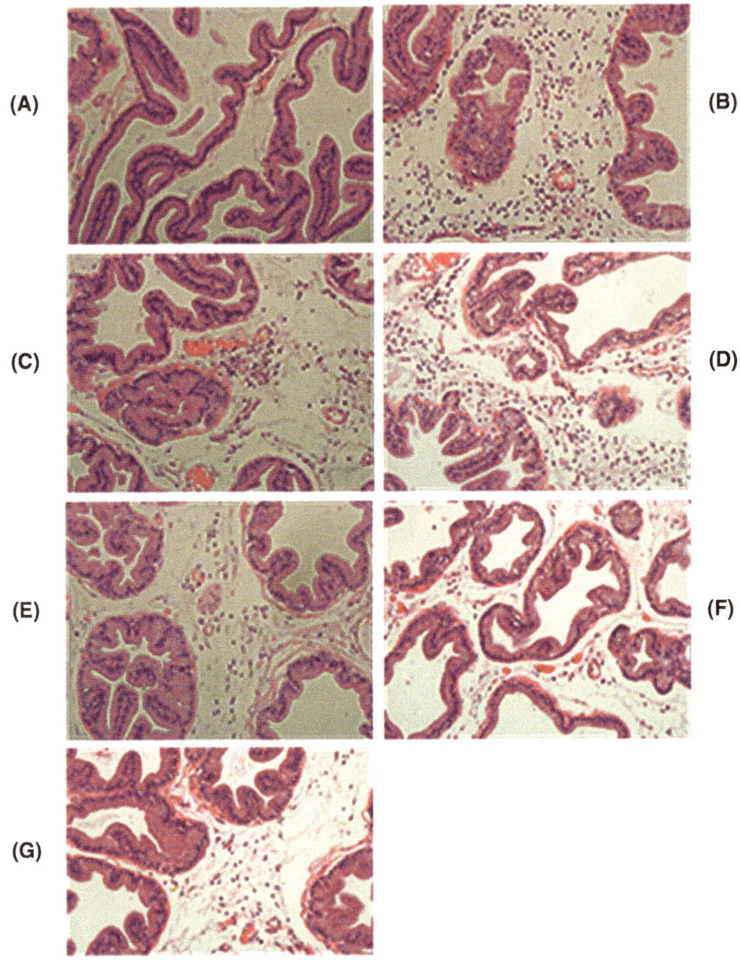

FIGURE 21.2 Representative photomicrographs of tissue sections stained with H&E (×200). (A) control group; (B) model group; (C) nexrutine group; (D) GHB; (E) berberine; (F) obaculactone; (G) berberine + obaculactone.

21.3.2 Plasma Protein Extravasation and the Analysis of Changes in Proinflammatory Cytokine Levels

As shown in Table 21.1, after capsaicin injection, Evans blue extravasation was significantly increased in the model group, which implied neurogenic plasma extravasation in the prostate. Administration with GHB, and its main components could effectively ameliorate, and recover to control level from the increased Evans blue extravasation. To investigate the effects of GHB on nonbacterial prostatitis, analyses of the inflammatory cytokines TNF-α was conducted. Rats in the capsaicin-induced NBP treatment group showed significant

Table 21.1 Expressions of TNF-α and Concentration of Protein Extravasation in all the Groups

Group	Plasma Protein Extravasation	TNF-α
SHAM	21.44 ± 1.61	1.00
MOD	29.49 ± 9.93*	3.39 ± 0.48**
M1	26.90 ± 2.07	1.98 ± 0.56##
M2	21.60 ± 2.73*	1.63 ± 0.79##
M3	20.83 ± 5.57#	1.46 ± 0.92##
M4	21.83 ± 2.23#	1.40 ± 0.48##
M5	19.43 ± 3.78##	1.34 ± 0.75##

*Significant difference from control at $p < 0.05$;
**Significant difference from control at $p < 0.001$;
#Significant difference from model at $p < 0.05$;
##Significant difference from model at $p < 0.001$.

increase in the proinflammatory cytokine TNF-α when compared to the control group ($p<0.01$). Similar to the inflammatory cell accumulation, these changes implied inflammatory processes in the prostate induced by capsaicin. Based on the aforementioned results, the NBP was considered to be successfully established. All these results indicated that the rats presented the typical pathological features of NBP. Of note, administration with GHB could effectively ameliorate these symptoms, as demonstrated by the proinflammatory cytokine level of TNF-α of the GHB, berberine, and obaculactone, berberine + obaculactone treatment groups, which showed significant decreases when compared to the model group ($p<0.01$).

21.3.3 Method Development and Validation

For the method validation study, 5 mL of urine from the samples from each group were pooled to get a quality control (QC) specimen, and the preparation of the QC specimen was the same as that of the samples. A number of consecutive injections of the QC sample were made to obtain a stable QTOF/MS system before experimental data acquisition. QC specimens were analyzed every 10 specimens throughout the whole analysis procedure. For the QC sample, eight selected characteristic ions (ESI^+:138.0914, 314.1236, 180.1032, 447.0922; ESI^+:124.0068, 312.1077, 173.0810, 407.2794) were picked out to examine the drift of the retention times, m/z, and peak areas. The results showed that variations in the retention times were less than 0.05 min, drift values of m/z were less than 5 ppm, and the relative standard deviation (RSD) of each peak area was below 8.5%, demonstrating that the system had excellent stability, and repeatability during the analysis procedure. BPI chromatograms of the urine samples of QCs based on UPLC-Q-TOF/HDMS were shown in Fig. 21.3. According to the optimized conditions of urine; the PCA of QCs urine samples

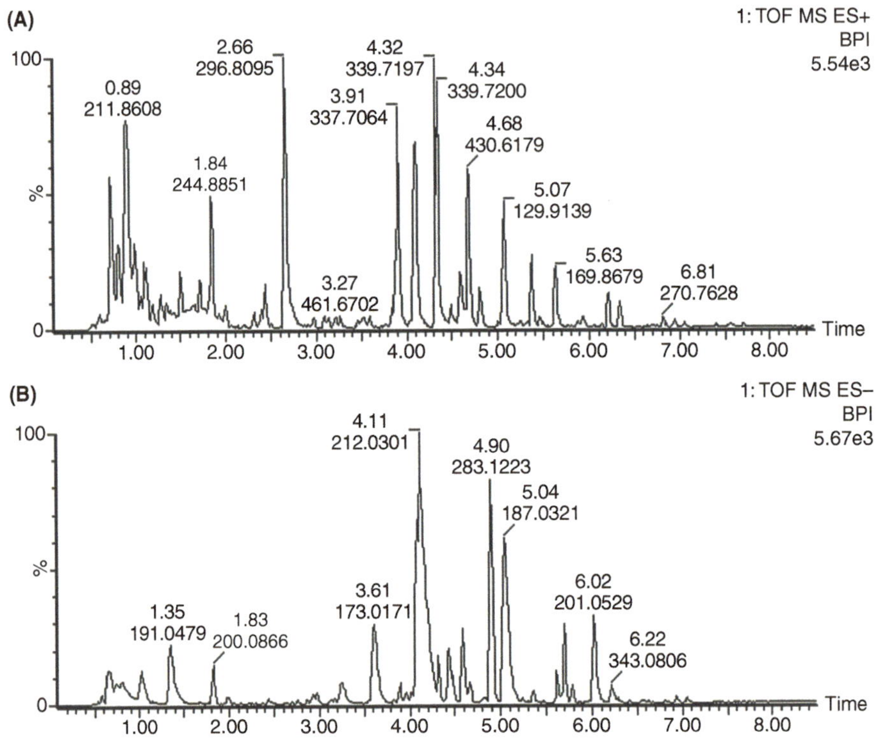

FIGURE 21.3 BPI chromatograms of urine samples of QCs based on UPLC-Q-TOF/HDMS.
(A) positive mode; (B) negative mode.

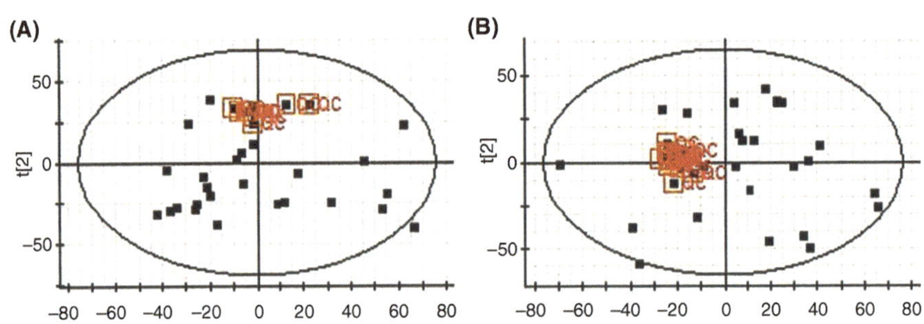

FIGURE 21.4
Principal component analysis of QCs urine samples from the prostatitis rat study, carried out with UPLC-Q-TOF/MS for urine metabolomic analysis: (A) positive mode; (B) negative mode. Black box represents the urine sample, and the sample in red box represents the QCs.

from NBP rats is shown in Fig. 21.4, the QC samples were gathered together to determine during the data collection, and the instrument was stable during the collected sample data, thus demonstrating that the system had excellent stability during the analysis procedure.

21.3.4 Metabolic Profile Analysis

The representative TIC of urine samples derived from control, model, and dosed groups in negative mode are presented in Fig. 21.5 by using the optimal UPLC-Q-TOF-HDMS conditions, as described already. Low molecular mass metabolites could be separated well in the short time of 10 min. A visual comparison of each group was made at day 9, and we found that slightly different contours, some ingredients may be hidden differences in the peaks, and in a visual comparison it is not easy to distinguish the thousands of signals of each group. In order to better visualize the subtle similarities, and differences among these complex data sets, multiple pattern recognition methods were employed to phenotype the urine metabolome of the rats.

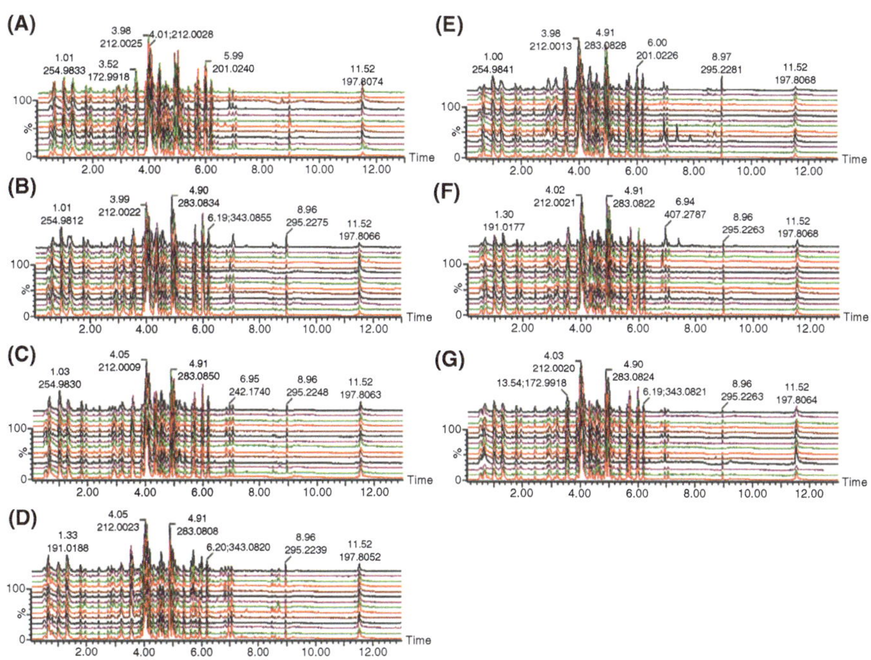

FIGURE 21.5

Metabolomic profilings analysis of all groups (×200): (A) control group; (B) model group; (C) nexrutine extra group; (D) GHB; (E) berberine; (F) obaculactone; (G) berberine + obaculactone.

21.3.5 Metabolomic Study of NBP

In order to identify the various different metabolic changes affecting the NBP rats, the MS spectrum were preprocessed in order to be able to carry out multivariate statistical analysis (PCA, PLS-DA, and OPLS-DA) which can give a comprehensive view of the clustering trend for complex metabolomics data. Here, PCA were used to classify the metabolic phenotypes, and identify the different metabolites. The PCA scores plot could divide the different urine samples into different blocks, respectively, suggesting that the metabolic profiles have changed. NBP model rats deviated from control rats in their metabolic profile, indicating that the metabolic networks of NBP rats were disordered. The differential metabolites associated with NBP were found according to the loading plot of OPLS, and the *p*-value of the Student's t-test.

With regard to the information analyst of PCA in our experiment (as shown in Fig. 21.6), the control, and the model groups were significantly divided into two classes, indicating that the model of NBP was successfully reproduced. We further used OPLS-DA with coefficient plots to directly visualize the results of the loadings, and correlation coefficients. The color-coded correlation coefficients indicate the significance of the metabolites in terms of their contribution to the separation between the different groups. The metabolites that are colored red are more significant than those that are colored green. Additionally, to exhibit how each of these variables is responsible for the separation more intuitively, we created an S-plot that is combined with a VIP plot, and a color coefficient scale bar. More subtle changes can be found by the pattern recognition approach score plots of OPLS. An OPLS score plot base on all 10,500 compounds showed an obvious separation of the control group, and the NBP model group (Fig. 21.7A). For the further analysis of feature ions, S-plot (Fig. 21.7B) from the OPLS was carried out to select distinct variables as potential biomarkers for distinguishing NBP from the controls. We generated

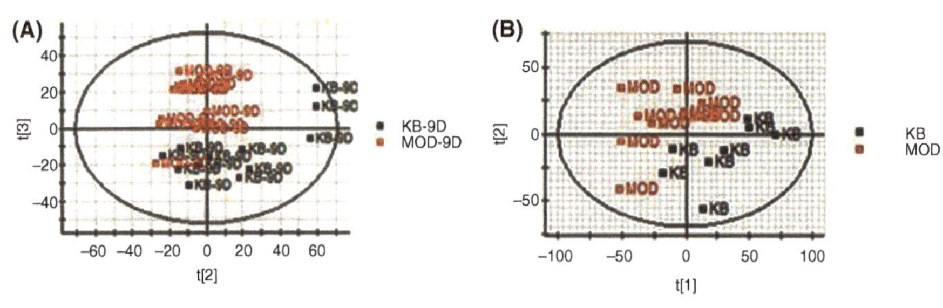

FIGURE 21.6 PCA score plots of urine samples in control (KB) vs. NBP groups at day 9.
(A) Positive ion mode; (B) negative ion mode. Note: Each colored point represents a sample.

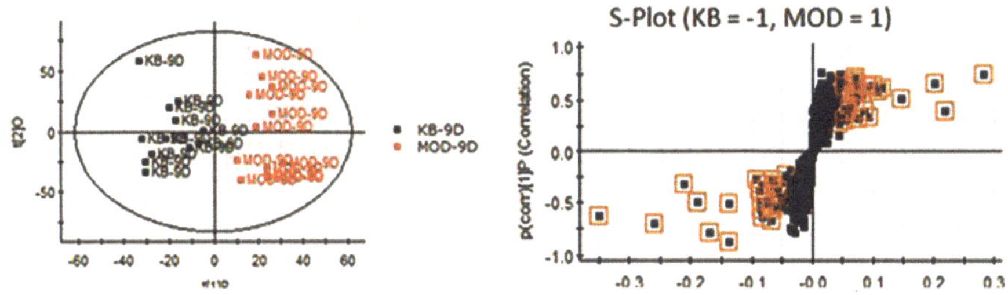

FIGURE 21.7 OPLS-DA results for control (KB) vs. NBP groups at day 9.
(A) Score plot; (B) S-plot.

VIP plots from the OPLS-DA with a threshold of three to identify the metabolites that significantly contribute to the clustering between groups.

21.3.6 Identification of Potential Biomarkers

We generated VIP plots from the OPLS-DA with a threshold of three to identify the metabolites that significantly contribute to the clustering between groups. 18 differentially expressed metabolites considered to be strong contributors were distinguished from those of the controls ($p < 0.05$, VIP > 3, 5 ions in the positive ion mode, and 13 ions in the negative ion mode). Of note, it was found that, among them, 5 ions were up-regulated, and 13 ions were down-regulated (Table 21.2). The small molecule metabolites of significant differences (t-test, $p < 0.05$) were searched. The potential markers were identified by using the "ID browser" to search in the Metlin database (http://metlin.scripps.edu/) and compared with the accurate mass charge ratio in some databases. We can know the probable name of potential biomarkers through the first step. In

Table 21.2 The Result of Stability of the Proposed Method in Urine Metabolomics

		Retention Time (min)		Mass Accuracy		Peak Area	
Mode	***m/z***	**Mean**	**CV (%)**	**Mean**	**CV (%)**	**Mean**	**CV (%)**
ESI^+	138.0914	1.2000	0.5893	138.0911	0.0002	7.0926	7.0204
	314.1236	1.1960	0.4580	314.1238	0.0002	19.4992	8.3213
	180.1032	2.9080	0.1538	180.1037	0.0004	6.2114	5.0650
	447.0922	5.3320	0.0839	447.0918	0.0001	29.6713	7.1428
ESI^-	124.0068	1.0600	0.0000	124.0069	0.0004	62.2240	2.7484
	312.1077	1.2160	0.4504	312.1085	0.0002	40.6800	5.5168
	173.0810	5.1680	0.0865	173.0799	0.0004	179.1556	2.8006
	407.2794	6.9420	0.0644	407.2792	0.0002	130.9644	5.3640

Table 21.3 Information of Potential Biomarkers Related With Prostatitis Identified in Positive and Negative Ion Mode

No.	Retention Time	Mass	Calc. Mass	[H−H]⁻/ [H+H]⁺	ppm	mDa	Formula	Compound	Trend
1	1.06	124.0068	124.0068	$[H-H]^-$	0.0	0.0	$C_2H_7NO_3S$	Taurine	↑
2	1.19	314.1236	314.1240	$[H+H]^+$	−1.3	−0.4	$C_{14}H_{19}NO_7$	Tyramine glucuronide	↓
3	1.27	167.0200	167.0205	$[H-H]^-$	−4.7	−0.8	$C_5H_4N_4O_3$	Uric acid	↓
4	1.97	194.0449	194.0453	$[H-H]^-$	−2.1	−0.4	$C_9H_9NO_4$	Salicylurate	↑
5	2.72	357.0817	357.0822	$[H-H]^-$	−1.4	−0.5	$C_{15}H_{18}O_{10}$	Dihydrocaffeic acid 3-O-glucuronide	↓
6	2.80	175.0241	175.0243	$[H-H]^-$	−1.1	−0.2	$C_6H_8O_6$	Ascorbic acid	↓
7	2.90	180.1032	180.1025	$[H+H]^+$	3.9	0.7	$C_{10}H_{13}NO_2$	Alpha-methylphenylal-anine	↓
8	2.83	283.0820	283.0818	$[H-H]^-$	0.7	0.2	$C_{13}H_{16}O_7$	*p*-Cresol glucuronide	↓
9	3.41	371.0981	371.0978	$[H-H]^-$	0.8	0.3	$C_{16}H_{20}O_{10}$	Dihydroferulic acid 4-O-glucuronide	↓
10	3.81	206.0450	206.0453	$[H-H]^-$	−1.5	1.5	$C_{10}H_9NO_4$	4-(2-Aminophenyl)-2,4-dioxobutanoic acid	↓
11	4.07	178.0503	178.0504	$[H-H]^-$	−0.6	n/a	C_9H8NO_3	Hippuric acid	↑
12	4.28	222.0775	222.0766	$[H-H]^-$	4.1	0.9	$C_{11}H_{13}NO_4$	Acetyl-L-tyrosine	↑
13	4.53	76.0396	76.0399	$[H+H]^+$	−3.9	−0.3	$C_2H_5NO_2$	Glycine	↓
14	4.56	256.0650	256.0657	$[H-H]^-$	−2.7	−0.7	$C_{12}H_{11}N_5S$	2-Amino-6-(benzylthio) purine	↑
15	4.57	192.0657	192.0661	$[H-H]^-$	−2.1	−0.4	$C_{10}H_{11}NO_3$	Methyl-hippuric acid	↓
16	4.61	431.0977	431.0978	$[H+H]^+$	−0.2	−0.1	$C_{21}H_{18}O_{10}$	5-L-Glutamyl-taurine′-O-glucuronide	↓
17	5.15	165.0546	165.0552	$[H-H]^-$	−3.6	−0.6	$C_9H_{10}O_3$	Phenyllactic acid	↓
18	5.59	170.0580	170.0582	$[H+H]^+$	−1.2	−0.2	$C_4H_{12}NO_4P$	Phosphodimethyletha-nolamine	↓

the present study, 18 potential biomarkers were identified (Table 21.3). The precise molecular mass of compounds with significant changes in the groups was determined within measurement errors (5 ppm) by Waters QTOF/MS, and meanwhile, the potential elemental composition, degree of unsaturation, and fractional isotope abundance of compounds were obtained. The presumed molecular formula was searched in Chemspider, HMDB, and other databases to identify the possible chemical constitutions, and the MS/MS data were screened to determine the potential structures of the ions. The primary and secondary mass spectrometry information was analyzed by Masslynx (v4.1, Waters) software, compared with the database.

Differential metabolites were identified according to their MS and MS/MS information. The m/z 178.0503 was taken to illustrate the process of differential metabolite identification. First, according to the data analysis results of differential metabolites, the relative content of m/z 178.0503 was significantly

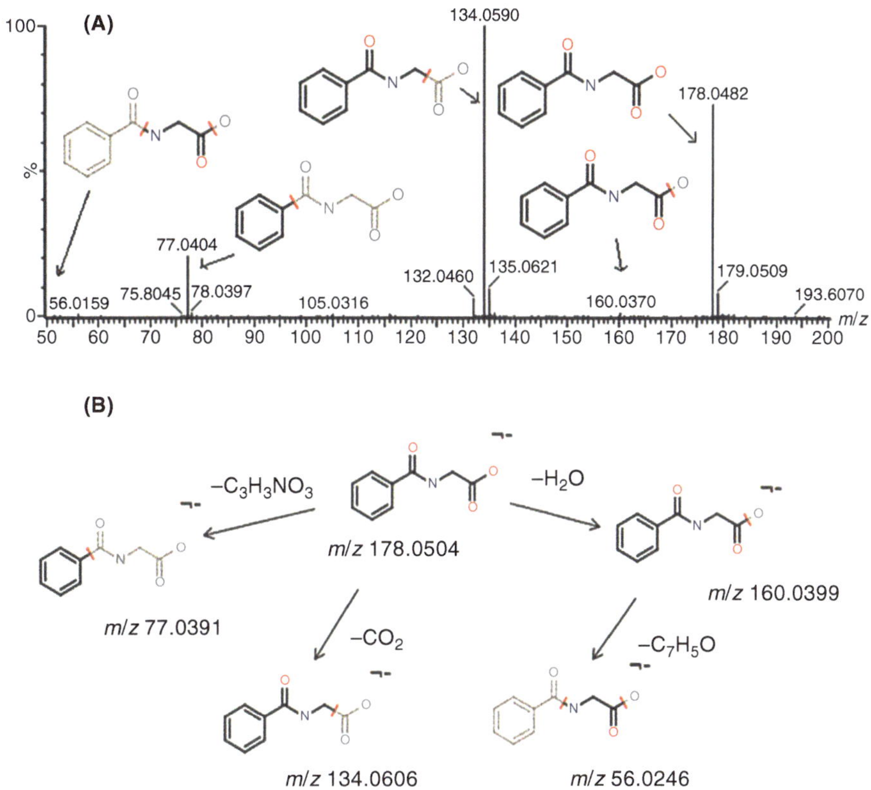

FIGURE 21.8

Mass spectra and fragmentation: (A) of t_R-m/z 4.07-178.0503 hippuric acid based on UPLC-Q-TOF/MS mass fragmentation pathway; (B) of t_R-m/z 4.07-178.0503 hippuric acid based on UPLC-Q-TOF/MS.

different, and was determined as an $[M\text{-}H]^-$ ion by its IFE, and MS information. Second, the MS/MS information about the fragmentation pattern of *m/z* 178.0503 was acquired from the Q-TOF system with 20 eV of collision energy. The main fragment ions analyzed by MS/MS screening were *m/z* 178, 134, and 77 (Fig. 21.8) which could correspond to lost $[M\text{-}H]^-$, $[M\text{-}H\text{-}CO_2]^-$, $[M\text{-}H\text{-}C_3H_3NO_3]^-$, respectively. Finally, the METLIN database was searched for the MS/MS information, and *m/z* 178.0503 was identified as hippuric acid. The identification process of other differential metabolites was similar to that of hippuric acid; the identification results are shown in Table 21.3.

21.3.7 Metabolic Pathway and Function Analysis

To gain insight into the metabolic mechanism of NBP, metabolic pathways of the significantly altered metabolites were analyzed using the "pathway analysis" module within the MetaboAnalyst software. The related pathway of the biomarker was identified by searching the KEGG PATHWAY Database. We

Table 21.4 Results From Ingenuity Pathway Analysis With Metabolyst Based on KEGG

No.	Pathway Name	Total	Hits	*p*	−log(*p*)	Impact
1	Taurine and hypotaurine metabolism	8	2	0.0025	5.9888	0.4285
2	Glycine, serine, and threonine metabolism	32	1	0.2773	1.2826	0.2919
3	Pentose and glucuronate interconversions	14	1	0.1316	2.0276	0.2727
4	Primary bile acid biosynthesis	46	2	0.0746	2.5954	0.0595
5	Glycerophospholipid metabolism	30	1	0.2623	1.3382	0.0444
6	Starch and sucrose metabolism	23	1	0.2075	1.5722	0.0377
7	Tyrosine metabolism	42	1	0.3480	1.0553	0.0269
8	Purine metabolism	68	1	0.5031	0.6869	0.0207
9	Glutathione metabolism	26	1	0.2314	1.4632	0.0057
10	Ascorbate and aldarate metabolism	9	1	0.0866	2.4464	0.000
11	Methane metabolism	9	1	0.0866	2.4464	0.000
12	Nitrogen metabolism	9	1	0.0866	2.4464	0.000
13	Cyanoamino acid metabolism	6	1	0.0585	2.8380	0.000
14	Porphyrin and chlorophyll metabolism	27	1	0.2393	1.4300	0.000
15	Tryptophan metabolism	41	1	0.3413	1.0750	0.000
16	Aminoacyl-tRNA biosynthesis	67	1	0.4978	0.6975	0.000

identified a total of nine distinct metabolic pathways that were significantly altered in the urine (Table 21.4) samples from model group. The detailed analysis of the most relevant pathways of NBP was performed by the MetaboAnalyst's tool. A total of feature compounds in nine pathways, which were identified together are important for the host response to NBP (Fig. 21.9). The predominant hits were pathways involved in taurine and hypotaurine metabolism, glycine, serine and threonine metabolism, pentose and glucuronate interconversions, etc. The detailed construction of the metabolism pathways with higher scores are shown in Fig. 21.9. The correlation network of all the potential biomarkers in response to NBP induced by capsicine, based on KEGG is shown in Fig. 21.10. It suggested that these target pathways showed the marked perturbations over the time course of NBP.

21.3.8 Effects of GHB and its Main Components on NBP Based on Metabolite Profile

The results of differential metabolite identification suggest that the development of NBP involves serious disorders of the metabolism of taurine and hypotaurine, glycine, serine and threonine, pentose and glucuronate interconversions. The dysfunction of the immune system in NBP rats may be caused by metabolic disorders. Using the presented UPLC-MS method, the urinary meta-

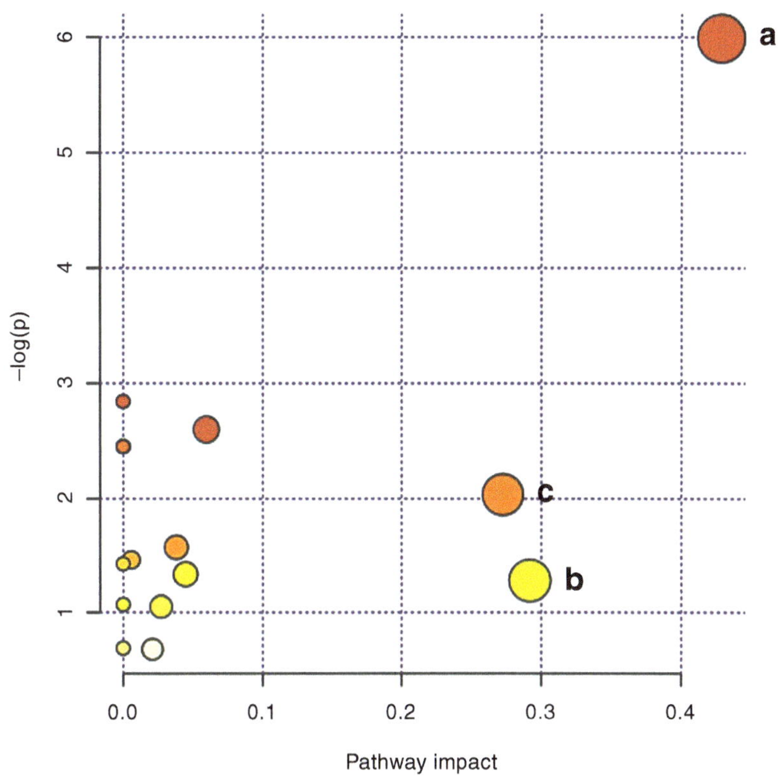

FIGURE 21.9

Summary of pathway analysis with MetPA online tool: (a) taurine and hypotaurine metabolism; (b) glycine, serine, and threonine metabolism; (c) pentose and glucuronate interconversions.

bolic profiles of GHB, berberine, and obaculactone, berberine + obaculactone-treated groups were obtained. As the 18 potential biomarkers from UPLC-MS have been found, it is reasonable to take them as the potential drug targets for further investigating the intervening mechanisms of GHB against NBP. Subtle changes could be found using the pattern recognition approach. Interestingly, it is important to note that the disturbed metabolic pathways can be partially reversed by GHB treatment. The score plot of the PCA showed that the control, the NBP, and the GHB treatment groups were separated clearly (Fig. 21.11A and B) and the GHB treatment group was closer to the control group than the NBP group, which suggested that GHB could reverse the pathological process of NBP. The urine PLS-DA model also clearly differentiates the three different groups based on the score plot result. Using the PLS score plot (Fig. 21.11C and D), the NBP group is clearly separated from the normal group in the direction of the component T1, which implies that the metabolic characteristics of the various small molecules are distinctively different. After treatment with GHB,

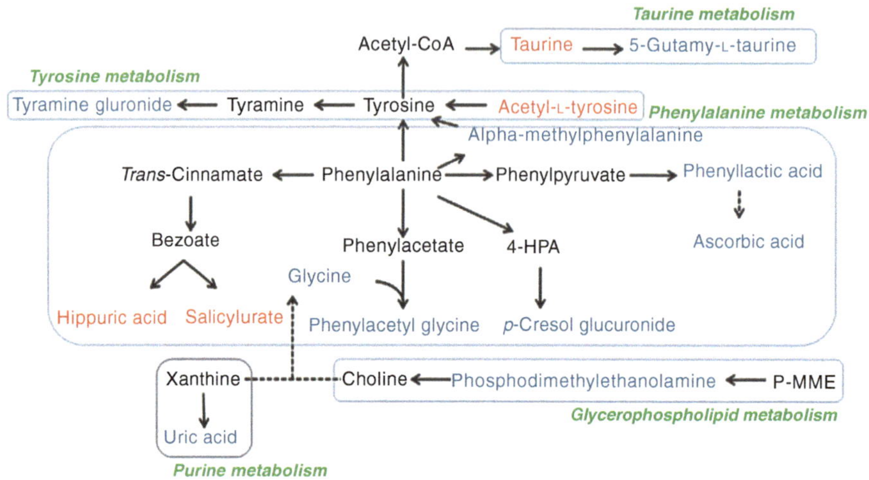

FIGURE 21.10 Correlation networks of the potential biomarkers, and disturbed metabolic pathways in response to prostatits induced by capsicine based on KEGG.
The multipoint and multipathway involve in a combined effect to the formation of NBP. Metabolites with a red dash represent a significant increase in the model group as compared to the normal group. Metabolites displayed with blue dash reppresent a significant decrease in the model group compared to the normal group.

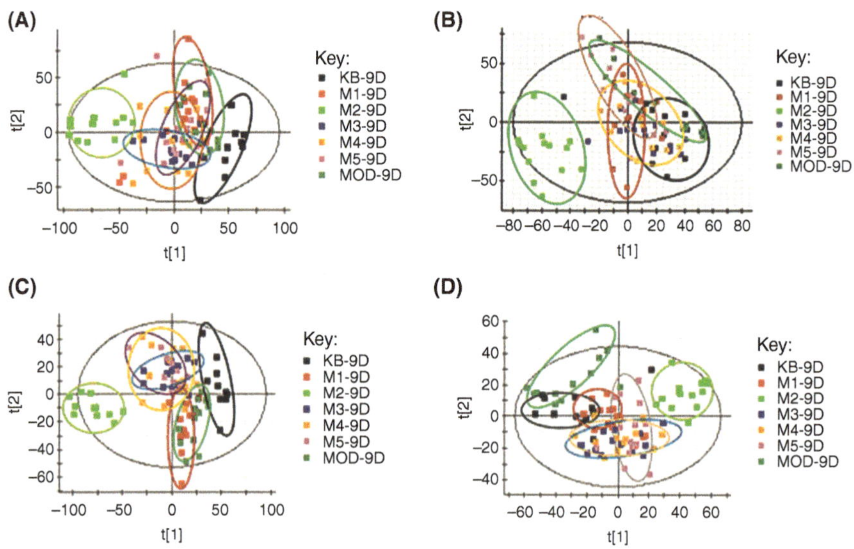

FIGURE 21.11
PCA results for all groups in day 9: (A) in the positive ion mode; (B) in the negative ion mode; (C) PLS-DA scores plots for all groups on, day 9 in negative mode; (D) negative mode; (E) KB, vehicle; MOD, model; M1, nexrutine extra; M2, GHB; M3, berberine; M4, obaculactone; M5, berberine + obaculactone.

or its main components, the score plot of OPLS showed that there was a significant difference in the metabolic profile of the NBP model group, GHB-treated group, berberine, and obaculactone, berberine + obaculactone treated group, and the control group in the direction of component T1. This difference confirmed that GHB, and its main components can play a role in regulating the abnormal metabolic network of the NBP rat.

The relative mean height intensity of different metabolites was graphed in Fig. 21.12. By comparing the identified biomarker level in the NBP group, 12 of the identified biomarkers, including taurine, slicylurate, tyramine glucuronide,

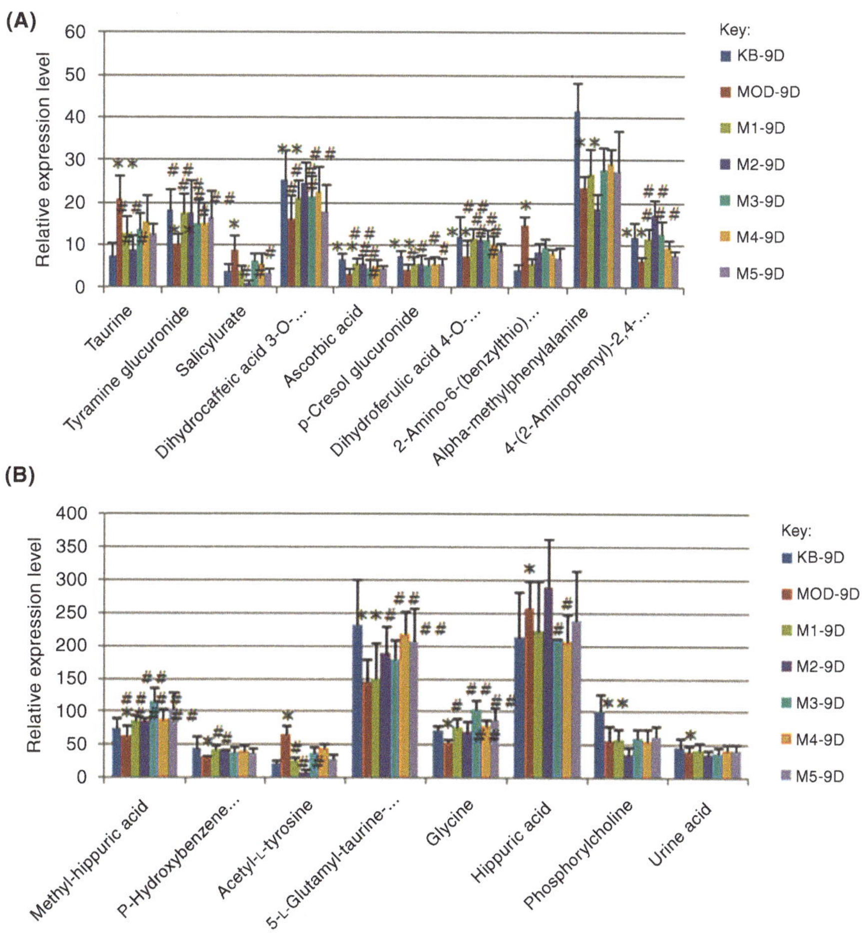

FIGURE 21.12 Differential expression levels of biomarkers in nonbacterial prosatatitis on intervention effects from all dosed groups.

KB, control; MOD, model; M1, nexrutine; M2, GHB; M3, berberine; M4, obaculactone; M5, berberine + obaculactone.

dihydrocaffeic acid 3-O-glucuronide, ascorbic acid, *p*-cresol glucuronide, dihydroferulic acid 4-O-glucuronide, 4-(2-aminophenyl)-2,4-dioxobutanoic acid, methyl-hippuric acid, phenyllactic acid, acetyl-L-tyrosine, and glycine are completely reversed by nexrutine-treated group; 12 of the identified biomarkers, including taurine, tyramine glucuronide, slicylurate, dihydrocaffeic acid 3-O-glucuronide, ascorbic acid, *p*-cresol glucuronide, dihydroferulic acid 4-O-glucuronide, 4-(2-aminophenyl)-2,4-dioxobutanoic acid, methyl-hippuric acid, phenyllactic acid, acetyl-L-tyrosine, and 5-L-Glutamyl-taurine are completely reversed by GHB (Fig.21.12); seven of the identified biomarkers, including tyramine glucuronide, dihydrocaffeic acid 3-O-glucuronide, dihydroferulic acid 4-O-glucuronide, 4-(2-aminophenyl)-2,4-dioxobutanoic acid, methyl-hippuric acid, glycine, and hippuric acid are completely reversed by berberine; nine of the identified biomarkers, including tyramine glucuronide, dihydrocaffeic acid 3-O-glucuronide, ascorbic acid, *p*-cresol glucuronide, dihydroferulic acid 4-O-glucuronide, methyl-hippuric acid, 5-L-glutamyl-taurin, glycine, and hippuric acid are completely reversed by obaculactone; six of the identified biomarkers, including tyramine glucuronid, slicylurate, *p*-cresol glucuronide, methyl-hippuric acid, 5-L-glutamyl-taurin, and glycine are completely reversed by berberine + obaculactone-treated group. Interestingly, according to the parametric t-test, we found that the relative concentration of these metabolites could be reversed after taking the GHB. Compared with the alterations of the NBP-related metabolites, most of them were reset to a normal state after GHB treated administration. GHB and its main components, which exert a multipathway, and a multitarget role in the body, effectively regulated the metabolic disorders of the rat with NBP. The relative content of the 12 biomarkers in the GHB-treated group were prominently regulated, and the content of these differential metabolites trended to the level of the control group (Fig. 21.12). GHB effectively repaired the metabolic network, thus alleviating the protective effects of the differential metabolites on immune response, and inflammation, ameliorating the dysfunction of the immune system, and the multiorgan inflammatory response during the development of NBP, thereby inhibiting the progression of NBP.

21.3.9 Hierarchical Clustering Analysis for GHB and its Main Components on NBP

To determine the variables that contribute to discrimination, the hierarchical clustering analysis (HCA), a supervised pattern recognition method was further employed for metabolic data. Heatmap plot (Fig. 21.13) of the 18 significantly changed metabolites (right side) among the cluster groups (top) among the control, and the NBP groups showed the cluster identification, and relative signal intensities. Hierarchical clustering analysis results showed distinct segregation between the control, the model group, and the berberine, and obaculactone, berberine + obaculactone dose groups

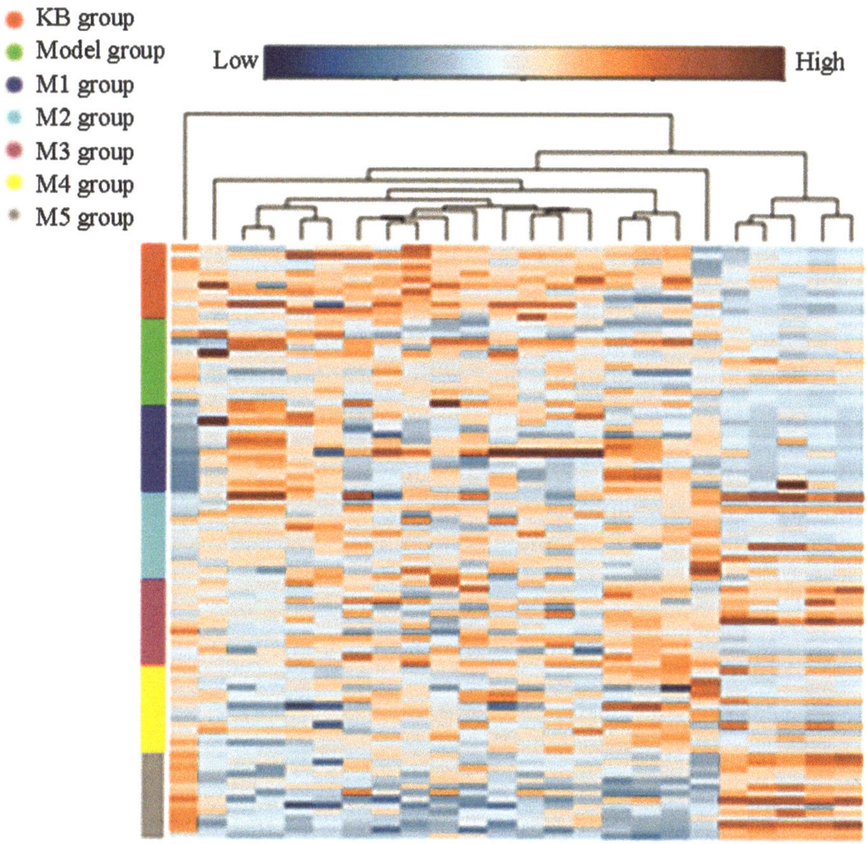

FIGURE 21.13 Heatmap clustering analysis visualization of the intervention effects of all dosed groups.

(Fig. 21.13). Combined with the results of the biological parameters assay, the HCA results display that the model group was far away from the remaining four groups, indicating that the changed metabolic pattern resulted from the NBP-induced may be significantly different from the others. The position of the treatment group was near to the control group, suggesting that the changed metabolic pattern was caused by the dose groups. The results showed that GHB, and its main components could change the abnormal metabolic status, and may have a different treatment mechanism of NBP. Metabolomics represents an emerging and powerful discipline that provides an accurate, and dynamic picture of the phenotype of biosystems through the study of the potential biomarkers of NBP that could be used for therapeutic targets, and the discovery of new drugs.

21.4 CONCLUSIONS

In this study, UPLC-Q-TOF/MS-based urine metabolomics analysis combined with multivariate statistical analysis was applied to evaluate the pathogenesis of NBP rats, and the mechanism of action of GHB, and its main components. As a result, 18 differential metabolites in the NBP model were identified by the MS, and MS/MS information. Compared with the alterations of NBP-related metabolites, most of them were reset to a healthier level after GHB, and its main components administration. Our findings also show that GHB exhibited preventive efficacy against NBP by adjusting these multiple metabolic pathways to their normal state. Particularly, GHB can effectively regulate the taurine and hypotaurine metabolism, glycine, serine and threonine metabolism, the pentose and glucuronate interconversions pathway, and exert a good therapeutic effect on NBP. These results provide a better understanding of the pathogenesis of NBP, and the mechanism of the action of GHB against NBP. It also proved that metabolomics is a powerful technology platform for studying the action mechanism of TCM, and investigating disease pathogenesis.

References

Chuang, Y.C., Yoshimura, N., Wu, M., Huang, C.C., Chiang, P.H., Tyagi, P., Chancellor, M.B., 2007. Intraprostatic capsaicin injection as a novel model for nonbacterial prostatitis and effects of botulinum toxin A. Eur. Urol. 51 (4), 1119–1127.

Chuang, Y.C., Yoshimura, N., Huang, C.C., Wu, M., Chiang, P.H., Chancellor, M.B., 2008. Intraprostatic botulinum toxin a injection inhibits cyclooxygenase-2 expression and suppresses prostatic pain on capsaicin induced prostatitis model in rat. J. Urol. 180 (2), 742–748.

Hu, Y., Xiong, L., Huang, W., Cai, H., Luo, Y., Zhang, Y., Lu, B., 2014. Anti-inflammatory effect and prostate gene expression profiling of steryl ferulate on experimental rats with non-bacterial prostatitis. Food Funct. 5 (6), 1150–1159.

Wang, H., Sun, H., Zhang, A., Li, Y., Wang, L., Shi, H., Dizou, X.L., Wang, X., 2013a. Rapid identification and comparative analysis of the chemical constituents and metabolites of Phellodendri amurensis cortex and Zhibai dihuang pill by ultra-performance liquid chromatography with quadrupole TOF-MS. J. Sep. Sci. 36 (24), 3874–3882.

Wang, H., Yan, G., Zhang, A., Li, Y., Wang, Y., Sun, H., Wu, X., Wang, X., 2013b. Rapid discovery and global characterization of chemical constituents and rats metabolites of Phellodendri amurensis cortex by ultra-performance liquid chromatography-electrospray ionization/quadrupole-time-of-flight mass spectrometry coupled with pattern recognition approach. Analyst 138 (11), 3303–3312.

Wang, L., Yan, G., Zhang, A., Shi, H., Sun, H., Wang, X., 2014. Fingerprinting and simultaneous determination of alkaloids and limonins in Phellodendri Amurensis Cortex from different locations by high-performance liquid chromatography with diode array detection. J. Chromatogr. Sci., 2015;53(1):161–166.

Xian, Y.F., Mao, Q.Q., Ip, S.P., Lin, Z.X., Che, C.T., 2011. Comparison on the anti-inflammatory effect of Cortex Phellodendri Chinensis and Cortex Phellodendri Amurensis in 12-O-tetradecanoyl-phorbol-13-acetate-induced ear edema in mice. J. Ethnopharmacol. 137 (3), 1425–1430.

CHAPTER 22

Metabolic Profiling Provides a System for the Understanding of Alzheimer's Disease in Rats Post-Treatment With Kaixin San

Xijun Wang, Ying Han, Aihua Zhang, Hui Sun

CONTENTS

22.1 INTRODUCTION

Alzheimer's disease (AD) is a neurodegenerative disorder that currently affects nearly 5% of people aged 65-years-old and over, and 30% of those aged 85-years-old (Langbaum et al., 2014). It is now estimated that there are 18–24 million people suffering from AD worldwide, and this number is expected to reach 34 million by 2025 (Scott et al., 2014). The economic burden of AD is monstrous. Even worse, at present there are only symptomatic therapies in nature, providing modest improvements in memory, without altering the progression of the disease pathology. Most previous studies have focused on the examination of Aβ and Tau in cerebrospinal fluid as markers in AD patients (Kumar and Nisha, 2014). Reports on other molecules of Aβ generation, transportation, and degradation as markers for the early diagnosis of AD are limited. New biomarkers are attracting increasing attention in research into early diagnosis, or as surrogate measures of AD. Today, there is no cure for this devastating disease, and therefore it is of great interest for researchers to find new drugs that can hinder the disease process. Drugs currently available on the market include different inhibitors of acetylcholine esteras, e and *N*-methyl D-aspartate-receptor antagonist (Ghezzi et al., 2013). Because of the side effects, safe and effective new drugs for AD are required. Herbal medicines, which have been used for thousands of years in many countries, provide an immense drug library for searching new drug leads, especially Traditional Chinese Medicine (TCM) that is regarded as natural and harmless, and has been accepted gradually in the world in recent years. TCM, a unique medical system with the significant characteristics of pursuing an overall therapeutic effect with a multitarget treatment, attempts to improve therapeutic efficacy, and to reduce the drug-related adverse effects at the same time (Zhang et al., 2014a).

Formulae – a combination of plant species – are designed on the basis of patients' symptoms, clinical experience, and TCM theories, and have been used

Chinmedomics. http://dx.doi.org/10.1016/B978-0-12-803117-9.00022-6

for centuries to enhance therapeutic efficacy, and reduce adverse effects (Wang et al., 2013). It is characterized by multicomponent, multitarget, and synergistic therapeutic efficacies, but the working mechanisms remain unknown. Unlike western medicine, TCM is a "system–to–system" treatment mode that focuses on both the pathological, and physiological changes. The systemic feature of metabolomics is consistent with the holistic characteristics of TCM (Cao et al., 2014). Metabolomics, monitors the entire pattern of low molecular weight compounds rather than simply focusing on individual metabolites, and focuses on the dynamic small molecule response in biological systems, and provides insights into the global metabolic status (which is coincident with the integrity, and systemic features of TCM) thus it is an effective approach to investigate how formulae works. In comparison with conventional research tools, ultra performance liquid chromatography coupled with quadrupole time-of-flight mass spectrometry (UPLC-Q-TOF/MS) has widely been used for metabolomic analysis, owing that UPLC possesses great peak capacity, and chromatographic resolution due to its special column with a stationary phase of 1.8 μm particles, and Q-TOF/MS can provide the chromatographic peak area, accurate mass value of m/z, and abundant structural information (Zhang et al., 2014b). To date, there are wide applications of the metabolomic approach in the field of TCM, which strive to illustrate the effective mechanism of TCM on diseases, and make further efforts to develop new drugs for complex diseases from the insights of TCM theory.

Kaixin San (KXS), is a famous classic Chinese herbal prescription for dementia, and was first recorded in the Chinese ancient medical prescription book *Qian-Jin-Yao-Fang*, written around 1300 years ago.. It is a TCM prescription consisting of four herbs: *Ginseng Radix*, *Polygalae Radix*, *Poria*, and *Acori Tatarinowii Rhizoma*. In modern times, KXS is still used continuously, and widely for the treatment of dementia, Alzheimer's Disease, Parkinson's Disease, etc., (Zhu et al., 2012, 2013; Mu et al., 2011). Recently, chemical analysis study of KXS has been reported that the following: polygalaxanthone III, ginsenoside Rb1, ginsenoside Rd, ginsenoside Re, and ginsenoside Rg1 were determined in the plasma of a rat postoral administration of KXS by ultrafast liquid chromatography with tandem mass spectrometry (Lv et al., 2014). Though many compounds have been isolated, and identified from KXS, evaluating the holistic efficacy, and the clarifying mechanism of the pharmacological action of KXS remains a difficult task, due to the mistiness of the active compounds, and the unknown synergistic actions of multiple components. New methods for activity evaluation, and molecular pathway identification of KXS are urgently required. Thus, in this study, an UPLC-Q-TOF-MS method was established to gain the global profiling of the endogenous metabolites in the urine of AD rats. The metabolite pathway changes posttreatment with KXS were investigated using pattern recognition approaches, adding a further understanding of how KXS could modify the AD symptom.

22.2 EXPERIMENTAL

22.2.1 Chemicals and Materials

Acetonitrile, HPLC grade, was obtained from Merck (Darmstadt, Germany); methanol, HPLC grade, was purchased from Fisher Scientific Corp. (USA); leucine enkephalin was purchased from Sigma–Aldrich (St. Louis, MO, USA). The deionized water (18.2 MΩ) was further purified using a Milli-Q system (Millipore, Billerica, USA); formic acid, HPLC grade, was purchased from Kermel Chemical Reagent Co., Ltd (Tianjin, China); $AlCl_3$ and D-galactose, analytical grade, was purchased from Tianjin Guangfu Fine Chemical Research Institute (Tianjin, PRC), Beijing Biosharp Biotechnology Company (Beijing, PRC). *Ginseng Radix, Poria, Polygalae Radix,* and *Acori Tatarinowii Rhizoma* were purchased from Harbin Tongrentang Drug Store (Harbin, China), and authenticated by Professor Xijun Wang, at the Department of Pharmacognosy of Heilongjiang University of Chinese Medicine. Voucher specimens were deposited at the authors' laboratory.

22.2.2 Preparation of KXS Samples

The KXS was prepared in the following procedure: *Ginseng Radix, Poria, Polygalae Radix,* and *Acori Tatarinowii Rhizoma* in the proportions 3:3:2:2 were ground into crude powders, mixed, and then a refluxed extraction in a rotary evaporator with six times the volume of 70% ethanol for 2 h twice, then the filtrate was dried under freeze-drying. The freeze-dried powder was dissolved in water and made concentrations of 1.08, 0.54, 0.27 g/mL solution, respectively.

22.2.3 Establishing Models of Rat With AD

$AlCl_3$ (Al) and D-galactose (D-gal) were used to induce the rat model for AD that was induced in the rats according to the method in a previous report with a slight modification (Luo et al., 2009). The rats were given Al orally (28 mg/kg/day) and injections of D-gal intraperitoneally (63 mg/kg/day) for 105 days to establish the AD animal model. To evaluate the learning/memory ability of the rats, a water maze behavioral test was performed at 0, 15, 30, 45, 60, 75, 90, and 105 days after the treatment. The daily administration of D-gal and Al was continued during the water maze performance tests.

22.2.3.1 Morris Water Maze

The Morris water maze test was applied to measure the cognitive functioning of the rats. The water maze model was used in a circular tank (diameter 1 m) filled with water. A platform was submerged below the water's surface in the center of the target quadrant. Swimming activity of the rats was monitored via a video camera mounted overhead, and automatically recorded via a video tracking system. At the beginning of each trial, the rats were placed into the water facing the wall of the pool at one of the four quadrants. Each rat was allowed 60 s to find, and mount the platform. When the rat found the platform, it was kept on the platform for 10 s. If the rat did not find the platform within 60 s, it was gently placed on it, and left there for 10 s. The latency to the platform, and the proportion of time spent in the target quadrant were recorded.

22.2.3.2 Place Navigation

The Morris water maze experiment lasted 5 days. The platform was set in the center of the maze in the northeast quadrant. The horizontal plane was designed to be 2 cm higher than the platform. The water temperature was kept at 25 ± 1°C. The rats in each group were trained for 5 days continuously. The model rats were set free from the west, and south points once a day; the longest swimming time was limited to 60 s. The time required to find the platform was the escape latency. We also observed the strategies used by the rats to get to the platform. The rats'activities within 60 s were analyzed using computer software.

22.2.3.3 Spatial Probe Test

The platform was removed on the day 6, and the model rats were placed in the water at the opposite quadrant. In this probe trial, the rats were put into the pool, and allowed to swim freely in the pool for 60 s. The number of crossings of the location of the platform was recorded. The number of times swimming across the original platform position was analyzed using computer software.

22.2.4 Animal Handling and Sample Preparation

Male Wistar rats (weighing 300 ± 20 g, 9 w) were supplied by the GLP Center of Heilongjiang University of Chinese Medicine (Harbin, China). The room temperature was regulated at 21 ± 5°C with 60 ± 5% humidity. A 12 h light–dark cycle was set, and there was free access to a standard diet and water. The animals were allowed to acclimatize in the metabolism cages for 7 days prior to dosing, and were put in metabolism cages during the urine collection period. After acclimatization, the animals were randomly divided into 5 groups with 10 rats in each group: control, model, KXS high dose group (1.08 g/mL, KXSH), KXS middle dose group (0.54 g/mL, KXSM), and KXS low dose group (0.27 g/mL, KXSL). Controls were subjected to oral gavage with distilled water, and injections of 0.9% saline solution intraperitoneally, whereas the AD-treated groups were orally administered Al solution, and injections of D-gal intraperitoneally at 14:00 p.m. for 105 days. The KXS treatment groups were orally administrated an accurate volume 10 mL/kg of KXS solution from day 90. After 15 days administration, all the rats were sacrificed, blood was collected for the following analysis. For each group, each rat was housed in a metabolism cage, and the urine samples were collected from each animal over the course. A standard diet was given to all animals, and free access to water throughout the studies. The urine samples were collected at 8:00 a.m. and centrifuged at 13,000 rpm for 5 min (−4°C), to remove solid debris, and the supernatant was stored at −80°C. Prior to UPLC–MS analysis, the urine samples were thawed at room temperature, and filtered through a 0.22 μm filter membrane before a UPLC-G2-Q-TOF/MS analysis. The experiments were carried out under the approval of the Committee of Experimental Animal Administration, of the Heilongjiang University of Chinese Medicine.

22.2.5 Metabolic Profiling

22.2.5.1 Chromatography

Chromatography analysis was performed on an Waters Acquity UPLC system coupled to a Xevo G2-Q-TOF (Waters MS Technologies, Manchester, UK) equipped with an electrospray ionization source operating in positive, and negative ion modes. An Acquity UPLC HSS T3 column (100 × 2.1 mm i.d., 1.8 μm, USA) equipped with a binary solvent delivery system, and an autosampler. The analysis was performed with a gradient elution using (A) acetonitrile with 0.1% formic acid, and (B) water with 0.1% formic acid as the mobile phase. The proportion of mobile phase A was optimized as follows: 0–2.5 min, linear increasing from 1–11% A; 2.5–4.5 min, 11–21% A; 4.5–7 min, 21–40% A; 7–8.5 min, 40–99% A. The column was maintained at 45°C and eluted at a flow rate of 0.40 mL/min. The sample injection volume was 2 μL. During the whole analysis, all the samples were maintained at 4°C. All the urine samples were mixed to get a quality control (QC) sample for method validation. A QC sample was injected every 10 samples during the analytical run to further monitor the system's stability. It was critical to acquire the QC data for assessing the changes in the analytical results, and the reliability of the metabolite fingerprint.

22.2.5.2 Mass Spectrometry

The ESI source in both positive and negative ion mode was used in the MS analysis. The source temperature was 110°C, and desolvation gas temperature was 450°C. Nitrogen was used as the cone and desolvation gas. The flow rates of cone and desolvation gas were set at 50 and 800 L/h, respectively. Capillary, cone, and extraction cone voltages were set at 3.0 kV, 20 V, and 5.0 V in positive ion mode, respectively; and at 2.0 kV, 30 V, and 5.0 V in negative ion mode, respectively. MS data was collected in full-scan mode in the mass range of 50–1000 Da. A lock-mass of leucine enkephalin (1 ng/μL) was used to set up mass spectrometer calibration via a lock spray interface at flow rate of 5 μL/min, monitoring in positive ion mode ($[M + H]^+ = 556.2771$) and in negative ion mode ($[M-H]^- = 554.2615$) to ensure accuracy during the MS analysis.

22.2.5.3 Data Processing and Multivariate Data Analysis

The UPLC-G2-Q-TOF/MS data of the urine samples were processed using Waters MassLynx (v4.1) and MarkerLynx software to transform the raw data into a single matrix containing aligned peaks with the *m/z* retention time pair along with the normalized peak intensities. The matrix was then introduced to EZinfo 2.0 software for principal component analysis (PCA), partial least-squares-discriminant analysis (PLS–DA), and orthogonal partial least-squared discriminant analysis (OPLS–DA). In order to gain an overview of the rat urine metabolic profiling; an unsupervised PCA method was used to give a comprehensive view of the clustering trends for the experimental groups. A supervised OPLS–DA analysis technique was used to search the biomarkers based on the variable importance in the projection (VIP) between the control and the model group. The integration areas of the detected metabolites with high VIP values were first tested for the normality of the distribution. If the distribution followed the normality assumption, a parametric Student's t-test was applied; otherwise, a nonparametric Mann–Whitney U test was performed to detect statistically significant metabolites that were increased or decreased between groups. Differences were considered significant at a value of $p < 0.05$.

22.2.5.4 Identification of Urine Biomarkers

The MassFragment™ application manager (Waters Corp., Milford, USA) was used to facilitate the MS/MS fragment ion analysis process by way of chemically intelligent peak-matching algorithms. For the identification of potential biomarkers, some available biochemical databases, such as HMDB (www.hmdb.ca/), KEGG (www.genome.jp/kegg/), METLIN (metlin.scripps.edu/), Massbank (www.massbank.jp), LIPIDMAPS (www.lipidmaps.org/) and Chemspider (www.chemspider.com) were used by comparing the accurate mass, and the fragments information obtained from the UPLC-Q-TOF/MS.

22.2.5.5 Pathway Analysis of Urine Metabolites

Metabolomic pathway analysis was performed with MetPA (metabolomics pathway analysis) based on potential metabolite biomarkers. MetPA is a user-friendly, web-based tool for pathway analysis, and the visualization of metabolomic data within the biological context of metabolic pathways. For the pathway analysis algorithms, the Hypergeometric Test was used for over-representation analysis, and relative-betweenness centrality was used for pathway topology analysis.

22.3 RESULTS AND DISCUSSIONS

22.3.1 Establishment of the AD Model

To validate the establishment of the animal model, Morris water maze testing (under the same conditions used in the place navigation test) were determined. Throughout the training, on day 105, the rat treated with Al and D-gal took longer to reach the platform than those of the control group. The escape latency from all four quadrants revealed significant differences between the control, and the model groups. There was no significant difference between the control and the model groups at 0, 15, 30, 45, 60, 75, and 90 days. From the statistics we found that administrating Al combined with D-gal for 105 days induced rat memory impairment, which mimic AD pathogenetics, and could therefore be used as an AD animal model.

22.3.2 KXS Improves Learning and Memory Ability in AD Rats

On the day 5, compared with the control group, the model group's escape latency was still prolonged. Compared with the model group, the KXSM, and the KXSL groups were shortened significantly. In the spatial probe test, compared with the model group, KXSH, KXSM, and KXSL group rats passed through the platform significantly fewer times. The administration of KXS can significantly improve learning ability, during the experimental period. Administration of KXS for a consecutive 15 days significantly decreased the latency to the platform, and increased the proportion of time spent in the target quadrant, as compared with the model group. These results showed that KXS had good anti-AD activity.

22.3.3 Urine Metabolomic Fingerprint Analysis

Using the optimal UPLC-G2-Q-TOF/MS condition described earlier, the base peak intensity (BPI) profiles of the urine samples of the control, the model, KXSH, KXSM, and KXSL in positive and negative ionization modes are shown in Figs. 22.1

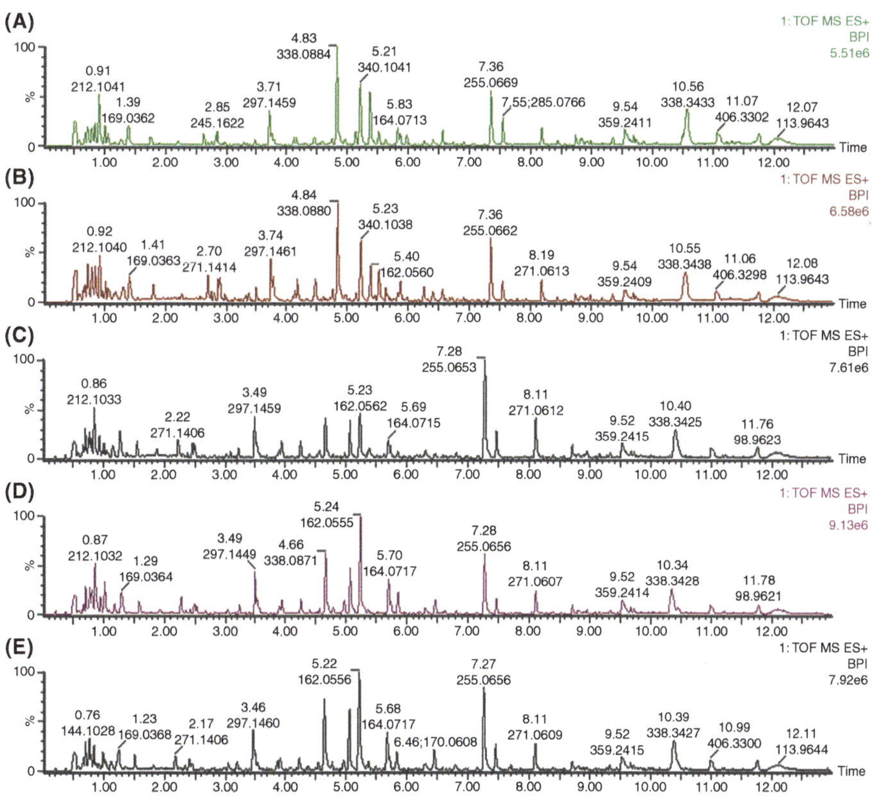

FIGURE 22.1 Typical urinary base peak chromatograms of: (A) control; (B) model; (C) KXSH; (D) KXSM; and (E) KXSL group, in positive mode.

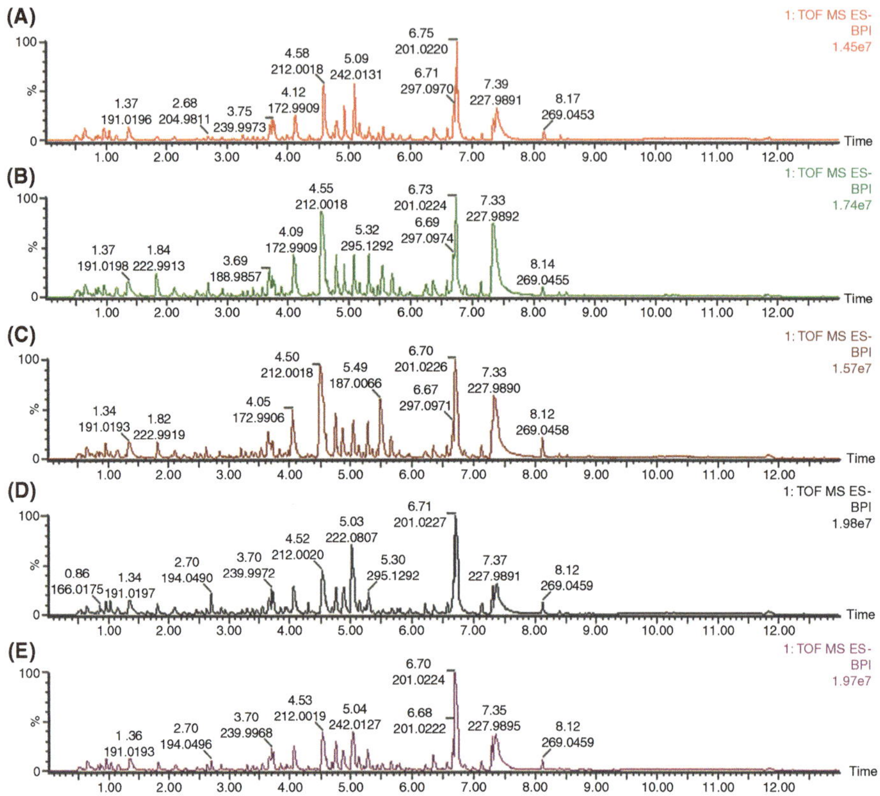

FIGURE 22.2 Typical urinary base peak chromatograms of: (A) control; (B) model; (C) KXSH; (D) KXSM; and (E) KXSL group, in negative mode.

and 22.2, respectively. To validate the stability of the established method, an inhouse QC sample was prepared by mixing an equal volume from each sample, and injecting every 10 samples during the whole sample analysis. A subset of peaks across the QC samples taking account of intensities, and retention time was examined. It was found that the relative standard deviation (RSD) was below 1% for the retention time, and below 5% for the peak area, in both ionization modes. To evaluate the repeatability, seven parallel samples obtained from a random urine sample were extracted by the same extraction method, and injected continuously. The same subset of peaks was tested, and the results showed that the RSD was below 0.5% for the retention time, and below 5% for the peak area, in both ionization modes. The results indicated that the method was reliable for the subsequent analysis.

22.3.4 Metabolic Profiling Analysis

22.3.4.1 Pattern Recognition Analysis

In spite of the differences in the BPI chromatogram that could be observed, the multivariate analysis method distinguished the experimental groups much better. PCA, PLS-DA, and OPLS-DA are frequently used pattern recognition approaches for metabolomics. Pareto scaling was used to reduce disturbances from noise, and artifacts in the models. To better understand the time-related trajectory of the metabolic pattern with Al and D-gal treatment from the very beginning of the established AD model to the end of treatment: a score plot of a PCA map from the AD group at various time points was illustrated, see Fig. 22.3. In the map, each spot represented a sample, and each assembly of samples indicated a particular metabolic pattern at different time points. The PCA plot of the model group deviated from the control group on day 105, and reached the maximum trend during day 75 to day 90; gradually keeping away to the normal state. The starting spots of the AD model group obviously deviated from those postadministration of Al and D-gal, suggesting that high fluctuations occurred in the metabolic regulatory network. The tracks of the metabolic profiles at different time points also clearly demonstrate the time-dependent changes in the urine metabolites. Typically, the trajectory analysis of the PCA score plots for the AD rats effectively reflects the differences between the model and the control groups.

22.3.4.2 Discovery and Identification of Urine Biomarkers Related to AD

All the collected samples were analyzed, and the low molecular weight metabolites were represented as the chromatographic peaks in the BPI chromatograms. To screen the statistically important variables (ions) related to the model rats, Progenesis QI was used to deconvolute, and align mass ions from the data files of the samples into a single data set. The result is a table that contains the m/z, and retention time coordinates, p-value, and the fold change for each feature, and the integrated feature intensities from all the aligned samples. Progenesis QI showed the detailed information for each individual feature including statistics, extracted ion chromatograms, spectrum details, and putative identifications. In this study, the metabolite profiling obtained from the model group at day 105 was performed by UPLC-MS in conjunction with Progenesis QI software analysis. The datasets from positive, and negative ionization modes contained 9434 and 9436 variables. In the OPLS-DA model, the variable importance in the projection (VIP) value was employed to identify the features contributing to group separation. The ions, with VIP values above 1.5, and p-value below 0.05, were considered as potential biomarkers. Following the criteria above, 11 metabolite ions were selected as potential biomarkers related with AD rats, and are listed in Table 22.1. Among them, 4 of the 11 biomarkers identified were up-regulated, and seven of them were down-regulated in the

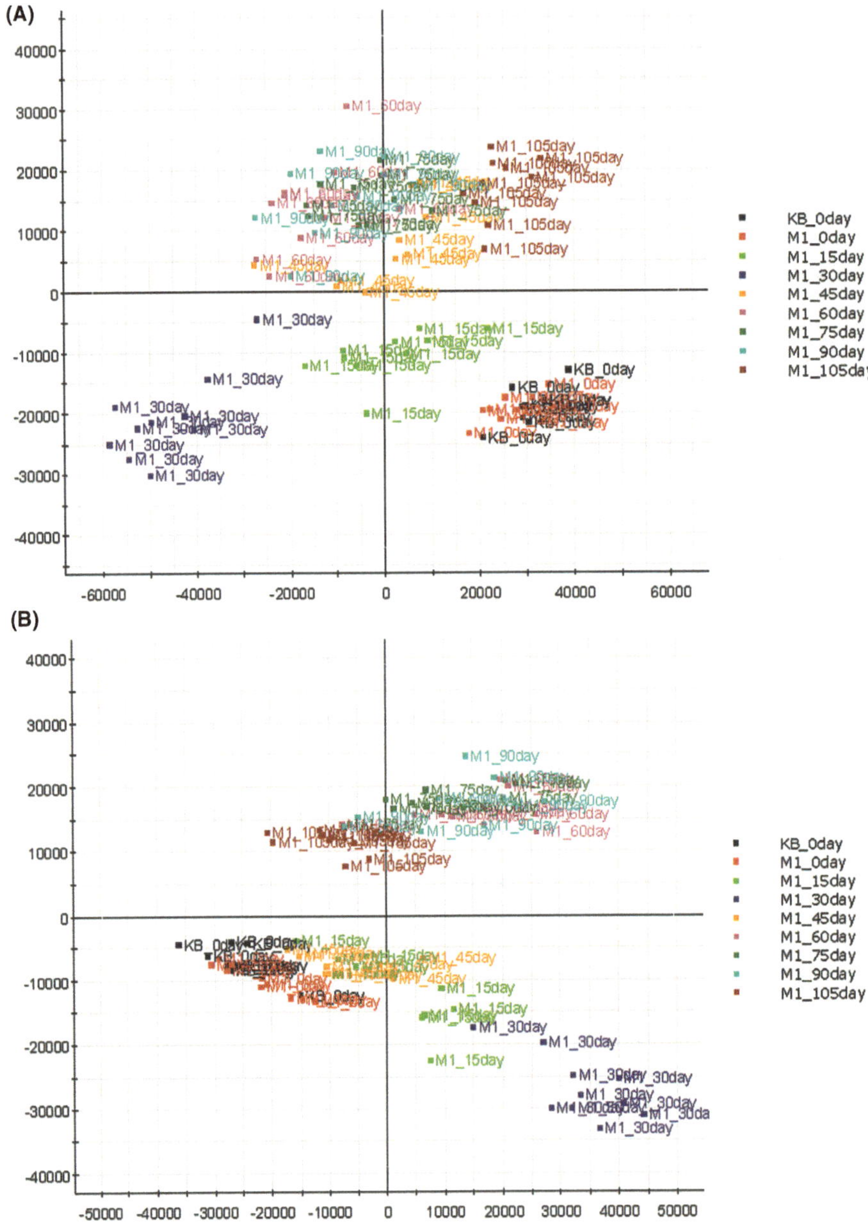

FIGURE 22.3 PCA score plots of trajectory analysis of urine metabolic profiling of AD rats. (A) positive ion mode; (B) negative ion mode.

Note: KB, control group; M1, model group.

Table 22.1 Information Relating to AD Biomarkers as Detected by UPLC-G2-Q-TOF/MS in Positive and Negative Ion Mode

No.	R_t min	*m/z* Deter-mined	*m/z* Calcu-lated	Scan Mode	Actual_M	Proposed Composition	Name	Trend	MS/MS	Losses
1	1.36	420.9714	420.9702	ESI^+	419.9624	$C_6H_{15}O_{15}P_3$	Inositol 1,4,5-trisphosphate	↓	382.9334 288.9878 179.0556	$-H_6O_2$ $-H_5O_6P$ $-H_5O_9P_3$
2	2.06	346.0558	346.0553	ESI^-	347.0631	$C_{10}H_{14}N_5O_7P$	Deoxyguanosine 5′-monophosphate (dGMP)	↑	267.0968 210.0879 134.0116	$-O_3P$ $-CHN_2O_4P$ $-C_4H_{11}N_3O_5P$
3	3.41	181.0971	181.0977	ESI^+	180.0898	$C_9H_{12}N_2O_2$	3-Hydroxykynurenamine	↑	137.0477 135.0320	$-C_2H_6N$ $-C_2H_8N$
4	3.69	178.0517	178.0504	ESI^-	179.0582	$C_9H_9NO_3$	Hippuric acid	↓	160.0399 136.0762 120.0449 107.0497	$-HO$ $-CO_2$ $-C_2H_2O_2$ $-C_2HNO_2$
5	3.71	192.1022	192.1025	ESI^+	191.0946	$C_{11}H_{13}NO_2$	5-Methoxytryptophol	↑	175.0633 165.0916 148.0762	$-CH_5$ $-CHN$ $-C_2H_4O$
6	3.87	191.0194	191.0192	ESI^-	192.0270	$C_6H_8O_7$	Citric acid	↓	174.0164 149.0450 130.9981 115.0031	$-HO$ $-CO_2$ $-C_2H_4O_2$ $-C_2H_4O_3$
7	4.93	191.0198	191.0192	ESI^-	192.0270	$C_6H_8O_7$	Isocitrate	↓	174.0164 149.0450 130.9981 115.0031	$-HO$ $-CO_2$ $-C_2H_4O_2$ $-C_2H_4O_3$
8	5.79	206.0811	206.0817	ESI^+	205.0738	$C_{11}H_{11}NO_3$	5-Methoxyindole-acetate	↑	188.0712 170.0242 148.0762	$-H_2O$ $-CH_8O$ $-C_2H_2O_2$
9	6.54	168.0448	168.0426	ESI^-	169.0504	$C_4H_{12}NO_4P$	Phosphorylcholine	↓	168.0426 153.0555 141.0191	$-CH_3$ $-CH_2O$ $-C_3H_6$
10	7.37	209.0633	209.0670	ESI^+	208.0592	$C_8H_{16}O_2S_2$	Lipoic acid	↓	156.0609 147.0480 85.0290	$-H_5OS$ $-C_2H_6S$ $-C_4H_{12}S_2$
11	8.77	300.2898	300.2903	ESI^+	299.2824	$C_{18}H_{37}NO_2$	Sphingosine	↓	257.2355 230.2120 212.1651	$-C_3H_7$ $-C_5H_{10}$ $-C_6H_{16}$

AD group that was shown in Fig. 22.4. To identify these metabolites, these variables were predicted by comparing the accurate MS, and MS/MS fragments with the online databases. The detailed method for the compound identification was described in our previous work. In brief, the corresponding quasi-molecular ion peak was found according to the accurate mass, and retention

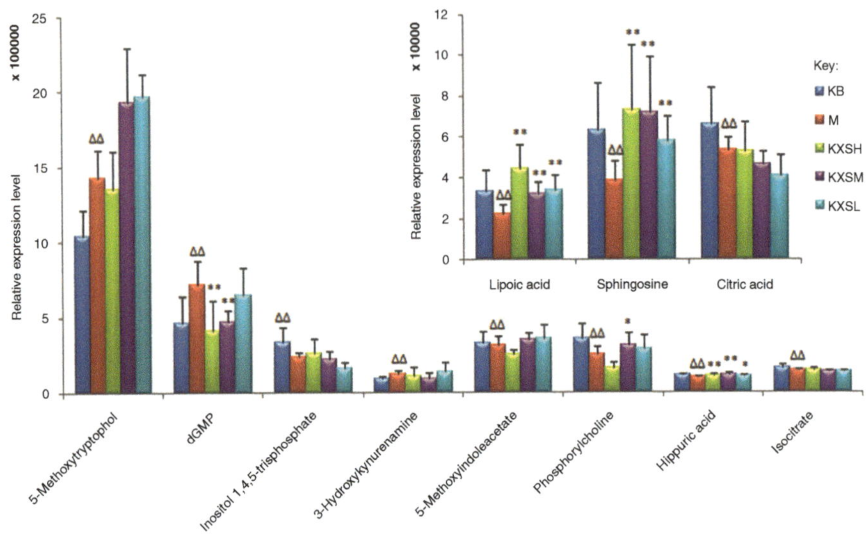

FIGURE 22.4 **Relative intensity of urine potential biomarkers identified by UPLC/MS.**

time in the extracted ion chromatogram, and speculated as the most probable molecular formula.

22.3.4.3 *Metabolic Pathway Analysis*

To determine whether our observations of changes in the metabolites in the setting of AD in fact reflected coordinate changes in the defined metabolic pathways, we used bioinformatics analysis MetPA software to identify the network pathway. We used the high-quality KEGG metabolic pathways as the backend knowledge base to identify the most relevant pathways (as listed in Table 22.2)

Table 22.2 Pathway Analysis Results According to the KEGG PATHWAY Tool

Pathway	Total	Expected	Hits	Raw *p*	Holm Adjust	FDR	Impact
Glyoxylate and dicarboxylate metabolism	16	0.1141	2	0.0052	0.42219	0.3291	0.2963
Inositol phosphate metabolism	26	0.1855	1	0.1712	1	1	0.1586
Citrate cycle (TCA cycle)	20	0.1427	2	0.0081	0.6502	0.3291	0.0949
Sphingolipid metabolism	21	0.1498	1	0.1405	1	1	0.0526
Glycerophospholipid metabolism	30	0.2134	1	0.1950	1	1	0.0444
Purine metabolism	68	0.4850	1	0.393	1	1	0.0076
Tryptophan metabolism	41	0.2924	2	0.0324	1	0.8740	0

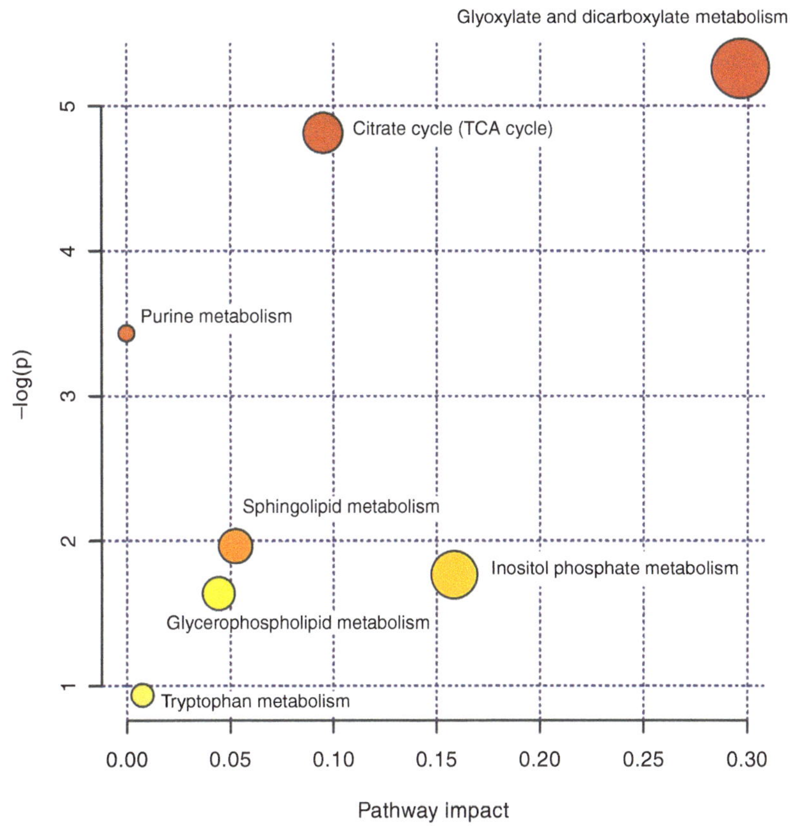

FIGURE 22.5 Summary of pathway analysis with the MetPA tool.
The map was gained by analyzing the known metabolic pathways.

for the model group was identified. Of the distinct metabolites identified from these pathways, many are in the various stages of progress of the AD. As shown in Fig. 22.5, the selected metabolites that were involved were glyoxylate and dicarboxylate metabolism, inositol phosphate metabolism, citrate cycle (TCA cycle), sphingolipid metabolism, suggesting that the target pathways might be the metabolic perturbations during AD, and might therefore be the targets for the treatment of AD. Our results showed that KXS could cause metabolite changes in the urine of AD rats. The content of potential biomarkers, such as lipoic acid, sphingosine, deoxyguanosine 5′-monophosphate, hippuric acid, and phosphorylcholine in KXS was higher or lower than in the AD rats. These metabolites were considered as biomarkers for the effect of KXS treatment. These changes may not occur immediately in response to the therapeutic effects of KXS for the AD rats.

22.3.5 Therapeutic Effects of KXS Based on Metabolite Profiling

The urinary metabolic profiles of the KXS-treated group were obtained using the presented UPLC–MS method. As shown in the PCA scores map as derived from the metabolic profile (Fig. 22.6), the KXSH, KXSM group is closer to the control group, which indicated the AD tended to recover. Furthermore,

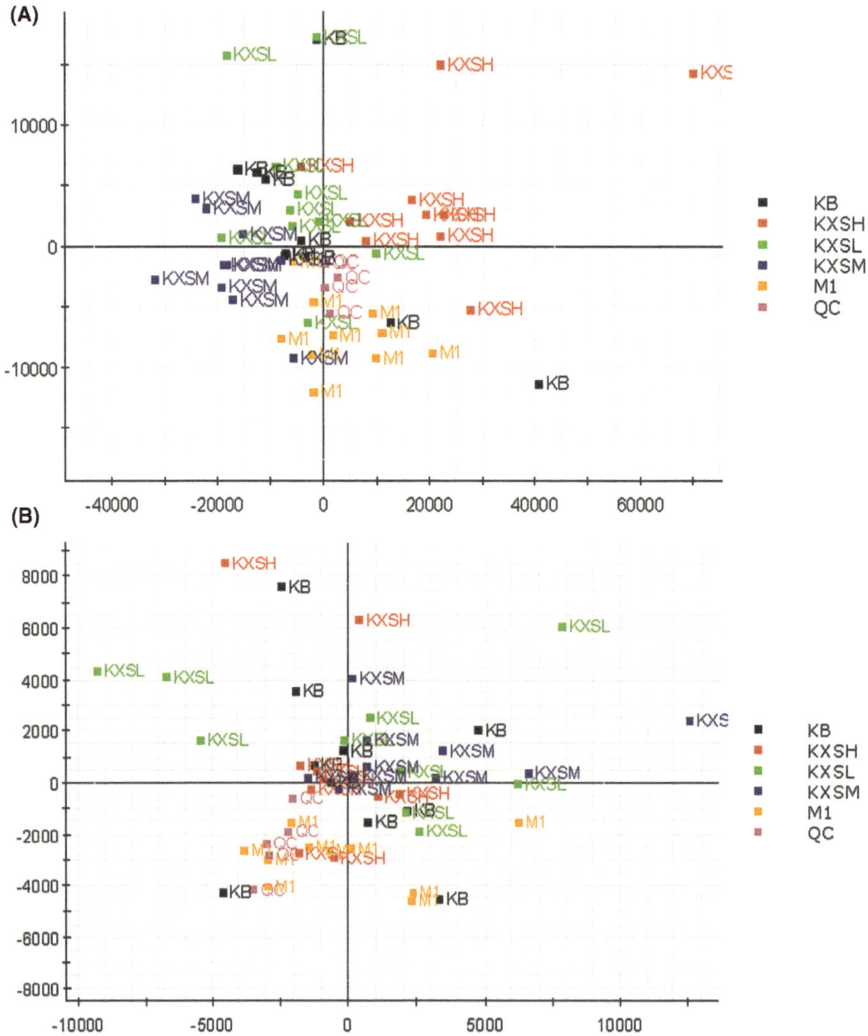

FIGURE 22.6 PCA score plots of urine samples among control, model, KXSH, KXSM, KXSL groups.

(A) Positive ion mode; (B) negative ion mode.

the spots of the KXS groups at day 105 treatment clustered near the control spots gradually, showing a backward tendency towards the normal status that might be an indication of the accumulated effect of KXS. To further evaluate the reversed condition of the potential biomarkers when KXS treated, the relative intensity of the 11 biomarkers (the relative peak areas of the 11 biomarkers to their respective total integrated area of the spectra) between the five groups were tested by ANOVA, and the Tukey posthoc test for comparisons of multiple groups. Similar to the results of the PCA score map change, 5 of the 11 detected metabolites also illustrated a return toward the control group values after 105 days of recovery. The other biomarkers were also regulated back to the relative normal status in varying degrees. Using the potential biomarkers found in this study as the criteria, we found that KXS had obvious recovery effects on AD through reversing the partial disturbed metabolic pathways. The KXS group at day 105 of the treatment ultimately approached the initial state, suggesting that KXS showed the recovering effects on AD to some extent. All the results above suggested that KXS treatment could effectively regulate some metabolic networks associated with their related metabolites in the AD rats.

22.4 CONCLUSIONS

In summary, we discovered a significant perturbance of metabolic profile in the urine of the AD rats, which could obviously be reversed by KXS treatment. Based on the metabolic pathway analysis, we concluded that the treating AD mechanism of KXS lay on correcting the perturbed metabolism; according to a variety of metabolic pathways. Eleven potential biomarkers had been identified, and five of them were considered to be correlated with the KXS therapeutic effects for the AD rats involved in glyoxylate, and dicarboxylate metabolism, inositol phosphate metabolism, citrate cycle (TCA cycle), and sphingolipid metabolism. The changes of the urinary endogenous metabolites showed that the AD rats were moving toward that of the healthy rats, posttreatment with KXS. These potential metabolites related with the efficacy of KXS in the AD model need to be further investigated to understand the mechanisms of KXS in treating AD. These results provide a better understanding of the pathogenesis of AD, and the action mechanism of KXS on AD. Taken together, the findings obtained in the current study can be used to predict some possible targets for AD, and KXS treatment. Furthermore, our study also highlights the importance of metabolomics as a potential tool for uncovering metabolic pathways to discover targets for understanding the holism of the TCM formulae.

References

Cao, H., Zhang, A., Zhang, H., Sun, H., Wang, X., 2014. The application of metabolomics in traditional Chinese medicine opens up a dialogue between Chinese and Western medicine. Phytother Res. 2015; 29 (2): 159–166.

Ghezzi, L., Scarpini, E., Galimberti, D., 2013. Disease-modifying drugs in Alzheimer's disease. Drug Des. Devel. Ther. 7, 1471–1478.

Kumar, N.S., Nisha, N., 2014. Phytomedicines as potential inhibitors of β amyloid aggregation: significance to Alzheimer's disease. Chin. J. Nat. Med. 12 (11), 801–818.

Langbaum, J.B., Hendrix, S.B., Ayutyanont, N., Chen, K., Fleisher, A.S., Shah, R.C., Barnes, L.L., Bennett, D.A., Tariot, P.N., Reiman, E.M., 2014. An empirically derived composite cognitive test score with improved power to track and evaluate treatments for preclinical Alzheimer's disease. Alzheimers Dement. 10 (6), 666–674.

Luo, Y., Niu, F., Sun, Z., Cao, W., Zhang, X., Guan, D., Lv, Z., Zhang, B., Xu, Y., 2009. Altered expression of Abeta metabolism-associated molecules from D-galactose/AlCl(3) induced mouse brain. Mech. Ageing Dev. 130 (4), 248–252.

Lv, C., Li, Q., Zhang, X., He, B., Xu, H., Yin, Y., Liu, R., Liu, J., Chen, X., Bi, K., 2014. Simultaneous quantitation of polygalaxanthone III and four ginsenosides by ultra-fast liquid chromatography with tandem mass spectrometry in rat and beagle dog plasma after oral administration of Kai-Xin-San: application to a comparative pharmacokinetic study. J. Sep. Sci. 37 (9–10), 1103–1110.

Mu, L.H., Huang, Z.X., Liu, P., Hu, Y., Gao, Y., 2011. Acute and subchronic oral toxicity assessment of the herbal formula Kai-Xin-San. J. Ethnopharmacol. 138 (2), 351–357.

Scott, T.J., O'Connor, A.C., Link, A.N., Beaulieu, T.J., 2014. Economic analysis of opportunities to accelerate Alzheimer's disease research and development. Ann. N. Y. Acad. Sci. 1313, 17–34.

Wang, X., Zhang, A., Wang, P., Sun, H., Wu, G., Sun, W., Lv, H., Jiao, G., Xu, H., Yuan, Y., Liu, L., Zou, D., Wu, Z., Han, Y., Yan, G., Dong, W., Wu, F., Dong, T., Yu, Y., Zhang, S., Wu, X., Tong, X., Meng, X., 2013. Metabolomics coupled with proteomics advancing drug discovery toward more agile development of targeted combination therapies. Mol. Cell. Proteomics 12 (5), 1226–1238.

Zhang, A., Sun, H., Wang, X., 2014a. Potentiating therapeutic effects by enhancing synergism based on active constituents from traditional medicine. Phytother. Res. 28 (4), 526–533.

Zhang, A., Zhou, X., Zhao, H., Guan, Y., Zhou, S., Yan, G.L., Ma, Z., Liu, Q., Wang, X., 2014b. Rapidly improved determination of metabolites from biological data sets using the high-efficient TransOmics tool. Mol. Biosyst. 10 (8), 2160–2165.

Zhu, K.Y., Mao, Q.Q., Ip, S.P., Choi, R.C., Dong, T.T., Lau, D.T., Tsim, K.W., 2012. A standardized Chinese herbal decoction, kai-xin-san, restores decreased levels of neurotransmitters and neurotrophic factors in the brain of chronic stress-induced depressive rats. Evid. Based Complement. Alternat. Med. 2012, 149256.

Zhu, K.Y., Xu, S.L., Choi, R.C., Yan, A.L., Dong, T.T., Tsim, K.W., 2013. Kai-xin-san, a Chinese herbal decoction containing ginseng radix et rhizoma, polygalae radix, acori tatarinowii rhizoma, and poria, stimulates the expression and secretion of neurotrophic factors in cultured astrocytes. Evid. Based Complement. Alternat. Med. 2013, 731385.

CHAPTER 23

Metabolic Profiles Delineate the Effect of Shengmai San on Alzheimer's Disease in Rats

Hui Sun, Ying Han, Aihua Zhang, Xijun Wang

CONTENTS

23.1 INTRODUCTION

Alzheimer's disease is the most common type of degenerative dementia, and is characterized by the progressive impairment of cognitive function, and changes in behavior and social adaptability (Choi et al., 2014). Growing evidence shows that herbal drugs can effectively treat AD. Shengmai San (SMS) which was first recorded in *Yi Xue Qi Yuan* which was written by Zhang Yuansu, a famous doctor in the Jin Dynasty, is one of the most famous Chinese herbal formulae. It consists of *Panax ginseng, Ophiopogon japonicas* and *Schisandra chinensis*, with a dosage proportion of 2:2:1 (Yao et al., 2008; Wu et al., 2011). According to the theory of Traditional Chinese Medicine (TCM), this formulae benefits "qi," nourishes "yin," prevents exhaustion, and replenishes body fluids. In China, it has long been used for the treatment of "qi–yin" deficiency, and is especially used to treat diverse symptoms including cardiovascular, and neurological disorders in modern times (Nishida et al., 2007a). Previous studies have demonstrated that SMS treatment of an ischemia–reperfusion model rat (Li et al., 2013) and PC-12 cells (Nishida et al., 2007b) exhibited decreased oxidative injury, such as the accumulation of lipid oxidation, and protein carbonyl in the brain and cells, respectively. SMS significantly decreased the AChE activity, and increased the ACh level in brain homogenates (Giridharan et al., 2011). However, the protective effects, and exact mechanism of SMS against AD are ambiguous. This has become the major obstacle for clinical application, and further pharmaceutical research. This dilemma calls upon the debut of metabolomics at the end of the last century. Metabolomics differs from the classical or traditional targeted phytochemical analysis in various fundamental aspects, such as being a data-driven approach with predictive power that aims to assess all measurable metabolites (Zhang et al., 2013a). Metabolomics reveals a dynamic metabolic fingerprint in urine or blood samples, and thus is able to provide important details in the metabolic spectrum with TCM intervention (Wang et al., 2011; Yang et al., 2012). This is very helpful to find new biomarkers related to

Chinmedomics. http://dx.doi.org/10.1016/B978-0-12-803117-9.00023-8

disease, and studying the targets, and mechanisms of drugs actions. In recent years, metabolomics was used to evaluate the therapeutic effects of many Chinese herbal medicines, and TCM prescriptions (Zhang et al., 2013b). Thus, in the present study, a UPLC-G2-Q-TOF/MS based urine metabolomic approach was employed to examine the metabolic changes in a rat model of AD, and to further explore the protective effects of SMS on AD.

23.2 EXPERIMENTAL

23.2.1 Chemicals and Materials

Methanol, HPLC grade, was purchased from Fisher Scientific Corp. (USA); acetonitrile, HPLC grade, was obtained from Merck (Darmstadt, Germany); leucine enkephalin was purchased from Sigma–Aldrich (St. Louis, MO, USA). The deionized water (18.2 MΩ) was further purified using a Milli-Q system (Millipore, Billerica, USA); formic acid, HPLC grade, was purchased from Kermel Chemical Reagent Co., Ltd (Tianjin, China); $AlCl_3$ (Al), and D-galactose (D-gal), analytical grade, was purchased from Tianjin Guangfu Fine Chemical Research Institute (Tianjin, PRC), Beijing Biosharp Biotechnology Co. (Beijing, PRC). *Panax ginseng, Ophiopogon japonicas* and *Schisandra chinensis* were purchased from Harbin Tongrentang Drug Store (Harbin, China).

23.2.2 Preparation of SMS Samples

The sample of SMS was prepared by combining *Panax ginseng, Ophiopogon japonicus,* and *Schisandra chinensis,* which were mixed in a ratio of 2:2:1. Then the mixture was extracted three times with 1000, 800, and 600 mL water for 1 h, respectively. The extracted solutions were combined, and evaporated by a to approximate 0.1 g/mL, then the filtrate was dried under freeze-drying. The freeze-dried powder was dissolved in water, and concentrations of 1.06, 0.53, 0.265 g/mL solution were made, respectively.

23.2.3 Establishing Models of Rat With AD

Al and D-gal were used to induce the rat model for AD. The rats were given Al orally (28 mg/kg/day) and injections of D-gal intraperitoneally (63 mg/kg/day) for 105 days to establish the AD animal model. A water maze behavioral test including place navigation and a spatial probe test were performed, to evaluate the learning/memory ability of the rats.

23.2.4 Animal Handling and Sample Preparation

After acclimatization, male Wistar rats (weighing 300 ± 20 g, 9 w) were randomly divided into 5 groups with 10 rats in each group: control, model, SMS high dose group (1.06 g/mL, SMSH), SMS middle dose group (0.53 g/mL, SMSM), and SMS low dose group (0.265 g/mL, SMSL). Controls were subjected to oral gavage with distilled water, and injections of 0.9% saline solution intraperitoneally, while the AD-treated groups were orally administered Al solution, and injections of D-gal intraperitoneally at 14:00 p.m. for 105 days. SMS treatment groups were orally administrated an accurate volume 10 mL/kg of SMS solution from day 1. After 15 days administration, all the rats were sacrificed. Each rat was housed in a metabolism cage, and urine samples were collected over the course. A standard diet was given to all the animals, and free access to water was provided throughout the studies. The urine samples were collected at 8:00 a.m. and centrifuged at 13,000 rpm for 5 min (−4 °C), to remove solid debris, and the supernatant was stored at −80 °C. Prior to UPLC/MS analysis, the urine samples were thawed at room temperature, and filtered through a 0.22 μm filter membrane before the UPLC-G2-Q-TOF/MS analysis.

23.2.5 UPLC-G2-Q-TOF/MS Conditions

The UPLC/MS analysis was carried out using a Waters ACQUITY UPLC system (Waters Corp., Milford, MA, USA) coupled with a Waters Micromass Xevo G2-Q-TOF Mass Spectrometer (Manchester, UK). The data were obtained using MassLynx (v4.1) workstation (Waters Corp., Milford, MA, USA), and the MarkerLynx aplication Manager (Waters Corp., Milford, MA, USA) was used for the peak detection. For the reversed-phase UPLC analysis, the ACQUITY UPLC™ HSS T3 column (100 × 2.1 mm i.d., 1.8 μm particles, USA) was used. The column temperature was maintained at 45 °C. The flow rate of the mobile phase was 0.4 mL/min. The injection volume was fixed at 2 μL. Mobile phase A consisted of acetonitrile and formic acid (100:0.1, v/v) while mobile phase B consisted of water and formic acid (100:0.1, v/v). The column was eluted with a linear gradient as follows: 0–2.5 min, linear increasing from 1–11% A; 2.5–4.5 min, 11–21% A; 4.5–7 min, 21–40% A; 7–8.5 min, 40–99% A. The eluate was directed to the mass spectrometer without splitting. All the urine samples were mixed to get a quality control (QC) sample for method validation. A QC sample was injected every 10 samples during the analytical run to further monitor the system stability.

The ESI source in both positive and negative ion mode was used in the MS analysis. The source temperature was 110 °C, and the desolvation gas temperature was 450 °C. Nitrogen was used as a cone and desolvation gas. The flow rates of cone and desolvation gas were set at 50 and 800 L/h, respectively. Capillary, cone, and extraction cone voltages were set at 3.0 kV, 20 V, and 5.0 V in positive ion mode, respectively; while at 2.0 kV, 30 V and 5.0 V in negative ion mode, respectively. MS data was collected in full-scan mode in the mass range of 50–1000 Da. A lock-mass of leucine enkephalin (1 ng/μL) was used to set up mass spectrometer calibration via a lock spray interface at a flow rate of 5 μL/min, monitoring in positive ion mode ($[M + H]^+$ =556.2771) and in negative ion mode ($[M-H]^-$ =554.2615) to ensure accuracy during the MS analysis.

23.2.6 Multivariate Data Analysis

UPLC/MS data of the urine samples were processed using Waters MarkerLynx software (v4.1) (Waters Corp., MA, USA), which used Apex-Track-peak detection package to integrate peaks in the UPLC/MS data. Peaks in each chromatogram were identified by *m/z*, retention time (R_t) as well as their associated height intensities. The three-dimensional data, peak number (R_t_*m/z* pair), sample name, and normalized ion intensity were introduced to EZinfo 2.0 software package for multivariate data analysis such as principal component analysis (PCA), partial least-squares-discriminant analysis (PLS–DA), and orthogonal partial least-squared discriminant analysis (OPLS–DA) analyses. PCA is an unsupervised multivariate statistical approach. It is used for variable reduction, and separation into classes. To maximize class discrimination and biomarkers, the data were further analyzed using the OPLS-DA method. S-plots were calculated to visualize the relationship between covariance, and correlation within the OPLS-DA results. Variables that had significant contributions to discrimination between groups were considered as potential biomarkers, and subjected to further identification of the molecular formula.

23.2.7 Identification of Biomarkers and Pathway Analysis

On the basis of determination, the biomarkers were scanned by tandem mass spectrometer according to their retention time and exact mass. The structures were identified according to their main daughter ions, and the splitting information, which was provided by mass fragment software in MassFragment™ application manager (Waters Corp., Milford, USA). Databases of HMDB (www.hmdb.ca/), KEGG (www.genome.jp/kegg/), METLIN (http://metlin.scripps.edu/), Massbank (www.massbank.jp), LIPIDMAPS (www.lipidmaps.org/) and Chemspider (www.chemspider.com) were used to identify the metabolite markers with tandem mass spectrometry. Metabolomic pathway analysis was performed with MetPA (metabolomics pathway analysis) based on potential metabolite biomarkers. For the pathway analysis algorithms, the Hypergeometric Test was used for over representation analysis, and relative-betweenness centrality was used for pathway topology analysis.

23.3 RESULTS AND DISCUSSION

23.3.1 SMS Improves Learning and Memory Ability in AD Rats

Morris water maze testing was determined to validate the establishment of the AD animal model. We found that administrating Al combined with D-gal for

105 days induced rat memory impairment, which mimics the AD pathogenetic, and could be used as an AD animal model. Administration of SMS for consecutive 15 days significantly decreased the latency to the platform, and increased the proportion of time spent in the target quadrant as compared with the model group (data not shown). These results indicated that SMS can significantly improve learning, and memory ability in AD rats.

23.3.2 Metabolomics on Efficacy Evaluation

Base peak intensity (BPI) chromatograms obtained from the control, model and SMSH, SMSM, SMSL in positive and negative ionization modes are shown in Figs 23.1 and 23.2, respectively. In spite of the differences in the BPI chromatogram that could be observed, the multivariate analysis method distinguished the experimental groups much better. PCA, PLS-DA and OPLS-DA are frequently used pattern recognition approaches for metabolomics. Metabolite

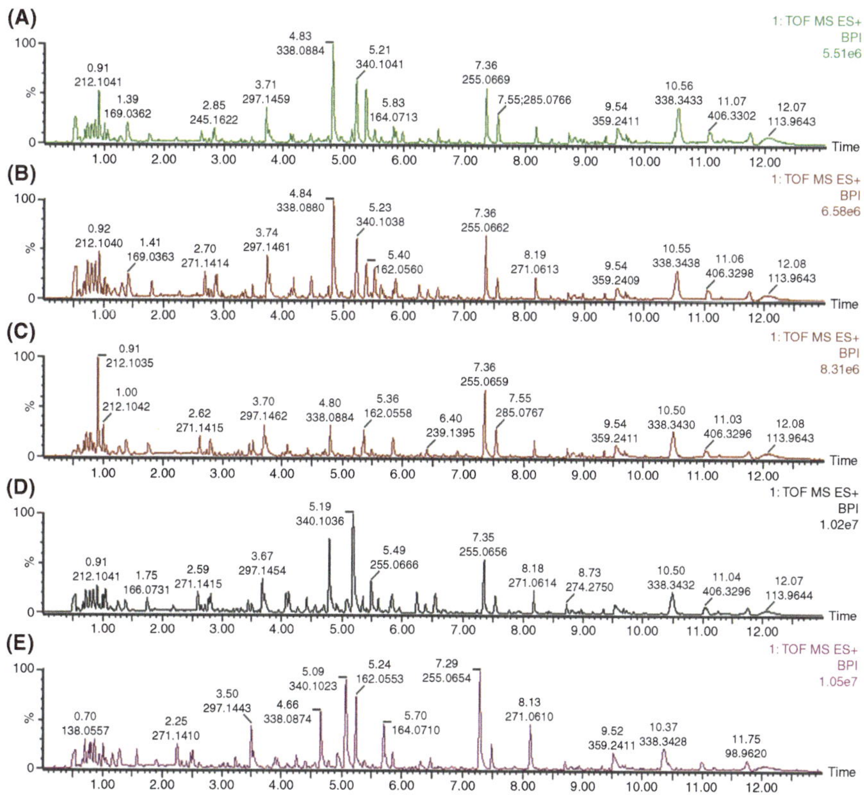

FIGURE 23.1

Base peak intensity (BPI) chromatograms obtained from the positive ion UPLC-MS analyses of control: (A) model; (B) SMSH; (C) SMSM; (D) SMSL group; (E) rats.

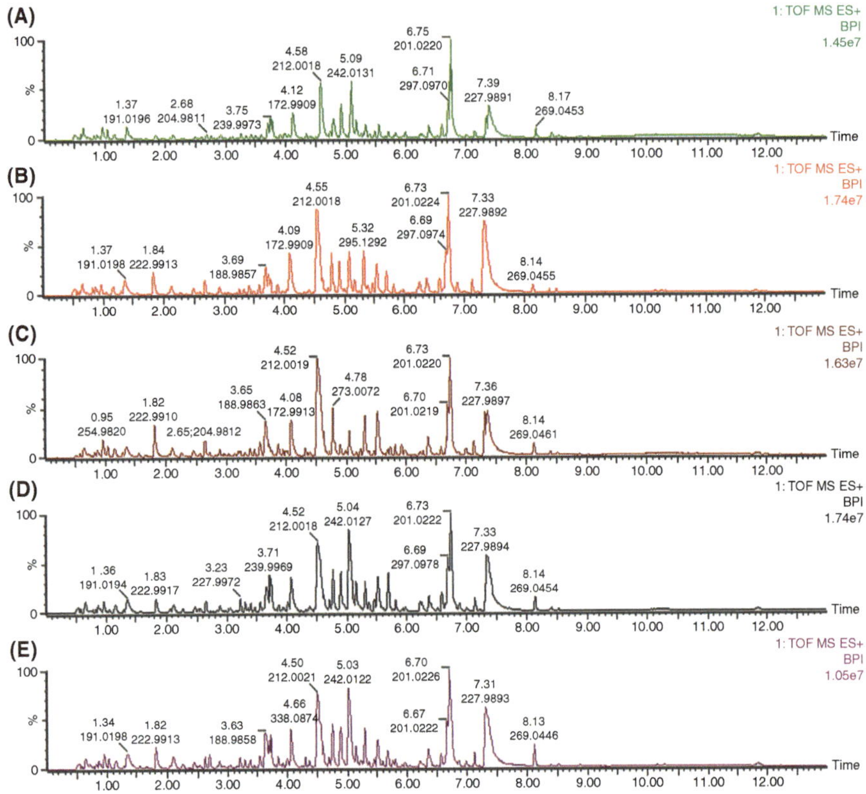

FIGURE 23.2

BPI chromatograms of rat urine samples of: (A) control; (B) model; (C) SMSH; (D) SMSM; and (E) SMSL group, in negative mode.

profiling obtained from model group at day 105 had been performed by UPLC-MS in conjunction with Progenesis QI software analysis. Following the AD experiments in the Chapter 22, 11 metabolite ions were selected as potential biomarkers related with AD rats. Among them, 4 of the 11 biomarkers were increased, and seven of them were decreased in the pathogenesis of the AD group. Further studies were focused on 11 metabolites, which were involved in altered glyoxylate, and dicarboxylate metabolism, inositol phosphate metabolism, citrate cycle, sphingolipid metabolism. As shown in the PCA scores plot (Fig. 23.3) the SMSH, SMSM group is closer to the control group, which indicated the AD toward the normal status. Similar to the results of PCA score plot change, 6 of the 11 marker metabolites also showed the same variation tendency. The content of changed metabolites, such as sphingosine, 5-methoxytryptophol, deoxyguanosine 5′-monophosphate, inositol 1,4,5-trisphosphate, phosphorylcholine, hippuric acid in SMS group were

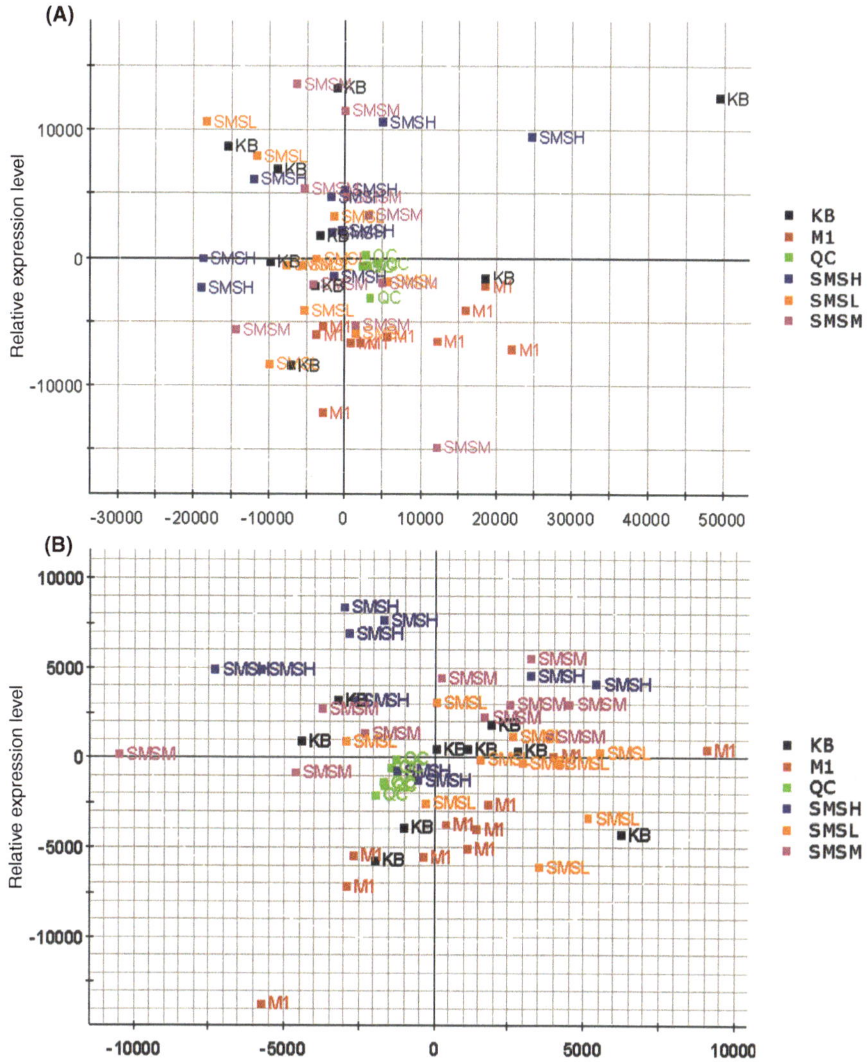

FIGURE 23.3 **Scores plot from PCA of the urinary metabolic profiling of the: control, model group, SMSH, SMSM, and SMSL treatment groups.**
(A) Positive ion mode; (B) negative ion mode.

significantly larger than in the AD model group (Fig. 23.4), and suggested that SMS acted on multitargets to relieve AD. These metabolites were considered as pharmacodynamic markers for SMS treatment on AD. It suggested that SMS could effectively regulate the changed metabolites, and the altered metabolic pathway in the AD toward the normal status.

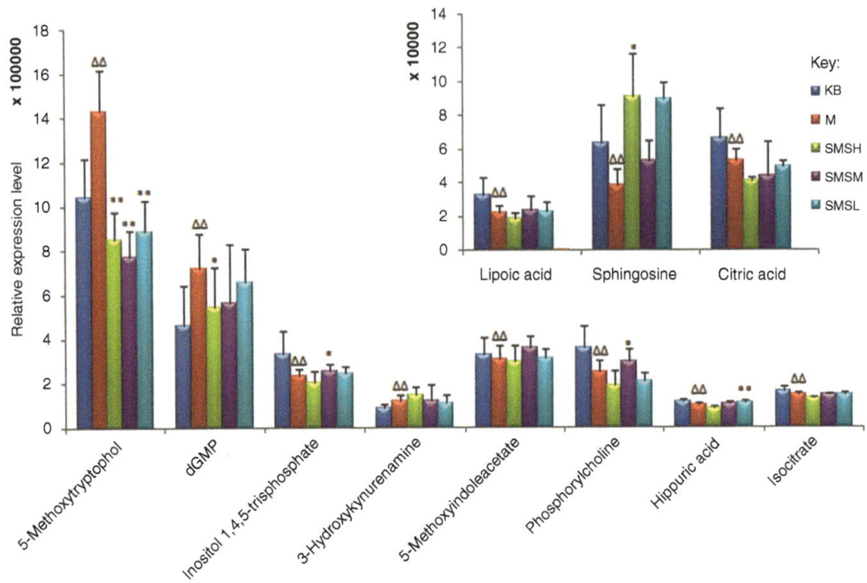

FIGURE 23.4 Comparison of the relative intensity of putative potential biomarkers in the control, model group, SMSH, SMSM, and SMSL treatment groups.

23.4 CONCLUSIONS

In this study, an UPLC-G2-Q-TOF/MS method with a novel multivariate statistical technique was established to gain the global profiling of the endogenous metabolites in the urine of AD rats. The metabolite changes after the treatment with SMS were investigated using pattern recognition approaches, and a further understanding of how SMS could modify the AD symptom. Six potential biomarkers were considered to be corelated with the SMS therapeutic effects for the AD rats involved in glyoxylate and dicarboxylate metabolism, inositol phosphate metabolism, citrate cycle, and sphingolipid metabolism. In conclusion, SMS showed significant effects on AD rats on the basis of the evaluation of entire metabolic changes, and the analysis of the variation of biomarkers. This research also demonstrated that UPLC-G2-Q-TOF/MS-based metabolomics was a promising tool to evaluate the efficacy of TCM.

References

Choi, S.H., Kim, Y.H., Hebisch, M., Sliwinski, C., Lee, S., D'Avanzo, C., Chen, H., Hooli, B., Asselin, C., Muffat, J., Klee, J.B., Zhang, C., Wainger, B.J., Peitz, M., Kovacs, D.M., Woolf, C.J., Wagner, S.L., Tanzi, R.E., Kim, D.Y., 2014. A three-dimensional human neural cell culture model of Alzheimer's disease. Nature 515 (7526), 274–278.

Giridharan, V.V., Thandavarayan, R.A., Konishi, T., 2011. Effect of Shengmai-san on cognitive performance and cerebral oxidative damage in BALB/c mice. J. Med. Food 14 (6), 601–609.

Li, L.H., Wang, J.S., Kong, L.Y., 2013. Protective effects of shengmai san and its three fractions on cerebral ischemia-reperfusion injury. Chin. J. Nat. Med. 11 (3), 222–230.

Nishida, H., Ichikawa, H., Konishi, T., 2007a. Shengmai-san enhances antioxidant potential in C2C12 myoblasts through the induction of intracellular glutathione peroxidase. J. Pharmacol. Sci. 105 (4), 342–352.

Nishida, H., Kushida, M., Nakajima, Y., Ogawa, Y., Tatewaki, N., Sato, S., Konishi, T., 2007b. Amyloid-beta-induced cytotoxicity of PC-12 cell was attenuated by Shengmai-san through redox regulation and outgrowth induction. J. Pharmacol. Sci. 104 (1), 73–81.

Wang, X., Sun, H., Zhang, A., Sun, W., Wang, P., Wang, Z., 2011. Potential role of metabolomics approaches in the area of traditional Chinese medicine: as pillars of the bridge between Chinese and Western medicine. J. Pharm. Biomed. Anal. 55 (5), 859–868.

Wu, F., Sun, H., Wei, W., Han, Y., Wang, P., Dong, T., Yan, G., Wang, X., 2011. Rapid and global detection and characterization of the constituents in ShengMai San by ultra-performance liquid chromatography-high-definition mass spectrometry. J. Sep. Sci. 34 (22), 3194–3199.

Yang, B., Zhang, A., Sun, H., Dong, W., Yan, G., Li, T., Wang, X., 2012. Metabolomic study of insomnia and intervention effects of Suanzaoren decoction using ultra-performance liquid-chromatography/electrospray-ionization synapt high-definition mass spectrometry. J. Pharm. Biomed. Anal. 58, 113–124.

Yao, H.T., Chang, Y.W., Chen, C.T., Chiang, M.T., Chang, L., Yeh, T.K., 2008. Shengmai San reduces hepatic lipids and lipid peroxidation in rats fed on a high-cholesterol diet. J. Ethnopharmacol. 116 (1), 49–57.

Zhang, A.H., Sun, H., Wang, X.J., 2013a. Recent advances in metabolomics in neurological disease, and future perspectives. Anal. Bioanal. Chem. 405 (25), 8143–8150.

Zhang, A.H., Sun, H., Qiu, S., Wang, X.J., 2013b. Recent highlights of metabolomics in Chinese medicine syndrome research. Evid. Based Complement. Alternat. Med. 2013, 402159.

Subject Index

Kolibri 360 GmbH
Ferdinand-Jühlke-Straße 7
99095 Erfurt, Deutschland
produktsicherheit@kolibri360.de